THEORY OF GEOSTATIONARY SATELLITES

THEORY OF GEOSTATIONARY SATELLITES

CHONG-HUNG ZEE
EBASCO SERVICES Inc., New York, U.S.A.

Kluwer Academic Publishers
Dordrecht / Boston / London

Library of Congress Cataloging in Publication Data

CIP

Zee, Chong-Hung, 1920-
Theory of geostationary satellites.

Includes index.
1. Geostationary satellites. I. Title.
TL796.6.E2Z4 1988 629.43'4 89-2453

ISBN 90-277-2636-1

Published by Kluwer Academic Publishers,
P.O. Box 17, 3300 AA Dordrecht, The Netherlands.

Kluwer Academic Publishers incorporates
the publishing programmes of
D. Reidel, Martinus Nijhoff, Dr W. Junk and MTP Press.

Sold and distributed in the U.S.A. and Canada
by Kluwer Academic Publishers,
101 Philip Drive, Norwell, MA 02061, U.S.A.

In all other countries, sold and distributed
by Kluwer Academic Publishers Group,
P.O. Box 322, 3300 AH Dordrecht, The Netherlands.

Printed in The Netherlands

To my wife – MEI-KUO

TABLE OF CONTENTS

CHAPTER 2

THE EARTH'S GRAVITATIONAL FIELD AND BASIC METHODOLOGY

CHAPTER 3

THE EFFECT OF EARTH'S EQUATORIAL ELLIPTICITY

CHAPTER 4

SUN AND MOON EFFECT

CHAPTER 5

COMBINED EFFECTS OF THE SUN, MOON, AND SOLAR RADIATION PRESSURE

CHAPTER 6

STATION-KEEPING

CHAPTER 7

VERIFICATION OF THE THEORY BY THE EARLY BIRD SYNCHRONOUS SATELLITE

INTELSAT V

Photo 1. INTELSAT V communications satellite.

(Courtesy of INTELSAT)

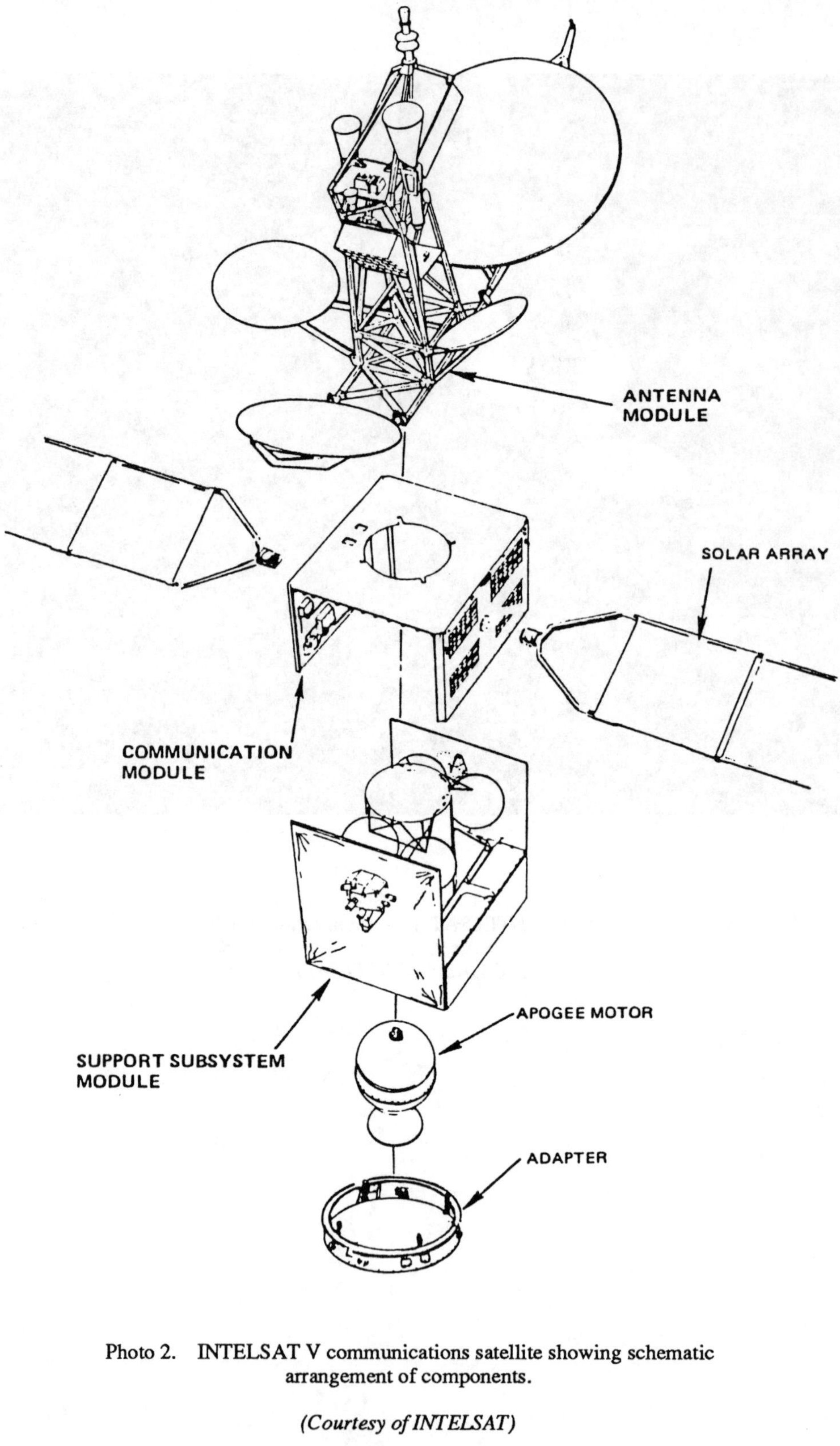

Photo 2. INTELSAT V communications satellite showing schematic arrangement of components.

(Courtesy of INTELSAT)

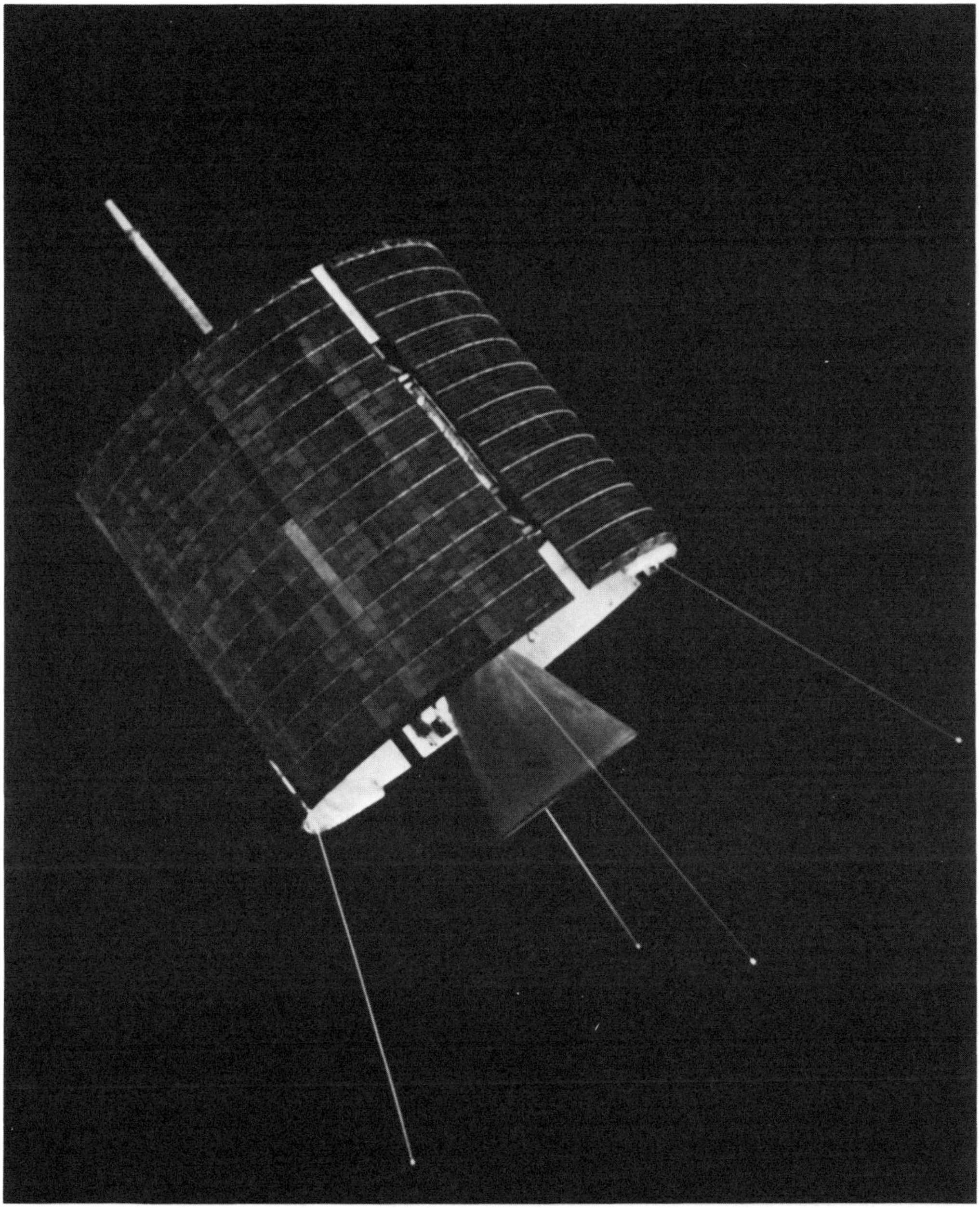

INTELSAT I

Photo 3. Early Bird communications satellite (INTELSAT I). (Communications antenna sticking out at the top and apogee motor rocket nozzle protruding from the bottom of solar cell covered drum.)

(Courtesy of INTELSAT)

PREFACE

Geostationary or equatorial synchronous satellites are a daily reminder of our space efforts during the past two decades. The nightly television satellite weather picture, the intercontinental telecommunications of television transmissions and telephone conversations, and the establishment of educational programs in remote regions on Earth are constant reminders of the presence of these satellites. As used here, the term 'geostationary' must be taken loosely because, in the long run, the satellites will not remain 'stationary' with respect to an Earth-fixed reference frame. This results from the fact that these satellites, as is true for all satellites, are incessantly subject to perturbations other than the central-body attraction of the Earth. Among the more predominant perturbations are: the ellipticity of the Earth's equator, the Sun and Moon, and solar radiation pressure. Higher harmonics of the Earth's potential and tidal effects also influence satellite motion, but they are of second-order when compared to the predominant perturbations.

This volume deals with the theory of geostationary satellites. It consists of seven chapters. Chapter 1 provides a general discussion including a brief history of geostationary satellites and their practical applications. Chapter 2 describes the Earth's gravitational potential field and the methodology of solving the geostationary satellite problem. Chapter 3 treats the effect of Earth's equatorial ellipticity (triaxiality) on a geostationary satellite. Chapter 4 deals with the effects of the Sun and Moon on the satellite's motion while Chapter 5 presents the combined influences of the Sun, Moon and solar radiation pressure. Chapter 6 describes various station-keeping techniques which may be used to make geostationary satellites practically stationary. Finally, Chapter 7 describes the verification of the theory developed in Chapters 3, 4, and 5 by utilizing the Early Bird synchronous satellite observed data as well as its numerically integrated results. Except for the first, second, and last chapters, each chapter is followed by a bibliographic review of the chapter's topic. It is by no means a thorough review of the literature but rather gives an overall view of previous investigations.

The presented approach to the problem is rather unique. The study of the motion of a satellite subject to the perturbations of various celestial bodies has been approached through Lagrange's equations for the general planetary motion with the aid of a disturbing function. In this volume, the problem is treated from the point of view of nonlinear mechanics. In particular, the problem is considered as one in nonlinear oscillation: all perturbing forces are expressed in terms of periodic functions, and the resulting differential equations are quasi-linear. As a consequence, these differential equations can be solved by the asymptotic method. This method of solution was originated in the 1930's by Kryloff and Bogoliuboff and was known as the averaging method. It was developed in its present form in the late fifties by Bogoliubov and Mitropolsky.

As might be anticipated, application of the asymptotic method requires increasingly more mathematical manipulation as the order of the solution increases. Fortunately, most of the problems in nonlinear mechanics require only low order solutions to reveal their essential characteristics. Bearing this in mind, the effects of the attractions of the Sun and Moon and also of the combined influences of the Sun, Moon, and solar radiation pressure are treated here by obtaining their improved first-order solution, while the effect of Earth's triaxiality is obtained by seeking an improved second-order solution. These solutions may serve as reference trajectories for a further refinement of the problem as, for example, when the Encke method is employed in celestial mechanics. In this process of refinement, additional higher order perturbing forces may be introduced into the problem.

As a matter of practical importance, each of the solutions presented here is discussed, along with graphs and tables which depict the principal characteristics. This should permit engineers more disposed toward vehicle design to use the numerical results in the design of geostationary satellites, while analysts may manipulate the analytical expressions for further general studies of these satellites. Moreover, in view of the various techniques involved in presenting the methods of solutions of non-linear differential equations, this book may also serve as a text and reference for students interested in nonlinear mechanics. In addition to the techniques presented in the main text, the reader will find additional techniques in the chapter concerning station-keeping, where references to various low thrust cases are reviewed.

Most of the work in this book was developed while the author was associated with the Grumman Aerospace Corporation. As a dynamicist working in the area of advanced development for twelve years, the author published a number of papers in the field of astrodynamics. These papers are referenced throughout the book. Special acknowledgement to the Pergamon Journals, Inc. for its permission to reproduce the materials in some of these papers is extended.

The author is indebted to the Grumman Aerospace Corporation for the opportunity to participate in systematic studies in the field of astro-dynamics. The kindness of the Ebasco Services Incorporated for providing all graphic works in this volume and of the International Telecommunications Satellite Organization in permitting three photographs of their satellites to be reproduced are greatly appreciated. The author also extends his thanks to

Vincent Sedita of the Grumman Data Systems without whose efforts the numerical results presented in this book would not have been obtained. Additionally, deep appreciation is extended to the late Dr. T. C. Fang and Eli Botkin, both of the Grumman Aerospace Corporation, and to Dr. Victor J. Slabinski of the Communications Satellite Corporation (now at INTELSAT) for their valuable review and comments. The author would like to convey special thanks to Dr. Slabinski for making available the Early Bird data and for his scholarly discussions and suggestions without which Chapter 7 could not have been written. Finally, I would like to express my appreciation to my wife, Mei-Kuo, and my two sons, Raymond and Roger, whose patience, support and kind understanding have made possible the completion of this book.

Melville, New York 1987 CHONG-HUNG ZEE

NOMENCLATURE

SUPERSCRIPTS

(.)	first derivative with respect to time, t
(..)	second derivative with respect to time, t
(-)	averaged value of a parameter over a 2π-period of θ

SUBSCRIPTS

A	point of application of an impulse
c	synchronous orbit
	circular orbit
D	drift angle and drift angle rate
E	the Earth
e	the Earth
	eccentricity
f	final value
L	long-period
M	the Moon
m	order of spherical harmonics
N	normal to the orbital plane direction
0	initial value
P	short-period
p	power-flight
R	radial direction
r	radial direction
S	the Sun
	single impulse
s	satellite
	secular
T	transverse direction
	truncated value
θ	transverse direction

ENGLISH SYMBOLS

A	Cross-sectional area
	Constant $\left(= \dfrac{\dot{m}\bar{c}}{M_0 g_0}\right)$, defined in a power-flight
	Constant
A_n	Constants defined in Appendix B (n = 1, ..., 44)

a	Acceleration
	Semimajor axis of an orbit
	Parameter in $u_k^{(0)}$ and $u_k^{(1)}$
	Dimensionless acceleration parameter $\left(= \frac{\dot{\theta}\bar{c}}{g_0}\right)$ in a power-flight
a_0	Constant
a_n	Taylor series coefficients
a_x, a_y, a_z	Direction cosines of $\vec{n}_r$
a_r, a_θ, a_ϕ	Acceleration in the directions of $\vec{r}$, $\vec{n}_\theta$, and $\vec{n}_\phi$
B	Constant $\left(= \frac{\dot{m}}{M_0}\left(\frac{r_0}{g_0}\right)^{1/2}\right)$, defined in a power-flight
	Constant
$B(\lambda_A)$	Function of λ_A
B_n	Constants defined in Appendix B (n = 1, ..., 46)
b_0	Constant
b_n	Taylor series coefficients
b_x, b_y, b_z	Direction cosines of $\vec{n}_\theta$
C	Constant $\left(= \frac{g_0}{r_0\dot{\theta}_0^2}\right)$ in a power-flight
	Constant $\left(= \frac{V_{\theta 0}}{(g_0 r_0)^{1/2}}\right)$ in a power-flight
	Constant $\left(= \left(\frac{d\rho}{d\tau}\right)_{\tau=0}\right)$ in a power-flight
$C_1(a)$, $C_2(a)$	Functions of parameter 'a'
c	0.9914825 (constant in solar perturbation on Moon's motion)
$\bar{c}$	Average jet velocity of a rocket
c_0	Constant
c_n	Taylor series coefficients
C_{nm}	Cosine coefficient of spherical harmonic potential term
c_x, c_y, c_z	Direction cosines of $\vec{n}_\phi$
D	Constant $\left(= \frac{V_{r0}}{(g_0 r_0)^{1/2}}\right)$ in a power-flight
	Constant $\left(= \left(\frac{d\theta}{d\tau}\right)_{\tau=0}\right)$ in a power-flight
	Constant $(= \varepsilon G)$
D_n	Constant defined in Appendix D (n = 1, ..., 49)
E	Elliptic integral of the second kind
E_n	Constants defined in Appendix D (n = 1, ..., 4)
e	Eccentricity of an orbit
F	Force
	Elliptic integral of the first kind

F	Force function
	Hamiltonian
	Rocket thrusting force
F_r, F_θ, F_ϕ	Gravitational attraction on a satellite due to the Earth in the r-, θ-, and ϕ-directions, respectively
F_H, F_Q, F_S	Force functions associated with a small parameter, ε, in the Laplace's form equations of motion related to the parameters H, Q, and S
f	Constant thrust acceleration
f_r, f_θ, f_ϕ	Force functions associated with a small parameter, ε, in the r-, θ- and ϕ-directions, respectively
G	Delaunay orbital element, angular momentum per unit mass or $[\mu_E a(1 - e^2)]^{1/2}$
G	Constant $\left(= \frac{1}{2\omega_{12}} (1 + \kappa) \sin 2\lambda \right)$
$G(\lambda_A)$	Function of λ_A
g	Delaunay orbital element for argument of perigee, ω
	1.003999175 (constant in solar perturbation on Moon's motion)
g_e	Gravitational acceleration at the Earth's surface
g_0	Gravitational acceleration at $r = r_0$
g_s	Gravitational acceleration at $r = r_c$
H	Delaunay orbital element, $[\mu_E a(1-e^2)]^{1/2} \cos i$
	Parameter in Laplace's form equations of motion
H, H_1, H_2, H_3	Hamiltonian Functions
h	Delaunay angular element for node, Ω
	Delaunay angular element, $\Omega - \theta_g$
	Dimensionless angular momentum parameter $\left(= \frac{\theta}{u^2 \dot{\theta}_0} \right)$
I	Mean inclination of Moon's orbit with respect to the ecliptic plane (5°09')
I_M	Inclination of Moon's orbit with respect to the equatorial plane of the Earth
I_{sp}	Specific impulse $(= \bar{c}/g_e)$
i	Inclination of osculating orbital plane
J_2	Earth oblateness coefficient
J_{22}	Coefficient of the Earth equator ellipticity term
J_{nm}	Harmonic coefficients of the Earth potential function
K	Constants defined in Table V
K_1	Constant defined in (4.80)
K_2	Constant defined in (4.80)
K_3	Constant defined in (4.80)
K_4	Constant defined in (4.80)
K_{22}	Constant defined in (3.77)
K_M, K_S	Delaunay orbital elements related to the Moon and Sun, respectively
k	Gravitational constant

k_1	Constant defined in (3.37)
k_2	Constant defined in (3.37)
k_5^2	Constant $(= 1/\sin^2 \tau_{max})$
k_6^2	Constant defined in (3.98)
k_M, k_S	Delaunay angular elements related to the Moon and Sun, respectively
L	Delaunay orbital element, $(\mu_E a)^{1/2}$
L^*	Delaunay orbital element, $(\mu_E r_s)^{1/2}$
ℓ	Direction cosine with respect to x-axis
	Delaunay angular element for mean anomaly, M
ℓ, ℓ_s, ℓ_L, ℓ_p	Semilatus rectum of osculating orbit and its secular, long-period and short-period values, respectively
M	Mass or satellite mass
	Mean anomaly
	Cos Ψ
M_x, M_y, M_z	Direction of cosines of $\vec{r}_{ME}$
M_p	Mass of fuel consumed, $(M_f - M_0)$
m	Mass
	Direction cosine with respect to y-axis
	$n_S/\bar{n}_M$
N	Sin Ψ
n	Mean motion
	Integer
	Direction cosine with respect to z-axis
$\bar{n}_M$	Observed mean value of Moon's mean motion
n_M'	$n_M/(1 - e_M^2)^3$
n_S'	$n_S'/(1 - e_S^2)^3$
$\vec{n}_r$, $\vec{n}_\theta$, $\vec{n}_\phi$	Unit vectors in the r, θ, and ϕ spherical coordinate system
P	Product of solar radiation pressure and the area-to-mass ratio of a satellite, $(P_r A/M)$
P_r	Solar radiation pressure intensity
p	Variable
p_i	Momentum variable
P_{nm}	Legendre function
Q	Parameter in Laplace's form equations of motion
q	Variable
q_i	Angle or position variable
R	Disturbing function
R_d	Disturbing function due to the Earth oblateness, the Sun and Moon
R	Reflectivity
R_E	Mean equatorial radius of the Earth
R_S	Disturbing function due to solar radiation pressure

R_{nm}	Disturbing function due to higher Earth harmonic potential terms
R_{22}	Disturbing function due to J_{22} Earth harmonic potential term
R_{ni}	Remainders of the Taylor series (i = 1, ..., 6)
R_{noi}	Remainders of the Taylor series (i = x, y, z)
r	Magnitude of radius vector in a orbit
r_c	Geostationary satellite orbital radius due to the Earth oblateness, $r_s \left(1 + \frac{1}{2}J_2 \frac{R_E^2}{r_s^2}\right)^{1/3}$
r_c'	Geostationary satellite orbital radius due to the Earth oblateness, the Sun and Moon, $r_s(1 + \frac{1}{6}\omega_4)$
r_s	Geostationary satellite orbital radius of a spherical Earth, $(\mu_E/\dot{\theta}_E^2)^{1/3}$
Δr_c	$r - r_c$
S	Determining function
	Generating function
	Parameter in Laplace's form equations of motion
	Parameter related to the eccentricity of an orbit in a power-flight
S_{nm}	Sine coefficient of spherical harmonic potential term
T	One half period of the station-keeping cycle
t	Time
t_T	Thrusting time
t_n	Time required to travel n half-periods
u	Dimensionless radius parameter (= r_0/r)
$u_k^{(0)}$, $u_k^{(1)}$	The zeroth-order and first-order solutions to X_k
v	True anomaly
V	Gravitational potential
	Satellite velocity
V_c	Stationary velocity (= $r_c\dot{\theta}_E$)
V_0	Circular orbital velocity (= $(g_0r_0)^{1/2}$)
V_r	Radial velocity
V_θ	Transverse velocity
ΔV	Impulsive velocity increment
w	Parameter
x	Variable
X	Dimensionless time parameter in a power-flight
X_n	Parameters; $X_1, \ldots, X_6$ defined in (2.89)-(2.91) and $X_2, \ldots, X_8$ defined in (3.10)-(3.12)
X, Y, Z	Dimensionless parameters defined as x/r_0, y/r_0 and z/r_0, respectively
$\dot{X}$, $\dot{Y}$, $\dot{Z}$	Dimensionless parameters defined as $\dot{x}/(g_0r_0)^{1/2}$, $\dot{y}/(g_0r_0)^{1/2}$ and $\dot{z}/(g_0r_0)^{1/2}$, respectively

x, y, z	Inertial rectangular coordinate system with origin at the center of the Earth and x-y plane being in the Earth equatorial plane, or in the initial orbital plane in the case of power-flight
$\vec{x}, \vec{y}, \vec{z}$	Unit vectors in the x, y, and z coordinate system
y	Variable $\dfrac{X_4^2}{\bar{\mu}_E r_c} - 1 \cong \dfrac{\Delta r_c}{r_c}$

GREEK SYMBOLS

α	Mean longitude of the Sun in the ecliptic plane
$\beta_1, \beta_2, \beta_3$	Constants
$\gamma_1, \gamma_2, \gamma_3$	Constants
δ	Radiation incident angle with respect to the normal of reflecting surface
	Parameter in Hamiltonian H_3
ε	Small parameter
	Constant $\left(= \dfrac{3}{4} \dfrac{K_1^2}{(1 - e_S^2)^{3/2}} \right)$
	Thrust to satellite weight ratio $(F/M_0 g_0)$
	Dimensionless acceleration parameter $(=f/g_0)$ in a power-flight
η	Variable
	Argument of latitude, angle measured from the ascending node to the radius vector $(\vec{r})$
θ	Spherical coordinate
θ_A	Angular position where an impulse is applied
θ_E	Angular position of the minor axis of the Earth's equator with respect to the vernal equinox
θ_g	Longitude of the Greenwich meridian from the vernal equinox in the ecliptic plane
θ_n	Spherical coordinate of a satellite at n half-periods
κ	Constant $\left(= \dfrac{K_2^2}{\sigma K_1^2} \left(\dfrac{1 - e_S^2}{1 - e_M^2} \right)^{3/2} \right)$
λ	Inclination of the equatorial plane of the Earth to the ecliptic plane (-23°27')
	Longitude
	Satellite's geographic longitude with respect to the Greenwich meridian (measured toward the east counted as positive)
	Mean longitude $(= M + \omega + \Omega)$
λ_{nm}	Harmonic coefficients of the Earth potential function
λ_{22}	Geographic longitude of the major axis of the equator ellipse of the Earth
λ_A	'Broken-legged' astronomical longitude $(= M + \omega + \Omega - \theta_g)$

μ	Gravitational constant
$\bar{\mu}$	Modified gravitational constant $\left(= \mu_E\left(1 + \frac{3}{2}J_2 \frac{R_E^2}{r_s^2}\right)\right)$
ν	True anomaly of the Moon in its own orbit
ξ	Variable $(= \omega_3(\theta - \theta_0) + \xi_0)$
	Variable
ρ	Position dependent mass density
	Dimensionless radius parameter $(= r/r_0)$ in a power-flight
σ	$(M_E + M_M)/M_M$
τ	Canonical transformation parameter
Γ	Satellite's angular position relative to the minor axis of the ellipse of the Earth equator (drift angle)
	Dimensionless time parameter $\left(= \left(\frac{g_0}{r_0}\right)^{1/2} t\right)$ in a power-flight
	Dimensionless time parameter $\left(= \frac{g_0 t}{\bar{c}}\right)$ in a power-flight
Γ_T, Γ_E, Γ_{SMR}	Total drift angle, drift angles due to the Earth's equator ellipticity effect and the effects of the Sun, Moon, and solar radiation pressure, respectively
Φ_0	Constant $(= \Psi_0 - \omega_1\theta_0)$
Φ	$(1 + \varepsilon^2)^{1/2}\omega_0\Delta t$
	Parameter
ϕ	Spherical coordinate
ϕ'	Colatitude
χ	Dimensionless solar radiation pressure parameter $(= 2P\bar{\bar{X}}_{60}^2/\varepsilon\mu_E^2)$
Ψ	Lead angle
	Fixed angle between a rocket thrust and a radius vector $\vec{r}$
	Variable $(= \omega_1(\theta - \theta_0) + \Psi_0)$
Ψ_n	Rotational angle of an orbital plane at n half-periods
Ψ_a	Required rotational angle of an orbital plane
ψ	True longitude of the Moon $(= \nu + \Omega_M + \omega_M)$
Ω	Longitude of ascending node
Ω_M'	Longitude of ascending node of Moon's orbit with respect to the vernal equinox in the Earth equatorial plane
ω	Argument of perigee of an orbit
$\tilde{\omega}$	$\omega + \Omega$
ω_0	Circular angular velocity $(= (g_0/r_0)^{1/2})$
ω_1	Mono-frequency $(= (\omega_{11}\omega_{12})^{1/2})$
ω_{11}	Constant defined in (4.86)
ω_{12}	Constant defined in (4.86)
ω_2	Constant defined in (3.42) and (3.43)
ω_3	Mono-frequency $(= \omega_{31}\omega_{32})^{1/2})$
ω_{31}	Constant defined in (4.86)

ω_{32}	Constant defined in (4.86)
ω_4	Constant defined in (4.125)

Chapter 1

INTRODUCTION

1.1. General

Telecommunication by means of satellites is by now a well-known application of space technology. Of the many orbits available for communication satellites, the 24-hr geostationary (equatorial, circular-synchronous) satellite offers a unique Earth-relative geometry and kinematics, thereby providing both economic and operational advantages for global communications. As shown in Figure 1.1, three geostationary satellites can give almost complete Earth coverage and hence, continuous world wide communication service. Of course, redundant units would be needed for a complete system in order to allow for communication outages.

Since geostationary satellites have nearly zero Earth-relative motion, the time-in-view for mutual visibility of any two ground stations is unlimited. That, in turn, results in multiple accessibility and inherent availability for a satellite communication system. The stationary feature also gives the following advantages to the operation and maintenance of the system:

- Only one ground antenna per terminal is required.
- The antenna steering system is simpler and less expensive, particularly for very large antennae.
- Frequency change-over from one relay to another as required by low altitude relays is avoided.
- Lower receiver system noise temperature and less variation of atmospheric alterations can be achieved, since the antennae do not have to be operated at low angles above the horizon.
- Doppler frequency shift is practically negligible, thus permitting simpler receivers and more efficient use of the frequency spectrum.
- Ease of acquisition and band width availability allow the satellite to be used for transmissions of time signals, navigation, weather, air traffic control, and many other types of data.
- Less interference is expected among various services such as teletype-continuous wave, telephone, television, broadcast, etc. because of the essentially fixed position of the satellite in space.

The applications of space technology to telecommunications were initiated during 1963-1964 by the SYNCOM satellite series of SYNCOM's I, II, and III. These satellites were put into the geostationary orbit for

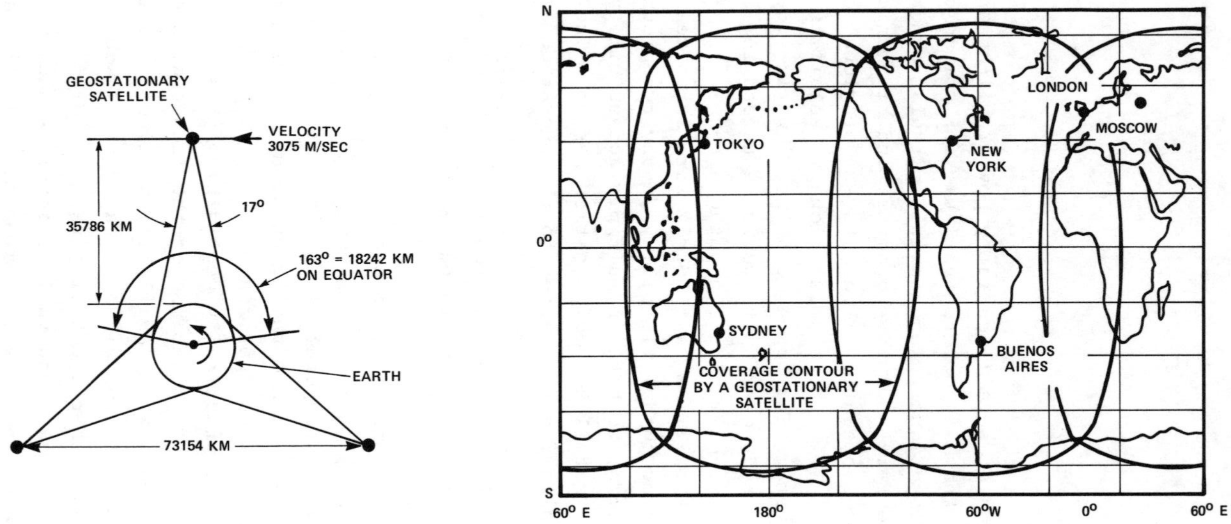

Fig. 1.1. Geometry and global coverage of geostationary satellites.

proving the feasibility of applying the spin-stabilization technique to active communications satellites. The success of SYNCOM led to the development of the INTELSAT (the International Telecommunications Satellite Organization) series. Spin-stabilization has been utilized in most of the subsequent communications satellites. However, the three-axis stabilization technique seems to be adopted for the next generation of communications satellites. Larger high-gain antenna and heavier spacecrafts due to the requirements of increased satellite radiated power and onboard processing will be possible as the space shuttle program is carried out in the 1990's. Consequently, the terminals on the Earth will be simpler and more economical to operate as the burden of processing is placed on the spacecraft. This is particularly important for Earth stations located in remote areas.

In addition to its early programs in experimental communication satellites such as Echo, Telstar, Relay, and Syncom, the Application Technology Satellite (ATS) program of the National Aeronautic and Space Administration (NASA) of the U.S. has led to the development of experimental communication satellites for the next generation. The ATS-F and ATS-G satellites launched in early 1973 and 1975, respectively, have accomplished their objectives in developing advanced technology and in performing scientific experiments, and their results have been applied to the fields of communications, navigation, and meteorology. The Lincoln Experimental Satellite (LES) series, developed under the US Air Force sponsorship, has also made a significant contribution to the communication satellite state of the art.

NASA is currently developing the Tracking and Data Relay Satellite System (TDRSS). When two TDRSS satellites are in geostationary orbit about 180° apart, NASA will have almost continuous coverage of its low orbiters, such as the shuttle and Landsat D, and spacecraft moving to transfer orbit. Consequently, the TDRSS satellite reduces NASA's dependence on overseas ground stations. The first Tracking and Data Relay Satellite (TDRS-1 or East) was launched from the space shuttle in April 1983. A malfunction of the second burn of the Inertial Upper Stage (IUS) booster put the spacecraft into an elliptical orbit with a perigee 8500 mi. short of synchronous altitude. However this satellite was later transferred to a circular geostationary orbit using its tiny altitude control thrusters.

NASA plans to launch Advanced Communications Technology Satellites (ACTS) in 1988 for communications experiments with:

- a new multibeam antenna on the satellite for communicating with small area on Earth of about 200 km in diameter;
- an on-board computer that receives signals from ground stations, sorts, groups, and stores them temporarily in its memory banks according to their destinations, and schedules their transmission to the appropriate ground terminals at proper time; and
- an on-board switch that will route high data rate message traffic over fixed spot beams to major communication centers.

Between 1963 and 1980, there were plans of launching a total of 92 24-hr geostationary satellites, the sponsors and the number of satellites (shown in the parentheses after the sponsor's name) are:

Canada (5), Europe (6), France/W.Germany (2), Indonesia (2), Italy (1), Japan (5), NATO (3), United Kingdom (2), U.S.A. (50), U.S.S.R. (15) and W.Germany (1).

Existing and planned programs for the applications of geostationary satellites are described in the following sections. In view of the dynamic world situation as well as the rapid advances of technology, additions and deletions of telecommunications satellite programs seems to be a 'fact of life'. The presentation of the following sections should be viewed as a general 'overview' rather than a 'log' of the satellite communications programs. With this in mind, readers will find themselves easily satisfied.

1.2. Global and Regional Communications Satellite Systems

1.2.1. Global Communications Satellite System

The International Telecommunications Satellite Organization (INTELSAT), having a current membership of 112 signatories (as of January 1987), was created in August 1964 as an international joint venture, with membership open to all members of the International Telecommunications Union (ITU). The INTELSAT organization owns the satellites and associated ground control equipment in a global system. The Communications Satellite Corporation (COMSAT) is the signatory for the United States.

Since INTELSAT inaugurated the commercial use of space technology and has been a major carrier of intercontinental telecommunications traffic today, a description of the development of the INTELSAT satellite series seems to be in order.

INTELSAT I (Early Bird), the world's first commercial communications satellite, was placed in service in June, 1965. It established the first satellite pathway between the United States and Europe and for the first time made live uninterrupted transoceanic TV programming possible. The satellite successfully provided answers to questions concerning the application of geostationary satellites to commercial telephone traffic. However, it provided service only to the northern hemisphere of the Earth and lacked battery storage power for operation during an eclipse.

The INTELSAT II satellites, placed over the Atlantic and Pacific Oceans during 1967, introduced a multi-point communications capability and extended satellite coverage to more than two-thirds of the world. They were placed into orbit following a specific request by NASA to provide communications in connection with the Apollo manned space flight program and also to provide additional capacity for commercial traffic. Unlike their predecessor, they could be operated during an eclipse.

The INTELSAT III satellites, placed in service over the Atlantic, Pacific, and Indian Oceans in 1969, established the global system. They also introduced the simultaneous transmission of all forms of communications - telephone, television, telex, data, and facsimile.

The INTELSAT IV satellites, placed in service between 1971 and 1975, further expanded the global system and introduced the spot beam concept: a concentration of satellite communications capabilities on small areas, particularly useful for colorful events in today's dynamic world.

The first INTELSAT IVA satellite was launched in September, 1975. A total of six were planned to replace the seven existing INTELSAT IV satellites, and five were successfully put into orbit. These satellites further extend the use of spot beam technology and apply it to frequency reuse by spatial separation of its east and west antenna beams to conserve the limited bands allocated for satellite and terrestrial communications. Today, a completely new generation of satellites called the INTELSAT V series handles requirements for communication satellites service into the mid-1980's. They employ advanced techniques to allow multiple use of the same frequences and operate in the 11/14-GHz bands, as well as in the 4/6-GHz bands. The first one was launched in December 1980, and there are eleven INTELSAT V's in orbit as of January 1987. Meanwhile, INTELSAT VI is already in the developing stage; each satellite will carry 120,000 circuits and three television channels over the C- and K-band portions of the frequency spectrum. The first launch date is scheduled for early 1989. The INTELSAT system is now relaying messages over the Atlantic, Pacific, and Indian Oceans to more than 165 countries, territories, and dependencies through its global network of 15 satellites.

1.2.2. The Intersputnik Organization

The Intersputnik Organization was proposed by the U.S.S.R. in 1968 as a counterpart to the INTELSAT. Eight countries (including Cuba) joined. Operation began in February 1974 with nine ground stations at the end of 1975. The introduction of the U.S.S.R.'s Statsionar series in the geostationary orbit in 1975 formed a global network that certainly competes with the INTELSAT. The Statsionar series is now known as Gorizont. The Intersputnik system is presently using Gorizont to serve 17 countries (U.S.S.R., Bulgaria, Hungary, East Germany, Cuba, Poland, Romania, Afghanistan, Vietnam, Iraq, Yemen, Mongolia, Czechoslovakia, Algeria, Laos, Nicaraqua, and Libya). During the period of 1983-1986, Gorizonts 7, 8, 9, 11, and 12 were put in the geostationary orbit. Gorizont 12 was launched on June 10, 1986 and stationed at 14° W longitude.

1.2.3. Regional Communications Satellite System

1.2.3.1. European Communications Satellite Program

The European communications satellite program had its first satellite

(OTS 1 - Orbital Test Satellite) launched in June 1977 and carried an 11/14-GHz payload to satisfy the needs of an experimental system for public communications and television exchanges in Europe. The second satellite (OTS 2 or MAROTS - maritime OTS) was launched at the end of 1977 and carried an L-band payload for maritime purpose. It was expected that the derivatives of the OTS would carry out the other missions such as TV broadcast and small terminal systems (data transmissions and conference links, mobil services, etc.).

In June 1983, the first of two European Communications Satellites (ECS) was placed in orbit. A new organization, Eutelsat (European Telecommunications Satellite Organization), linked 20 member nations from the European Confederation of Postal, Telephone, and Telegraph together in the first regional European satellite system. The second ECS satellite was put in orbit in August 1984. The demand for ECS transponder leases has already exceeded the supply available on the two satellites. The European Space Agency (ESA) plans to build L-Sat (Olympus) in anticipating the increasing demand of communications satellite service in the late 1980's and early 1990's.

1.2.3.2. French and West German Telecommunications Satellite Program

France and West Germany have a joint effort in developing a communications satellite system. In 1967, they signed an agreement for developing and launching two experimental (Symphonie) telecommunications satellites. Launched in 1974 and 1975, these satellites demonstrated that Europe has the ability to design and build a modern spacecraft. Symphonie has an impressive record of testing, including service to the People's Republic of China.

In 1979, they again undertook a cooperative program for developing, fabricating, and launching two DBS (Direct Broadcasting Satellite) satellites for television broadcasting over French and West German territories. These two satellites, TV-Sat for West Germany and TDF-1 for France, are to be collocated in geostationary orbit.

1.2.3.3. Arab Satellite Communications Organization

The Arab Satellite Communications Organization owns the Arabsat, which provides enhanced radio, telephone, data, and television links across Africa and the Middle East. The organization is a consortium that is headquartered in Riyadh, Saudi Arabia. The consortium consists of 22 members: Algeria, Bahrain, Djibouti, Egypt, Iraq, Jordan, Kuwait, Lebanon, Libya, Mauritania, Morocco, Oman, Qatar, Saudi Arabia, Somalia, Sudan, Syria, Tunisia, the United Arab Emirates, Yemen, South Yemen, and the PLO (Palestine Liberation Organization). Three Arabsats were constructed by Aerospatiale, a French company. The first Arabsat was launched in February 1985 by an European Ariane Rocket, but its stabilizing gyroscope started to fail soon after the launching. The second one (Arabsat 1-B) was put in orbit by space shuttle Discovery on June 18, 1985 and the third is a spare that may be launched later.

1.2.3.4. Nordic Telecommunications Satellite System
A Swedish-Norwegian-Finnish telecommunications satellite, known as Tele-x, will carry out telecommunications in the Nordic region with shared costs of 82% by Sweden, 15% by Norway and 3% by Finland. The satellite has several missions, viz.:

- direct broadcasting of TV and sound (DBS);
- data and video services for business communications;
- permanent and temporary link services for TV and sound program reduction;
- closed-circuit TV, tele-education etc.

Launch into a geostationary orbit was expected by the middle of 1987.

1.3. Domestic Communications Satellite System

1.3.1. U.S. Domestic System

There are many potential domestic satellite system operators, all of them are required to obtain authorization from the Federal Communications Commission (FCC) to proceed with their plans.

1.3.1.1. Comsat General/American Telephone & Telegraph Company (AT & T)
During 1976 and 1978, Comsat General placed three high-capacity Comstar (D1, D2, D3) satellites in the geostationary orbit and they have been capable of providing reliable communications to the 48 contiguous states, Alaska, Hawaii, and Puerto Rico. AT & T and General Telephone & Electronics (GT & E) have leased the entire capacity of the three in-orbit Comstars, and have built their own Earth station network and integrated satellite communications with their vast terrestial facilities to bring an added dimension and flexibility to their domestic U.S. service. In February 1981, Comstar D4 was launched to share the role of D1, which was then collocated with D2. As a result, the number of active transponders on the D1/D2 composite was reduced. This reduction will certainly extend Comstar's life.

In July 1983, AT & T launched the first in its new series of three wholly owned and maintained Telstar-3 satellites. The second and third were launched in September 1984 and January 1985, respectively. It was anticipated that the new satellites would occupy the geosynchronous orbital positions at 87° W, 95° W, and 128° W longitudes as these positions were taken by the then existing AT & T leased satellites. New features in the Telstar-3 satellites included a spare transponder for each group of four working transponders, solid state power amplifiers in 12 of the 24 working transponders and in all six of the spare transponders, combined continental-U.S. off-shore coverage beams to Alaska, Hawaii or Puerto Rico, and sufficient battery capacity and station-keeping fuel to ensure a 10-yr in-orbit life.

With the U.S. Postal Service, Comsat (the parent company of Comsat General) initiated the Intelpost experiment to send information internationally in facsimile form via satellite in 1979. A standard business letter can now be sent in six seconds from Intelpost Service Centers in New York and Washington D.C. to Argentina, Belgium, West Germany, France, Iran, The Netherlands, and the United Kingdom. This 'electronic mail' has its counterpart in the existing 'Mailgram' system used by the Western Union.

1.3.1.2. Western Union Telegraph Company

Western Union has five Westar (I - V) satellites in geostationary orbit, with their primary coverage of the contiguous 48 states. They also serve Hawaii and Alaska at reduced levels. The ground stations for the Westars directly augment the long-distance Western Union terrestrial transmission system. In 1979, the then five-year-old Westars I and II were approaching a full traffic load, and Westar III was put in orbit. With the advanced designs in both solar power and antenna systems to improve service and area coverage, Westars IV and V were developed to have 24-channels C-band taking advantage of cross-polarization isolation and frequency interleaving to reuse the available spectrum. Both Westars IV and V were put into orbit in 1982. Westar VI failed in attaining its orbit in February 1984.

1.3.1.3. American Satellite Company (ASC)

The American Satellite Company has leased three transponders from the Western Union system for service to New York, Los Angeles, and Dallas since July 1974. Service to San Francisco and Chicago was instituted by leased terrestrial facilities initially and later upgraded to full satellite service with ASC-owned and -operated ground stations. The ASC will own three dual-band commercial communications satellites. Each satellite has 18 transponders at C-band and six transponders at Ku-band for coverage of the 50 states plus Puerto Rico. This hybrid satellite features on-board command encryption, a first for domestic commercial spacecraft. The first one was launched on August 27, 1985 from the shuttle Discovery.

1.3.1.4. General Telephone & Electronics (GT & E)

The General Telephone & Electronics, through its subsidiary GTE Satellite Corporation (GSAT), plans to have a domestic satellite system providing a trunking capability between areas it serves. It will construct five ground stations in Hawaii, California, Pennsylvania, Indiana, and Florida. The four-satellite 'GSTAR' communication system of GSAT consists of four Ku-band satellites, each utilizing 16 channels operating at the 14/12-GHz band. The satellites provide coverage of the 48 contiguous states by shaped beam, and Alaska and Hawaii are covered by spot beam. The first launch was carried out in May 1985, the second in March 1986. With the acquisition of the Southern Pacific Communications Company in 1983, GT & E may become the largest U.S. communications satellite operator.

1.3.1.5. RCA Global Communications

In January 1974, RCA Global Communications, Inc. and RCA Alaska Communications, Inc. began domestic satellite communications service (RCA Satcom). Initially, Satcom was using transponders in the Canadian Telesat system to link four ground stations, one each in Pennsylvania and California and two in Alaska. The second phase involved four RCA satellites (Satcom), each containing 24 transponders and a network of stations to serve 50 states. Satcom I was launched in December 1975, and Satcom II in March 1976. Satcom III-R was placed in orbit in November 1981 and Satcom IV in January 1982. All four Satcom's are operational, handling television, voice, and high-speed data transmission.

The second generation of the Satcom, Advanced RCA Satcom, was put into operation in 1982. These satellites provided up to 50% in voice/data capacity. RCA Satcom V was launched on October 27, 1982 and has been dedicated to Alaskan service. RCA Satcom I-R was placed into orbit on April 11, 1983, while RCA Satcom VII was launched in September 1983, Satcom K2 in November 1985, and Satcom K1 in January 1986.

1.3.1.6. Satellite Business System (SBS)

This system was formed in 1975 by subsidiaries of COMSAT General, Aetna Life & Casualty, and IBM Corporation to provide an unique, all-digital domestic service, principally to business and government agencies with large communications requirements. Four satellites were planned for use in the SBS system; the first one, SBS-1, was launched in November 1980, the second in September 1981, the third launched by the space shuttle in the fall of 1982, and the fourth one launched from the shuttle Discovery in August 1984.

The SBS system involved a combination of fixed and moving spot beams to serve both high- and low-traffic areas. The satellite linked the high-traffic areas with fixed spot beams. For a corporation with widely separated facilities scattered over a large area, SBS would link television, facsimile and computer equipments installed at different localities for its daily operations.

1.3.1.7. Satellite Television Corporation (STC)

The Satellite Television Corporation will develop a direct broadcast satellite system to handle direct-to-home television broadcasting, which will cover the entire nation with four operating and two spare in-orbit satellites. The first two launches are scheduled for 1986. Each operating satellite will serve an area corresponding to a mainland time zone. The satellite located in the Pacific time zone will also serve Hawaii and Alaska.

1.3.1.8. Southern Pacific Communications Company

The Southern Pacific Communications Company has a four hybrid C/Ku-band satellite 'Spacenet' communications system. The satellites provide coverage of the 48 contiguous states and either Alaska and Hawaii or Puerto Rico and the Virgin Islands. The first two (Spacenet 1 and 2) were launched in May and November 1984. The satellite system is capable of relaying a wide variety of communications signals.

1.3.1.9. Hughes Communications
Hughes Communications has three Galaxy satellites (a commercial domestic system) and sold 18 of the 24 transponders on the first satellite, Galaxy I, much as a condominium sells apartments. The company operates and maintains the satellites for their nine-year lives. Galaxy I, launched in 1983, distributes video signals in the continental U.S., Alaska, and Hawaii. The second, launched in September 1983, and the third in September 1984 relay video, voice, data, and facsimile communications for the business community.

1.3.2. U.S.S.R. Communications Satellite System

The U.S.S.R. domestic communications program is known as the Molniya System. This system was placed into service in 1967. In 1965, a series of Molniya 1 satellites were placed in highly elliptical orbits, at an inclination of about 64° to the equatorial plane with apogee of 40,000 km in the northern hemisphere and perigee on the order of 500 km in the southern hemisphere. By proper spacing of four of these satellites in 12-hr-series-synchronous orbits in orbital planes 90° apart, each satellite is in view of major Soviet ground terminals for about eight hours a day, including those at near polar latitudes, hard to reach with an equatorial geostationary satellite. These satellites are heavier than the U.S. communications satellites and have 10 times the power output of the Early Bird's. Their higher power has permitted the establishment of relatively low-cost terminals to distribute TV programs from Moscow to outlying points, as well as to carry telephone and computer data traffic. The Molniya system not only provided the Russian domestic communications service but also the connections to certain overseas points. One such overseas connection was between Moscow and Washington to establish direct communications between the chiefs of states. The counterpart of the U.S. side is provided through the INTELSAT system.

Currently the Molniya 1's play a secondary role and new Molniya 2 and Molniya 3 satellites operating at higher frequencies provide the prime service. They are arranged in the same orbital planes as their earlier counterpart.

In 1974, for the first time, Russia put Cosmo 637 and then Molniya 1S-1 into the 24-hr synchronous equatorial orbit. The Statsionar plan of 24-hr synchronous satellites was announced in 1975, and a total of ten or eleven satellites was intended to be put into orbit before 1980. Roduga, the first Statsionar, was launched in December 1975, and thereafter practically every year a new Roduga was put into orbit. Roduga 18 was in orbit on

January 17, 1986 stationed at 15° W longitude.

In spite of the three Molniya systems and the Roduga group to encompass the four TV channels carrying programs originating in Moscow, one-fifth of the population isolated in the vast territories of central Siberia still does not have television. On October 26, 1976, Statsionar-T (T for television) was put into geostationary orbit at 99° E longitude, and this satellite (now named Ekran) carried a system for broadcasting television directly to local antennae. Ekran is now directly transmitting TV programs to those previously isolated regions. The latest Ekran (Ekran 15) was launched on May 24, 1986.

1.3.3. Canadian Telesat System

Canada has established its own satellite (Telesat) system for domestic communications. Two ANIK-A satellites, launched in November 1972 and April 1973, respectively, were stationed at 114° W and 109° W longitudes. The third ANIK-A satellite was launched in May 1975 stationed at 104° W longitude. These satellites illuminated most of Canada and served the Canadian Broadcasting Corporation, the Trans-Canada Telephone system, and the Canadian National and Canadian Pacific railways. Of particular importance, the Telesat system provided reliable service to the remote northern regions of the country. Each satellite had 12 transponders and each transponder was used for 480 duplex voice channels or one color television channel.

In January 1976, an experimental communications satellite known as Hermes, was launched. This satellite demonstrated both technology and applications in the 14/12-GHz range.

ANIK-B, an advanced dual-band, domestic communications satellite, was launched in December 1978. It provided high capacity TV, telephone, and data transmission including high power coverage for improved quality service. On ANIK-B, on-board antennae provided coverage in the 6/4-GHz band shaped to match the outline of Canada as seen from the geostationary orbit and 14/12-GHz band spot beams to serve four major regions in Canada. The first ANIK-D was put into orbit in August 1982, and the second was launched in November 1984; these two satellites replaced the three ANIK-A satellites. ANIK-C3, Canada's first dedicated communications satellite at 14/12-GHz in the Ku-band and having tripled the power output of ANIK-A, was injected into orbit in November 1982 and ANIK-C2 was launched from the space shuttle in June 1983.

1.3.4. French Telecommunications Satellite System

In addition to the joint venture with West Germany in the development of communications satellites, France has committed to the development of the Telecom communications satellite, which is basically a business satellite and has transponders at C-band, X-band, and Ku-band. The X-band is designed to interconnect the French navy. Telecom-A and -B were launched in August 1984 and May 1985, respectively. A later version, Telecom-2, in the 1990 time frame will use 30/20-GHz transponders to handle the growing need. Another proposed project, Athos satellite, is intended to use 6/4-GHz and 30/20-GHz transponders. The former will be connected to an antenna system radiating four beams, three of them illuminating France, central Africa, and northern Africa, while the latter will illuminate only France.

1.3.5. West German Domestic Satellite System

West Germany began development in 1983 of a national domsat program. This domsat, Deutsche Fernmedie Satellite (DFS), also called Copernicus, has transponders at C-band, Ku-band, and Ka-band, and will provide broadband services such as video and high-speed data transmissions for the West German terrestrial networks.

1.3.6. Italian Communications Satellite System

In 1977, Italy launched the high frequency Sirio satellite to the geostationary orbit at 150° W longitude. Five years later, another satellite Sirio-Z was put in orbit. Tests with these satellites for television distribution and on the propagation characteristics at both 12- and 18-GHz continued through 1983, and now include links between Italy and the People's Republic of China.

In 1979, Italy decided to build Italsat, a 30/20-GHz experimental/preoperational satellite designed for flexible communication within the Italian territory. It will consist of a single-beam system covering all of Italy and a spot-beam system using six spot beams to cover Italy with a regenerative repeater driving each spot. Italsat was scheduled to be launched in 1987.

1.3.7. The United Kingdom Communications Satellite System

In addition to the Skynet series in her military communications satellites, the United Kingdom participated in the Telstar experiment in 1963 and has played a major role in the developments of ECS, OTS, and L-Sat for the European Space Agency and the Eutelsat.

Recently, the U.K. decided to build Unisat as a combined direct-broadcasting and business service satellite. Three Unisat satellites have been procured, and the first one was scheduled for launching in 1986.

1.3.8. Japanese Domestic System

Late in 1977, Japan launched her first communications satellite, Sakura, to provide communication service in the 30/20-GHz range. In March 1978, the Japanese Broadcast Satellite Experiment (BSE) was placed in the geostationary orbit at 110° E longitude. It was intended to be used for three years to conduct direct-broadcast experiments in Japan and its outlying islands. In 1983, two operational satellites using 30/20-GHz (Ka)-band, CS-2A and CS-2B, were successfully orbited from the Tanigashima launching base. A series of CS-3's is in the process of being built and is intended to be launched in 1987 and 1988. Nippon Telephone & Telegraph is considering designs of INSS-1 and INSS-2 satellites for the 1990's. INSS-2 would have 30 spot beams and weigh 2½ times more than INTELSAT V.

1.3.9. Indian National Satellite System

Insat, the domestic communications satellite of India, is a multipurpose satellite system for telecommunications, direct television broadcasting, and meteorological data gathering. In 1982, Insat-1A was damaged during deployment causing depletion of its altitude control fuel supply and making the satellite inoperable. Insat-1B was injected into its orbital slot in September 1983 following the shuttle's first night launch. India's Department of Space controls the communications, DBS-TV, and meteorological payloads from its mission control center in Hassen after three months of operational testing. Two additional satellites, Insat-1C and Insat-1D are expected to be launched in January 1988 and October 1990, respectively.

1.3.10. Australian National Satellite Communications System (AUSSAT)

Aussat provides a wide range of domestic services to the entire continent, its offshore islands, and Papua New Guinea. This includes direct television broadcast to homes and remote communities, high quality television relays between major cities, digital data transmission for both telecommunications and business use, voice applications for urban and remote areas, centralized air traffic control services, and maritime radio coverage. The first two Australian satellites, Aussat 1 and Aussat 2 were launched in August and November 1985, respectively. The third will be launched at a future unspecified date.

1.3.11. People's Republic of China Communications Satellite System

In April 1984, China launched her first experimental communications satellite into the geostationary orbit. Two years later, in February 1986, a second communications satellite was put into geostationary orbit stationed at 103° E longitude.

In 1974, China imported INTELSAT standard-A ground stations and established international telecommunication links above both the Pacific and Indian Oceans. New types of communications ground stations with 15- and 10-m-diameter antennae were developed and used in joint experiments with Germany and France via Symphonie from May 1976 to July 1979. These experiments included news and format facsimile, sound and video transmissions, digital and analog telephone transmissions, etc. Also a co-operation project for microwave transmission experiments between China and Italy was carried out from June 1983 to October 1984 using Italian Sirio-1 satellite.

At present, China has signed an agreement with INTELSAT to lease relay capacity corresponding to 1/4 of a transponder and will install more than 30 ground stations to link coastal cities with remote regions and the mainland with off-shore islands.

1.3.12. Indonesian Domestic System

The Indonesian domestic satellite system is known as PALAPA, having two Palapa-A satellites with 50 ground stations. These satellites were launched in July 1976 and March 1977, respectively, and both are in the synchronous orbit. The first one is located at 83° E longitude over the Indian Ocean, and the second one is at 77° E longitude. The system carries out normal telecommunications, national and educational television, and communications for national security. It may be possibly extended to all countries of the Association of South-East Asian Nations for their telecommunications needs. The establishment of the system enables the remote islands and the remote areas of the large island of the country to have direct access to the outside world, and the consequence of this will be an acceleration of the development of her natural resources such as oil, mineral deposits, etc.

Palapa-A has been replaced by Palapa-B, which is twice as large as, and has twice the capacity and about four times the power of its predecessor. Palapa-B1 was injected into orbit by the space shuttle in June 1983, while Palapa-B2 failed in achieving its orbit in February 1984.

1.3.13. Algerian Domestic System

The Algerian domestic satellite system is designed to operate with any of the INTELSAT satellites over the Atlantic and Indian Oceans. Its operation is based on the principle to utilize the spare capacity of the INTELSAT system on a preemptible basis in exchange for a reduced rate of the transponder leasing service offered by the INTELSAT. The system has 14 Earth stations and each station has an 11-m antenna.

1.3.14. The Philippines' Domestic System

With a possible later implementation of a regional dedicated satellite system, the Philippines is now leasing the space segment of communications satellites from the INTELSAT. In January 1976, the system had five Earth stations with an ultimate goal to reach fifteen stations. The system serves both rebroadcastable color television for educational and commercial purposes and also for ordinary telecommunications.

1.3.15. Other Countries

The telecommunications satellites are especially beneficial to countries having broad territorial expanses of water or land over which a land-based system of telecommunications lines would be extremely expensive to maintain. Mexico has Morelos 1 and 2 (launched in June and November 1985) providing wire, video data, and facsimile communications. Brazil has her national communications satellites, Brazilsat 1 and 2 (launched in February 1985 and March 1986). Pakistan will have her own satellite - Pakisat. Other countries such as Malaysia, Nigeria, Spain, Zaire, etc. either have leased or plan to use the INTELSAT transponder capacity for their educational and domestic communications.

1.4. Military Communications Satellite System

Since military communications satellite systems are related to the interest of national security, the descriptions in the sections followed are limited to the available information in open literature.

1.4.1. U.S. Defense Satellite Communications System (DSCS)

The U.S. Department of Defense (DOD) has directed communications satellite programs for over a decade. The Defense Satellite Communications System Phase I (DSCS I) consisted of 26 simple spin-stabilized satellites launched between June 1966 and June 1968 into quasi-synchronous equatorial orbits (18,300 n. mi.) chosen to produce a drift rate of the satellite over the equator of approximately 30° per day. Of the 26 placed into orbit, seven still performed satisfactorily (as of September 1975). The DSCS Phase II (DSCS II) satellite is a spin-stabilized design with a capability for station-keeping in the geostationary orbit. The first two were launched in November 1971, and two more in December 1973. The former two had failure in their despin control systems in 1972 and 1973, and the latter two were still operational (as of September 1975). In May 1975, two more satellites were intended to be put into orbit; however, these satellites were lost due to failure of the launch vehicle.

The DOD has also been developing satellite communications for mobile and tactical users, primary through systems using the 225-400-MHz band. The Tactical Communications Satellite (TACSAT) was launched into the geostationary orbit in 1969. Its design control system failed in 1972. Further development was followed by the Lincoln Experimental Satellite (LES) series and the FLEETSAT/AFSATCOM. The latter provides service to ships at sea, submarines, and the Air Force (in particular, the Strategic Air Command). The U.S. Navy has launched five FLEETSAT communications satellites since 1978. These satellites employ one SHF and 12 UHF transmitters and contain a total over 5,000 integrated circuits and over 50,000 discrete electronics parts with full redundancy for reliability. The DOD relies on FLEETSAT for its tactical satellite communications needs until Leasat becomes available.

Leasat, scheduled for launch in this decade, has an internal propulsion system that takes it to the geosynchronous orbit after it is spin-stabilized and ejected from the shuttle bay. Each of the four operational Leasat (and a fifth ground spare) will have an end-of-life power capacity of 1,200 W. Leasat provides world-wide communications satellite services to the DOD. Users include mobil air, surface, subsurface, and fixed Earth stations of the Navy, Marine Corps, Air Force, and Army.

The first DSCS III satellite was put into operation in 1982. The key features of the satellite are its radiation hardness and multibeam antennae, which give selective coverage and broadband nulling on receive. The X-band system on this satellite will be the first with commandable-contour antennae, using arrays with RF lenses to form transmit and receive beams in the desired direction. The latest launch of DSCS III satellite was in October 1985. The next generation of military communications satellite will be the Milstar, which will provide highly survivable narrowband communications to Army, Navy, and Air Force strategic and tactical forces.

1.4.2. United Kingdom Defense Satellite Communications System

The U.K. entered into a cooperative effort with the U.S. during DSCS I for its military communications satellite system. Two SKYNET I satellites were launched into the geostationary orbit in November 1969 and August 1970. The first one had a failure in its communications system in late 1970, while the second was lost during burn of the apogee motor. Again, the first launch of SKYNET II in January 1974 resulted in loss of the satellite due to a malfunction of the booster. The second SKYNET II launched in November 1974 was successfully orbitting in the geostationary orbit, and SKYNET IV was intended to be launched in 1985. In addition to SKYNET satellites, the U.K. defense satellite communications system includes large fixed ground stations and shipborne and land mobil terminals.

1.4.3. NATO Satellite Communications Program

With the assistance from the U.S., NATO launched two satellites (NATO II) into the geostationary orbit in March 1970 and February 1971 (the U.S. DSCS I satellites were used for NATO I). The operation system includes 12 ground stations near the capitals of the 12 participating countries. It is expected to have 10 more ground stations and two large mobil terminals. The operational command of NATO directs the NATO fleets through this system. NATO III satellites were launched into the geostationary synchronous orbit in 1977 and 1978. These satellites have one transponder connected to a wide beam antenna to provide coverage of the entire NATO area and another transponder connected to a narrow-beam antenna to provide coverage throughout Europe. The latest launch of NATO III (III D) was in November 1984 and NATO III E was scheduled to be launched in 1986.

1.5. Special Service Satellite Programs

There are many satellites which are put into the geostationary orbit to serve a particular single purpose. They are described in the following sections.

1.5.1. Synchronous Weather Satellites

Synchronous weather satellites provide our daily TV weather picture and serve as a basic source for weather forecasting and storm warning. The first experimental synchronous weather satellite was launched in late 1966 to take photographs of the weather patterns around the Earth. A successor, called ATS 3 (Application Technological Satellite) launched in November 1967, provided daily weather photographs for television weather forecasters until January 1975. In May 1974, the SMS 1 (Synchronous Meteorological Satellite) went aloft to provide time-lapse film strips to television stations and to show the movement of weather systems such as tropical hurricanes, etc. After another experimental satellite was placed in the geostationary orbit in February 1975, the GOES 1 (Geostationary Operational Environmental Satellite) was launched in October 1975 to replace the SMS 1. Three additional GOES satellites (B, C and 2) were launched in 1977 and 1978 as part of a global-girdling international weather satellite network. This network was completed late in the nineteen-seventies by a Japanese satellite (GMS-1) placed over the Western Pacific, an European satellite (METEORSAT) over the eastern Atlantic, and a Soviet satellite (GOMS) over the Indian Oceans. As a further strengthening of the projected world weather watch program, GOES-D, -E, and -F were added in 1980, 1981, and 1983, respectively, together with the Japanese GMS-2 and the European Space Agency's METEORSAT 2 in 1981. These satellites are utilized by meteorologists to have the most detailed study of the Earth atmosphere. The SMS 1 and GOES satellites are operated by the National Oceanic and Atmospheric Administration (NOAA). In August 1984, GMS-3 was launched into orbit to replace the GMS-2 because of motor failure.

The National Weather Service of the NOAA is currently working on an experimental project that monitors GOES's photos every five minutes for tornado tracking, since the occurrence of tornados can frequently be predicated by noting the intersection of their convection lines on satellite photos.

In 1985, NASA announced the building of the next generation GOES series (GOES I, J, K, L, and M); the first launch is expected in late 1989. The new satellites will advance the capability for measuring the Earth's atmosphere, its surface, cloud cover, and electromagnetic environment.

Particular emphasis will be placed on improving short-term forecasts over local areas.

Data from weather satellite have been used for such diverse purposes as helping free a Coast Guard ice breaker, the USS Glacier, from pack ice in the Antarctic in 1975 and in aiding the eradiction of screwworms along the Mexican border. In the latter case, satellite temperature and moisture data determine the best time for releasing sterile insects that breed with fertile screwworms. Pollution in the New York bight also is being monitored by GOES.

1.5.2. Maritime Satellite System (MARISAT) and INMARSAT

MARISAT provides maritime communications services via satellites over the oceans to the U.S. Navy, commercial shipping interests, and the offshore industries. It comprises three multi-frequency satellites, positioned over the Atlantic, Pacific, and Indian Oceans. Each satellite operates in three frequency bands: UHF for U.S. Navy service, L-band between commercial vessels and satellite, and C-band between shore stations and satellite. The design lifetime of these satellites is five years. They began their services in March 1976, June 1976 and January 1977, respectively. MARISAT is the predecessor of the global satellite system that was established on July 16, 1979 by the International Maritime Satellite Organization (INMARSAT) of 45 (as of July 1986) country-members based in London; Comsat Corporation was designated as the U.S. participant in the INMARSAT. Marecs 1, launched on December 20, 1981, provided high quality voice and data transmission between ship and shore stations. In addition to telephone and telex calls, and distress alerts, Marecs can also be used for transmitting data on ship engine performance, fuel consumption, position reporting, weather condition, etc. Marecs-B failed to reach the orbit and Marecs-B2 was launched in November 1984. In 1982, the first INMARSAT maritime communications payload by INTELSAT V was put into orbit, and three more INTELSAT V's will carry the maritime communications system. INMARSAT is currently planning for the second-generation Inmarsat 2 satellites for launching in 1988, which will also include the aeronautical communication services.

1.5.3. Satellites for Education

Satellites can be used for educational purposes. The ATS-1 (Applications Technology Satellite) initiated educational experiments in the Pacific region of the U.S.; the ATS-3 did a similar role for Brazil. The ATS-6 over the Indian Ocean began in 1975 for a period of one year to perform the Satellite Instructional Television Experiment (SITE) in India. The satellite transmitted televised instruction in farming and family-planning to thousands of remote villages in India. The experiment is not only a success in education but also raises important questions about the potential impact of televised propaganda on illiterate people in remote areas of the world. It clearly indicates that the seeds have been sown for a radical change in their life styles.

In 1977, Canada started satellite educational experiments by its Communications Technology Satellite (CTS).

On September 3, 1985, the Association for Media-Based Continuing Education for Engineers (AMCEE) in the United States presented its first satellite-relayed programming as it began providing six hours a day of National Technology University noncredit short courses, seminars, and special events. Most transmissions originated from a new TV studio at Georgia Institute of Technology in Atlanta, Georgia and offered one-hour presentations two or three times a week for the duration of a course. Others were three- or four-hour, one-time events.

1.5.4. Direct Broadcasting Satellite (DBS)

With the success of television distribution to cable networks by U.S. commercial satellites in the 4-GHz fixed satellite service (FSS) band, the Direct Broadcasting Satellite (DBS) system has generated a great deal of interest in direct-to-home TV broadcasting by satellite. In 1982, the U.S. FCC approved permits for nine DBS applicants in the 12-GHz band: Satellite Television Corporation (STC, a Comsat's subsidiary), Western Union, Direct Broadcast Corporation, RCA Americom, CBS, Graphic Scanning Corporation, Focus Broadcast Satellite Company, U.S. Satellite Broadcasting Company, and Video Satellite Systems. These companies have each proposed systems of one to six satellites. The first satellite was proposed for launch in 1985-1986 by STC.

The DBS satellites are and also will be developed in other countries such as Australia (AUSSAT), Austria (joint Austria-Switzerland project), Canada (ANIK's B and C), Italy (Italsat), Japan (BS-2), France (TDF-1), Germany (TV-Sat), India (Insat), Luxembourg, Scandinavian countries (Nordsat), Sweden (Tele-x), Switzerland (Tel-Sat), United Kingdom (Satellite Broadcasting-SBC, Satellite Television-STV, Unisat), etc. as the U.S.S.R already has its Ekran group.

Table I lists all communications satellites described previously in this chapter.

1.6. Other Applications

Geostationary satellites need not be limited in their applications to the above programs, they may also be used for other purposes. The following are some proposals for their utilizations.

1.6.1. Satellite Solar Power Station (SSPS)

An equatorial synchronous satellite can be used as a source of power by

Table I. SUMMARY OF GEOSTATIONARY SATELLITES
(as of January 1987)

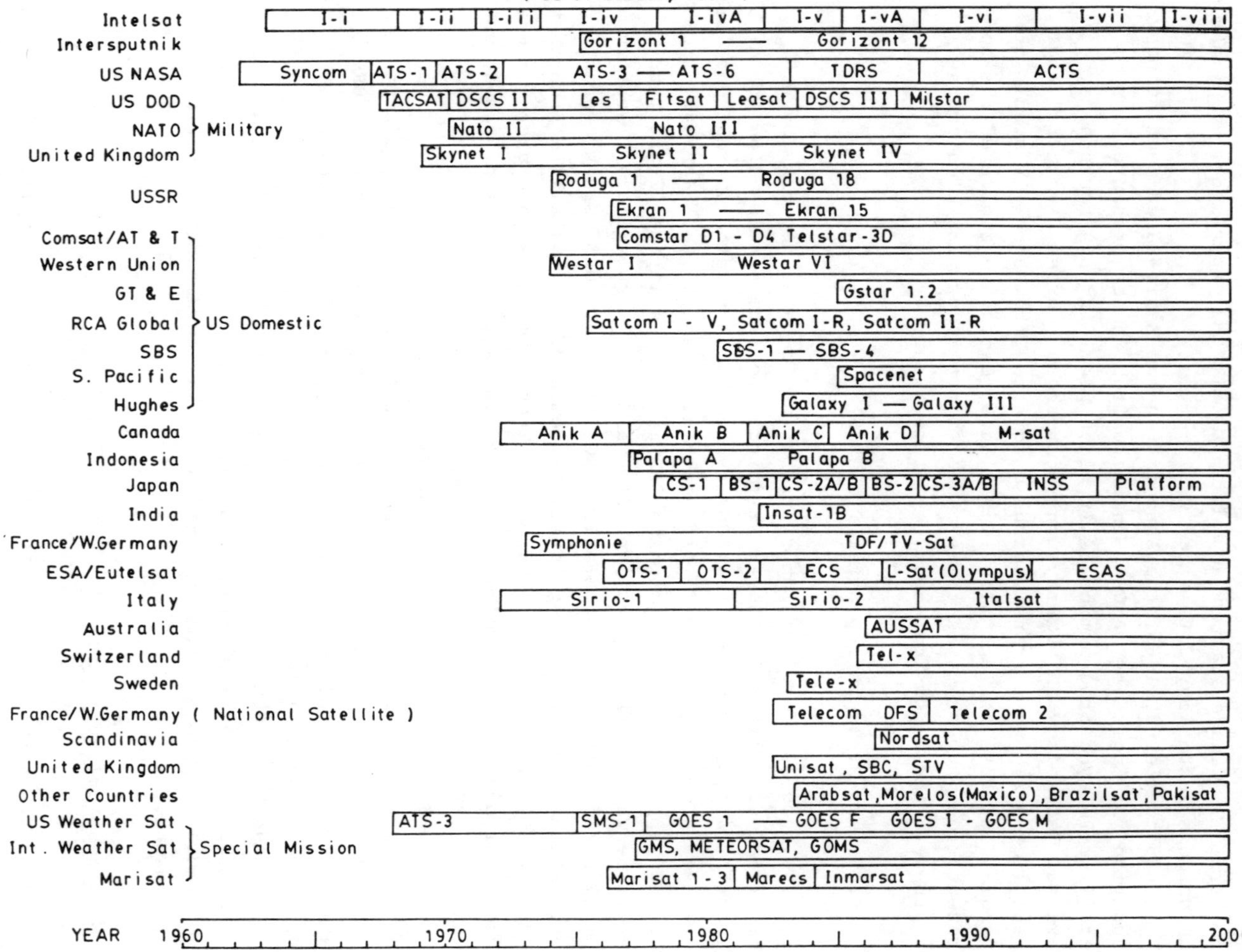

equipping large solar panels to absorb solar energy. This energy is collected by an array of photovoltaic cells, converted into electric energy, and transformed into microwave power. The beam of microwave power is then transmitted to the Earth and reconverted into electrical power by an antenna-rectifier array. To obtain a useful amount of electrical power, the area of photovoltaic cells will probably measure in square kilometers. Again the stationary feature of the satellite facilitates the transmission of microwave power. There is a potential of delivering 2,000 to 15,000 MW of power to the Earth.

1.6.2. Power-Relay Satellite

The power generated from large solar-thermal, ocean-thermal, or nuclear-thermal generating facilities located in remote areas can be transmitted to the populated areas by means of microwave beams to the load center via a geostationary satellite. Electricity generated at a source is converted into microwave energy and transmitted to the geostationary satellite, which reflects the beam to a rectenna at another location on the Earth. The antenna on Earth would be considerably larger than those of the relay satellite, and they are designed so that the beam leaving the transmitting antenna on the Earth would be considerably larger than those of the relay satellite will diverge as it approaches the receiving antenna; thus, the microwave densities at the Earth's surface are not high enough to create an environmental or safety hazard. Since a nuclear power plant can be located in a remote area, more large nuclear plants near reprocessing facilities in remote areas can be constructed and maximum security could be maintained to prevent the theft of fissionable material and the hazards of nuclear power to the general public are reduced.

A geostationary satellite can also be used as a power-relay satellite when the power-generating facility is located on another satellite which collects solar energy. This satellite may be in a low-altitude Earth orbit and Sun-synchronous, so that the satellite is always illuminated. The transmission of power will be by means of laser beams, which are narrower in comparison with microwave beams and need small aperture area for beaming and receiving. As a result, energy can be beamed to a geostationary satellite and back down to wherever it is needed on Earth.

1.6.3. Others

In addition to the above proposed applications, many proposals have been mentioned that a geostationary satellite may be used as a space parking station for further interplanatary exploration, or as a communication park as it is established in the geostationary orbit. The current Earth observation program of NASA includes two geostationary Earth observatory satellites (SEOS) to be launched in the 1980's. They are to be used for viewing short-lived events either when they occur or at the first cloud-free opportunity, or for monitoring continuous time-variable environmental phenomena. Among the more important short-lived events are transient hazardous occurrences such as fires, floods, volcanic eruption, earthquakes and storms, etc., against which precautionary action could be taken, or where emergency aid or public health action might be required. These kinds of tasks are usually beyond the capability of low altitude satellites.

1.7. The Needs for Geostationary Satellite Theory

In practice, equatorial geostationary satellites will not maintain their stationary position because they are subject to various perturbing forces. The existence of longitude-dependent terms (triaxiality) in the Earth's potential function will tend to induce long-term large scale oscillation in longitude. Solar and lunar gravitational attractions will cause the change of orbital plane inclination. Solar radiation pressure acting on satellites, such as the SSPS equipped with large solar panels or on communications satellites, will bring about variations in the orbit eccentricity. As a result, the desired condition of zero Earth-relative motion is disturbed, and orbital correction maneuvers are necessary to minimize the satellite's relative motion with respect to a fixed point on the equator of the Earth.

Existing geostationary communications satellites have employed longitudinal (east-west) station-keeping to reduce the perturbing effect of Earth's triaxiality. The SYNCOM and INTELSAT I satellites were allowed to oscillate through bands of 10 or more degrees in longitude; INTELSAT II satellites have been maintained in longitude bands on the order of one degree; SYNCOM II was placed in longitude of low triaxial acceleration over the Pacific Ocean, thereby reducing longitudinal motion. Although no attempt has been made at latitudinal (north-south) station-keeping, the INTELSAT II orbits were given an initial inclination biases of 1 to 2 degrees to counter the effect of the Sun and Moon attractions, thus reducing the maximum latitudinal variation over a period of several years.

For a communication system requiring a narrow beam, high gain, fixed ground antenna, the satellites of the system would require much more precise station-keeping than has thus far been necessary. This is also true for geostationary satellites that may be used for solar power stations, meteorological observation, orbital tracking station for navigation control, etc. Reduction of the relative satellite motion to the order of milliradian may be required. For example, SBS requires station-keeping to $\pm\ 0.03°$ in both latitude and longitude. This refined requirement needs precise orbital corrections. To determine the proper correction, it is necessary to analyze in detail the perturbed motion of the geostationary satellite. A detailed analysis of the motion will constitute the main purpose of this book.

In the following chapter, we shall discuss the Earth's gravitational field, and the standard methods and the basic methodology of our approach to study the motion of a geostationary satellite in general. In the subsequent three chapters, these descriptions will lay the foundation of analysis for our separate studies of the effects of the Earth's equator ellipticity, the perturbations of the Sun, Moon and solar radiation pressure on a geostationary satellite. The complexity of the problem rules out the possibility of obtaining an integrated, approximate analytical result encompassing all these effects; even different forms of parameters will have to be used in the process of seeking separate solutions. As will be seen later, although the satellite is under the action of various perturbing forces, each perturbing force plays a dominant role only on certain orbital elements. This fact allows the analytical expressions of the separate effects of various perturbing forces to serve our purposes of implementing the short-period station-keeping as well as of studying the long-period orbital behavior, and they are treated in the remaining two chapters.

References

1. Bargellini, P.L.; Communications Satellite Technology, AIAA Progress in Astronautics & Aeronautics **33** (1974).
2. Bargellini, P.L.; Communications Satellite System, AIAA Progress in Astronautics & Aeronautics **32** (1974).
3. deMendona, F.; 'Education through Satellite in Brazil', 27th International Astronautical Congress, IAF-76-137, Anaheim, California (1976).
4. Feldman, N.E. and Kelly, C.M.; Communications Satellite for the 70's: System, AIAA Progress in Astronautics & Aeronautics **26** (1971).
5. Feldman, N.E. and Kelly, C.M.; Communications Satellite for the 70's: Technology, AIAA Progress in Astronautics & Aeronautics **25** (1971).
6. Glaser, P.E.; 'Evolution of the Satellite Solar Power Station (SSPS) Concept', J. of Spacecraft and Rockets **13**, 573 (1976).
7. Hawkridge, D.G.; 'Satellites and Education: Prospects and Problem', J. of the British Interplanetary Society **30**, 147-154 (1977).
8. Hogwood, P.; 'Palapa - Indonesia to the Fore', J. of the British Interplanetary Society **30**, 127-130 (1977).
9. Jarett, D.; Satellite Communications: Future Systems, AIAA Progress in Astronautics & Aeronautics **54** (1977).
10. Jones, W.S. and Hunter, M.W.; 'Solar Power Satellite: The Laser Option', Astronautics & Aeronautics **17**, **59**, and **67** (1979).
11. La Vean, G.E. and Martin, E.J.; 'Communications Satellites: The Second Decade', Astronautics & Aeronautics **12**, 54-61 (1974).
12. La Vean, G.E. and Schmidt, W.G.; Communications Satellite Developments: System, AIAA Progress in Astronautics & Aeronautics **41** (1976).
13. La Vean, G.E. and Schmidt, W.G.; Communications Satellite Developments: Technology, AIAA Progress in Astronautics & Aeronautics **42** (1976).

14. Lagrade, J.B.; 'Setting Up of a Worldwide Maritime Satellite System', J. of the British Interplanetary Society 30, 123-126 (1977).

15. Marsten, R.B.; Communications Satellite System Technology, AIAA Progress in Astronautics & Aeronautics 19 (1966).

16. Roa, B.S., Karnik, K.S., and Gupta, V.D.; 'A Study of Satellite Television Broadcasting System in India', J. of Spacecraft and Rockets 10, 537-540 (1973).

17. Rosen, H.A. and Jones, C.R.; 'CTS (Space Transportation System) - Optimized Satellite Concept', Astronautics & Aeronautics 15, 48-53 (1977).

18. Ruppe, H.O.; 'A History of the Communications Satellite', J. of the British Interplanetary Society 30, 131-134 (1977).

19. Shrivastava, S.K.; 'Orbital Perturbations and Stationkeeping of Communication Satellites', J. of Spacecraft and Rockets 15, 67-78 (1978).

20. Williams, J.R.; 'Geosynchronous Satellite Solar Power', Astronautics & Aeronautics 13, 46-52 (1975).

21. Report to the President and the Congress, Communications Satellite Corporation, Washington D.C., Nov. 30, 1978.

22. Astronautics & Aeronautics, monthly issues (1974-1983).

23. Spaceflight, monthly issues (1974-1986).

24. Aerospace America, monthly issues (1984-1986).

Chapter 2

THE EARTH'S GRAVITATIONAL FIELD AND BASIC METHODOLOGY

2.1. General

A geostationary satellite revolving around the Earth is subject to the attractions of all planets in the solar system. Among these planets, the Earth constitutes the basic center of attraction, while the Sun and Moon play a minor role in determining the motion of the satellite. The effects of other planets can be ignored because of their large distances from the satellite. Thus,the motion of a geostationary satellite can be studied in an Earth-Sun-Moon system and constitutes a four-body problem in Celestial Mechanics. The four bodies involved are the three constituents of the system and the satellite itself.

In the Earth-Sun-Moon system, the Sun and Moon are considered to be spherical body masses, while the potential of the Earth can mathematically be described in terms of spherical harmonics. The attraction of the Earth on a satellite is then obtained by expressing an acceleration as a gradient of the Earth's potential. In other words, the mass of the Earth forms a potential field (or gravitational field) in its vicinity as usually discussed in classical mechanics. In the following section, we shall briefly explore the development of this potential field for our use.

Classically, a four-body problem is solved by setting up a system of Lagrangian equations with six orbital elements as dependent variables and time as the independent variable. Disturbing functions are introduced into the Lagrangian equations to relate the attractions (or perturbations) of the Sun, Moon, and higher harmonics of the Earth's potential while a spherical Earth is considered as the center of attraction to establish the basic (unperturbed) elliptical orbit. We shall present the canonical approach and its variations of solving the problem after the section that deals with the Earth's potential field.

Our approach to the problem of the motion of a geostationary satellite in the Earth-Sun-Moon system is from a point of view of nonlinear mechanics. We start the development of the solution of the problem with the basic equations of motion in spherical coordinates. This approach will help scientists and engineers to understand the problem without involving training in the field of Astronomy or Celestial Mechanics. However, our approach also provides the scientists with astronomical background a different avenue to solve the problem. A description of our basic methodology dealing with the solution of the problem and the transformation between our variables and orbital elements constitute the section following those of the canonical approach and its variations. And finally, a comparison of various methods of solution and the proposed scheme of solutions are presented. All these preparatory works will pave the way for our developments of the theory of geostationary satellites in the following chapters.

2.2. Earth's Gravitational Field - Spherical Harmonics Model

All of the irregularities of mass distribution in the body of the Earth are manifest as small contribution to the total gravitational potential of the Earth. To fully model these, a formulation is required which makes use of an infinite sequence of orthogonal functions. The Earth's geometry dictates the representation of its surface in spherical coordinates which, in turn, suggests the formulation of the Earth's gravitational potential in terms of spherical harmonic expansion. We shall present a brief discussion of the derivation of the spherical harmonics model.

Newton's universal law of gravitation gives the force, F, of attraction between two particles of masses m and M at a distance r from each other as

$$F = k \frac{mM}{r^2} , \tag{2.1}$$

where k is the gravitational constant. If M >> m, the acceleration, a, of mass m with respect to mass M is

$$a = k \frac{M}{r^2} \tag{2.2}$$

as a result of interpreting (2.1) by means of Newton's second law.

The acceleration, a, in (2.2) may be expressed as the gradient of a scalar, called a potential, V. Thus

$$a = \frac{\delta V}{\delta r} , \tag{2.3}$$

where

$$V = - \frac{kM}{r} . \tag{2.4}$$

For a large body consisting of masses, M_i's, the resulting potential at mass m is

$$\Sigma V_i = - \Sigma \frac{kM_i}{r_i} \tag{2.5}$$

which, in turn, may be expressed in integral form as

$$V = \Sigma V_i = - \int \frac{k\rho \; d\tau}{r} , \tag{2.6}$$

where ρ may be the position dependent mass density associated with the differential volume $d\tau$.

The potential function V in (2.6) is characterized by satisfying the Laplace equation when the mass m is located outside of the large body ΣM_i; while the mass m is situated inside the large body ΣM_i, the partial differential equation of Poisson will be satisfied (Wills, 1931). For the development of the Earth's gravitational potential model, we shall study the case of the Laplace equation because our interest is the Earth's attraction on satellites which are revolving around the Earth.

The Laplace equation in spherical coordinates is

$$r^2\nabla^2 V = \frac{\partial}{\partial r}\left(r^2 \frac{\partial V}{\partial r}\right) + \frac{1}{\sin\phi'}\frac{\partial}{\partial\phi'}\left(\cos\phi' \frac{\partial V}{\partial\phi'}\right) + \frac{1}{\cos^2\phi'}\frac{\partial^2 V}{\partial\lambda^2} = 0 \tag{2.7}$$

where

r = radial distance from the center of mass,

ϕ' = colatitude ($= \frac{\pi}{2}$ - latitude),

and

λ = longitude (measures eastward as looking from the positive end of the z-axis toward the origin in the counterclockwise direction).

A trial solution to (2.7) is approached by separating the variables r, ϕ' and λ as

$$V(r, \phi', \lambda) = \Phi(r)Y(\phi', \lambda), \tag{2.8}$$

where Φ is a function of r only and Y is a function of ϕ' and λ. Substituting (2.8) into (2.7) yields two equations:

$$2r\frac{d\Phi}{dr} + r^2\frac{d^2\Phi}{dr^2} = n(n+1)\Phi, \tag{2.9a}$$

and

$$\frac{\delta^2 Y}{\delta^2\phi'} + \cot\phi' \frac{\delta Y}{\delta\phi'} + \frac{1}{\sin^2\phi'}\frac{\delta^2 Y}{\delta^2\lambda} = -n(n+1)Y, \tag{2.9b}$$

where n is an integer and the constant n(n + 1) is written in a convenient form. The solution to (2.9a) is

$$\Phi(r) = A_n r^n + B_n r^{-(n+1)}, \tag{2.10}$$

where A_n and B_n are constants to be determined. If $Y_n(\phi', \lambda)$ is a particular solution to (2.9b), then

$$V(r, \phi', \lambda) = A_n r^n Y_n(\phi', \lambda) + B_n r^{-(n+1)} Y_n(\phi', \lambda) \tag{2.11}$$

are particular solutions to V with n taken integer values 0, 1, 2, ... A property of the Laplace equation is that any linear combination of particular solutions is also a solution. Thus

$$V = \sum_{n=0}^{\infty}[A_n r^n Y_n(\phi', \lambda) + B_n r^{-(n+1)} Y_n(\phi', \lambda)]. \tag{2.12}$$

When the boundary condition, $V \to 0$ as $r \to \infty$, is applied to (2.12), it can be seen that all A_n's are zero; therefore

$$V = \sum_{n=0}^{\infty} B_n r^{-(n+1)} Y_n(\phi', \lambda). \tag{2.13}$$

A further separation of variables is accomplished by setting $Y(\phi', \lambda) = g(\phi')h(\lambda)$. Substituting this expression into (2.9b) results in

$$\sin\phi' \frac{d^2 g}{d\phi'} + \cos\phi' \frac{dg}{d\phi'} + \left[n(n+1)\sin\phi' - \frac{m^2}{\sin\phi'}\right]g = 0, \tag{2.14}$$

and

$$\frac{d^2 h}{d\lambda^2} + m^2 h = 0, \tag{2.15}$$

where m^2 is a constant written in convenient form.

The solution to (2.15) is

$$h(\lambda) = C_m \cos m\lambda + S_m \sin m\lambda, \tag{2.16}$$

where C_m and S_m are constants to be determined.

The solution to (2.14) is the Legendre Function $P_{nm}(\cos \phi')$; its expression and characteristics can be found elsewhere (Abramowitz and Stegun, 1964). Consequently,

$$Y_n(\phi', \lambda) = (C_m \cos m\lambda + S_m \sin m\lambda) P_{nm}(\cos \phi'), \tag{2.17}$$

where m = 0, 1, 2, ..., n.

The linear combination of the solutions in (2.17) is also a solution; thus

$$Y(\phi', \lambda) = \sum_{m=0}^{n} (C_m \cos m\lambda + S_m \sin m\lambda) P_{nm}(\cos \phi'). \tag{2.18}$$

This expression combining with (2.13) results in

$$V(r, \phi', \lambda) = - \sum_{n=0}^{\infty} \frac{1}{r^{n+1}} \sum_{m=0}^{n} (C_{nm} \cos m\lambda + S_{nm} \sin m\lambda) \times \times P_{nm}(\cos \phi'), \tag{2.19}$$

where the constants, $C_{nm} = B_n C_m$ and $S_{nm} = B_n S_m$, are to be determined. A minus sign is added to (2.19) to conform with our definition of the potential function, V, in (2.6).

Equation (2.19) is the surface spherical harmonics of the gravitational potential, V. Its individual components, $C_{nm} \cos m\lambda \, P_{nm}(\cos \phi')$ and $S_{nm} \sin m\lambda \, P_{nm}(\cos \phi')$, are orthogonal. This property makes the spherical harmonics a natural means of representing a function over a spherical surface (analogous to the use of Fourier series for functions in rectilinear space).

Since the Earth is of a spherical shape in general, equation (2.19) is a natural candidate to represent its gravitational potential. In order to make C_{nm} and S_{nm} nondimensional, we shall apply the factor $\mu_E R_E^n$ to (2.19), where μ_E is the gravitational constant of the Earth and R_E is the mean equatorial radius of the Earth; consequently, the Earth's gravitational potential can be represented by

$$V_E(r, \phi, \lambda) = - \mu_E \sum_{n=0}^{\infty} \frac{R_E^n}{r^{n+1}} \sum_{m=0}^{n} (C_{nm} \cos m\lambda + S_{nm} \sin m\lambda) \times \times P_{nm}(\sin \phi), \tag{2.20}$$

where $\cos \phi'$ in the Legendre function is replaced by $\sin \phi$ because ϕ is the latitude and ϕ' is the colatitude.

The Earth's gravitational potential may be written in another form,

$$V_E(r, \phi, \lambda) = - \mu_E \sum_{n=0}^{\infty} \frac{R_E^n}{r^{n+1}} \sum_{m=0}^{n} J_{nm} P_{nm}(\sin \phi) \cos(\lambda - \lambda_{nm}), \tag{2.20a}$$

and the constants C_{nm}, S_{nm}, J_{nm}, and λ_{nm} are related by

$$C_{nm} = J_{nm} \cos \lambda_{nm} \quad \text{and} \quad S_{nm} = J_{nm} \sin \lambda_{nm}.$$

The constants, C_{nm} and S_{nm} (or J_{nm} and λ_{nm}), are determined from actual Earth satellite's tracking data. There are many Earth potential models existing in open literature, among the better known models are the SAO-66-M1 model (Gaposchkin, 1966), the SAO-69 model (Gaposchkin and Lambeck, 1970; Kozai, 1969), Wagner's model (1969), Gaposchkin's model (1974), and Gem 6 (Smith et al., 1976). The truncated fourth-order Earth gravitational potential field of Gem 6 is presented in Appendix A.

2.3. Canonical Method of Solution

We shall present the canonical method of studying the movement of geostationary satellites by introducing the classical Lagrange's equations for the general planetary motion with associated disturbing functions of the problem. The Lagrange's equations are then transformed to the Delaunay equations by means of the Delaunay variables, and finally the Delaunay equations are solved with the aid of the von Zeipel method. We shall explore the above procedures in detail in the following sections.

2.3.1. Lagrange's Equations

The classical Lagrange's equations for the general planetary motion are (Brouwer and Clemence, 1961):

$$\begin{aligned}
\frac{da}{dt} &= \frac{2}{na}\frac{\partial R}{\partial M}, \\
\frac{de}{dt} &= \frac{1-e^2}{na^2 e}\frac{\partial R}{\partial M} - \frac{(1-e^2)^{1/2}}{na^2 e}\frac{\partial R}{\partial \omega}, \\
\frac{di}{dt} &= \frac{\cos i}{na^2(1-e^2)^{1/2}\sin i}\frac{\partial R}{\partial \omega} - \\
&\quad - \frac{1}{na^2(1-e^2)^{1/2}\sin i}\frac{\partial R}{\partial \Omega}, \\
\frac{d\omega}{dt} &= -\frac{\cos i}{na^2(1-e^2)^{1/2}\sin i}\frac{\partial R}{\partial i} + \frac{(1-e^2)^{1/2}}{na^2 e}\frac{\partial R}{\partial e}, \\
\frac{d\Omega}{dt} &= \frac{1}{na^2(1-e^2)^{1/2}\sin i}\frac{\partial R}{\partial i},
\end{aligned} \tag{2.21}$$

and

$$\frac{dM}{dt} = n - \frac{1-e^2}{na^2 e}\frac{\partial R}{\partial e} - \frac{2}{na}\frac{\partial R}{\partial a},$$

where a (semimajor axis), e (eccentricity), ω (argument of perigee), i (inclination), Ω (longitude of ascending node), and M (mean anomaly) are the six basic orbital elements, n is the mean motion, t is the time, and R is the disturbing function.

2.3.2. Disturbing Functions

For the case of geostationary satellites, the disturbing function, R, consists of three components:

$$\begin{aligned}
R = \;& R_d \text{ (perturbations of the Earth oblateness, } J_2\text{, the Sun and the Moon)} \\
&+ R_{nm} \text{ (perturbations of the higher Earth harmonics potential terms)} \\
&+ R_S \text{ (perturbation of solar radiation pressure).}
\end{aligned}$$

Because both the eccentricity and the inclination are very small for the geostationary satellite orbit, Kozai (1973) derived the main term of the disturbing function (R_d) due to the perturbations of the Earth oblateness, the Sun, and the Moon as

$$R_d = n^2a^2J_2\left(\frac{R_E}{a}\right)^2(\tfrac{1}{2} - \tfrac{3}{4}\sin^2 i)(1 + \tfrac{3}{2}e^2) + n_M^2a^2\frac{M_M}{M_M + M_E} \times$$
$$\times (1 + \tfrac{3}{2}e^2)(1 - e_M^2)^{-3/2}[\tfrac{1}{4}(1 - \tfrac{3}{2}\sin^2 i)(1 - \tfrac{3}{2}\sin^2 i_M) +$$
$$+ \tfrac{3}{16}\sin 2i \sin 2i_M \cos(\Omega - \Omega_M') +$$
$$+ \tfrac{3}{16}\sin^2 i \sin^2 i_M \cos 2(\Omega - \Omega_M')] + n_S^2a^2(1 + \tfrac{3}{2}e^2) \times$$
$$\times (1 - e_S^2)^{-3/2}[\tfrac{1}{4}(1 - \tfrac{3}{2}\sin^2 i)(1 - \tfrac{3}{2}\sin^2 \lambda) -$$
$$- \tfrac{3}{16}\sin 2i \sin 2\lambda \cos\Omega + \tfrac{3}{16}\sin^2 i \sin^2\lambda \cos 2\Omega], \quad (2.22)$$

where M_E and M_M are the masses of the Earth and Moon, respectively, n_S and n_M are the respective mean motion of the Sun and Moon with respect to the Earth with e_S and e_M being the corresponding eccentricity of their orbits, i_M and λ are the respective inclination of the Moon's orbit and the ecliptic plane with respect to the equatorial plane of the Earth, and Ω_M' is the longitude of the ascending node of the Moon's orbit in the equatorial plane of the Earth measured from the vernal equinox (♈).

It should be noted that the perturbation terms in R_d due to the Earth oblateness (J_2), the Sun, and the Moon are of the same order of magnitude for the case of geostationary satellites:

$$\left.\begin{aligned} &J_2\frac{R_E^2}{a^2} \cong 2.46 \times 10^{-5} \text{ - Earth oblateness } (J_2), \\ &\frac{M_M}{M_M + M_E}\left(\frac{n_M}{n}\right)^2 \cong 1.63 \times 10^{-5} \text{ - Moon}, \\ \text{and } &\left(\frac{n_S}{n}\right)^2 \cong 0.75 \times 10^{-5} \text{ - Sun}. \end{aligned}\right\} \quad (2.22a)$$

Among the higher Earth harmonics potential terms, the J_{22}-term (having an order of magnitude of J_2^2) is of the most interest to the movement of the geostationary satellites because of its magnitude, as well as the angular arguments in its expression. The disturbing function of this term has a form of (Kaula, 1966; Morando, 1962a)

$$R_{nm} \cong R_{22} = \frac{3}{4}\frac{\mu_E}{a}J_{22}\left(\frac{R_E}{a}\right)^2(1 + \cos i)^2(1 - \tfrac{5}{2}e^2 + \tfrac{3}{16}e^4 + \ldots) \times$$
$$\times \cos 2(\omega + M + \Omega - \lambda_{22} - \theta_g), \quad (2.23)$$

where $J_{22} > 0$, θ_g is the longitude of the Greenwich meridian in the ecliptic plane measured from the vernal equinox (♈), and $\lambda_{22} + \theta_g$ depicts the angular position of the major axis of the elliptical Earth equator. Since geostationary satellite's orbits have a small inclination ($\cos i \cong 1$) and near zero eccentricity ($e \cong 0$), (2.23) may be simplified as

$$R_{22} = 3\frac{\mu_E}{a}J_{22}\left(\frac{R_E}{a}\right)^2 \cos 2(\omega + M + \Omega - \lambda_{22} - \theta_g). \quad (2.24)$$

As geostationary satellites revolving around the Earth have a mean secular rate of

$$\dot{\omega} + \dot{M} + \dot{\Omega} - \dot{\theta}_g \cong 0, \tag{2.25}$$

R_{22} thus becomes a resonant disturbing function (Kaula, 1966; Morando, 1962b). It is well known that the usual first approximation for the orbit of a secularly precessing ellipse in the method of solution of satellite motion breaks down in the vicinity of a resonance, and a different avenue will be chosen to solve the problem. In fact, geostationary satellites have a longitudinal motion of long-period libration about the minor axis of the elliptical Earth equator because of the resonance. We shall discuss the libration of geostationary satellites in detail in Chapter 3.

The disturbing function R_S of solar radiation pressure can be written as (Mello, 1972)

$$\begin{aligned} R_S = P{\times}r{\times}[&- \tfrac{1}{4}(1 + \cos i)(1 - \cos \lambda)\cos(v + \omega + \Omega + \lambda_S) - \\ &- \tfrac{1}{4}(1 + \cos i)(1 + \cos \lambda)\cos(v + \omega + \Omega - \lambda_S) - \\ &- \tfrac{1}{4}(1 - \cos i)(1 + \cos \lambda)\cos(v + \omega - \Omega + \lambda_S) - \\ &- \tfrac{1}{4}(1 - \cos i)(1 - \cos \lambda)\cos(v + \omega - \Omega - \lambda_S) - \\ &- \tfrac{1}{2}\sin i \sin \lambda \cos(v + \omega + \lambda_S) - \\ &- \tfrac{1}{2}\sin i \sin \lambda \cos(v + \omega - \lambda_S)], \end{aligned} \tag{2.26}$$

where P is the product of solar radiation pressure (P_r) and the 'area-to-mass' (A/M) of a satellite, λ_S is the mean longitude of the Sun, r and v are the radial position and the true anomaly of the satellite in the orbit, respectively, and all other parameters have been defined previously. For the case of geostationary satellites, approximations $r \cong a\ (1 - e \cos M)$ and $v \cong M$, can be made because they are in a near-circular orbit.

The solar radiation pressure (P_r) on an Earth satellite (Seifert, 1959) can be computed from

$$\begin{aligned} P_r &= 4.86 \times 10^{-7}\ (1 + R)\cos^2 \delta \quad (\mathrm{kg/m^2}) \\ &= 1.00 \times 10^{-7}\ (1 + R)\cos^2 \delta \quad (\mathrm{lb/ft^2}), \end{aligned}$$

in which R is the reflectivity and δ is the radiation incidence angle with respect to the normal of the reflecting surface. A perfect reflector has an R equal to unity.

2.3.3. Delaunay Variables and Delaunay Equations

The expressions in (2.21) are rather complicated; in the canonical method of solution, the Delaunay variables are introduced to replace the six orbital elements in (2.21) and the resulting expressions become extremely simple. The Delaunay variables are defined as (Brouwer and Clemence, 1961)

$$\begin{aligned} &L \equiv (\mu_E a)^{1/2}, \quad G \equiv L(1 - e^2)^{1/2}, \quad H \equiv G \cos i\ , \\ &\ell \equiv M, \qquad g \equiv \omega, \quad \text{and} \quad h \equiv \Omega. \end{aligned} \tag{2.27}$$

With these variables, (2.21) has the form of

$$\begin{aligned} &\frac{dL}{dt} = \frac{\partial R}{\partial \ell}, \quad \frac{d\ell}{dt} = n - \frac{\partial R}{\partial L}, \quad \frac{dG}{dt} = \frac{\partial R}{\partial g}, \quad \frac{dg}{dt} = -\frac{\partial R}{\partial G}, \\ &\frac{dH}{dt} = \frac{\partial R}{\partial h}, \quad \text{and} \quad \frac{dh}{dt} = -\frac{\partial R}{\partial H}, \end{aligned} \tag{2.28}$$

where R is considered to be expressed in terms of the Delaunay variables and is assumed to be independent of t explicitly.

An alternative form of (2.28) is

$$\frac{dL}{dt} = \frac{\partial F}{\partial \ell}\,,\ \frac{d\ell}{dt} = -\frac{\partial F}{\partial L}\,,\ \frac{dG}{dt} = \frac{\partial F}{\partial g}\,,\ \frac{dg}{dt} = -\frac{\partial F}{\partial G}\,,\ \frac{dH}{dt} = \frac{\partial F}{\partial h}\,,$$

and

$$\frac{dh}{dt} = -\frac{\partial F}{\partial H} \tag{2.29}$$

with

$$F = \frac{\mu_E^2}{2L^2} + R, \tag{2.30}$$

and F is known as the force function.

Equation (2.29) are the canonical equations of motion in Delaunay variables, and they are known as the Delaunay equations.

The disturbing function R_d in (2.22) depends explicitly on t because i_M and Ω_M' are related to Ω_M by (Kozai, 1973)

$$\begin{aligned} \sin^2 i_M &= 0.164 + 0.068 \cos \Omega_M\,, \\ \sin i_M \cos \Omega_M' &= 0.396 + 0.086 \cos \Omega_M\,, \\ \sin i_M \sin \Omega_M' &= 0.086 \sin \Omega_M\,, \end{aligned} \tag{2.31}$$

and Ω_M is a linear function of t as Ω_M is the longitude of the ascending node of the Moon's orbit in the ecliptic plane measured from the vernal equinox (♈).

The disturbing function R_S in (2.26) also depends on t, since the mean longitude of the Sun, $\lambda_S \equiv \alpha$, is a linear function of t.

When the disturbing function R is explicitly dependent on t, the force function in (2.30) requires some modification. For the cases of R_d and R_S, it is convenient to introduce k_M, k_S, K_M, and K_S as additional variables by putting

$$\frac{dk_M}{dt} = \dot{\Omega}_M = -\frac{\partial F}{\partial K_M}\,,\quad \frac{dk_S}{dt} = \dot{\alpha} = -\frac{\partial F}{\partial K_S}\,, \tag{2.32}$$

where $\dot{\Omega}_M$ is the rate of regression of the ascending node of the Moon's orbit in the ecliptic plane, $\dot{\alpha}$ is the angular rate of the Sun revolving around the Earth, and both are constant.

The addition of the variables requires the addition of $-\dot{\Omega}_M K_M$ and $-\dot{\alpha} K_S$ to F, i.e.,

$$F = \frac{\mu_E^2}{2L^2} - \dot{\Omega}_M K_M - \dot{\alpha} K_S + R\,. \tag{2.33}$$

The Delaunay equations then become

$$\frac{dL}{dt} = \frac{\partial F}{\partial \ell}\,,\ \frac{d\ell}{dt} = -\frac{\partial F}{\partial L}\,,\ \frac{dG}{dt} = \frac{\partial F}{\partial g}\,,\ \frac{dg}{dt} = -\frac{\partial F}{\partial G}\,,\ \frac{dH}{dt} = \frac{\partial F}{\partial h}\,,\ \frac{dh}{dt} = -\frac{\partial F}{\partial H}$$

$$\frac{dK_M}{dt} = \frac{\partial F}{\partial K_M}\,,\ \frac{dk_M}{dt} = -\frac{\partial F}{\partial K_M} = \dot{\Omega}_M,\quad \frac{dK_S}{dt} = \frac{\partial F}{\partial k_S}\,, \text{ and}$$

$$\frac{dk_S}{dt} = -\frac{\partial F}{\partial K_S} = \dot{\alpha}, \tag{2.34}$$

where F is defined in (2.33).

2.3.4. von Zeipel Method

The process of solving the Delaunay equations basically involves a succession of transformations of the force function F, such that

$$\begin{aligned} &F\,(L, G, H, K_M, K_S, \ell, g, h, k_M, k_S) \\ &\rightarrow F'\ \ (L', G', H', K_M', K_S', -, g', h', -, -) \\ &\rightarrow F''\ \ (L'', G'', H'', K_M'', K_S'', -, -, h'', -, -) \\ &\rightarrow F'''\ \ (L''', G''', H''', K_M''', K_S''', -, -, -, -, -). \end{aligned} \tag{2.35}$$

It can be seen that the first transform eliminates the short-period mean anomaly (ℓ) and the explicit time dependent variables, k_M and k_S, and the second and third transformations eliminate the remaining two angular variables, g and h, respectively. The last transformation is obviously solvable as

$$\begin{aligned} &L''' = L_0''', \ G''' = G_0''', \ H''' = H_0''', \ K_M''' = K_{M0}''', \ K_S''' = K_{S0}''', \\ &\ell''' = \ell_0''' + \text{constant}_{(\ell)}(t - t_0), \\ &g''' = g_0''' + \text{constant}_{(g)}(t - t_0), \\ &h''' = h_0''' + \text{constant}_{(h)}(t - t_0), \\ &k_M''' = k_{M0}''' + \dot{\Omega}_M(t - t_0), \end{aligned} \tag{2.36}$$

and

$$k_S''' = k_{S0}''' + \dot{\alpha}(t - t_0),$$

where L_0''', ℓ_0''', etc. are integration constants.

In transforming the force function from F to F', a determining function S is introduced, which is a function of L', G', H', K_M', K_S', ℓ, g, h, k_M, and k_S, and can be developed in terms of the order of magnitude of the small parameter in the disturbing function; thus

$$S\,(L', G', H', K_M', K_S', \ell, g, h, k_M, k_S) = S_0 + S_1 + S_2 + \ldots\,. \tag{2.37}$$

Let

$$L = \frac{\partial S}{\partial \ell}, \ G = \frac{\partial S}{\partial g}, \ H = \frac{\partial S}{\partial h}, \ K_M = \frac{\partial S}{\partial k_M}, \ K_S = \frac{\partial S}{\partial k_S},$$

$$\ell' = \frac{\partial S}{\partial L'}, \ g' = \frac{\partial S}{\partial G'}, \ h' = \frac{\partial S}{\partial H'}, \tag{2.38}$$

and

$$S_0 = L'\ell + G'g + H'h + K_M'k_M + K_S'k_S, \tag{2.39}$$

where S_0 is defined so that restricting S to S_0 will simply yield the identical transformations:

$$L = \frac{\partial S_0}{\partial \ell} = L', \quad \ell' = \frac{\partial S_0}{\partial L'} = \ell, \text{ etc.}$$

We are not interested in k_M' and k_S' because k_M and k_S can be obtained directly from (2.32) by integration.

Similarly, the transformation from F' to F'' requires a determining function S^* $(L'', G'', H'', K_M'', K_S'', g', h')$, and subsequently $S^{**}(L''', G''', H''', K_M''', K_S''', h'')$ is introduced for transforming F'' to F'''.

The use of a determining function to effect a transformation in accordance with (2.35) is known as the von Zeipel method.

We shall briefly explore the von Zeipel method by examining the role of S in transforming F to F', while the interested reader may find S^* in transforming F' to F'' and S^{**} for transforming F'' to F''' by similar procedures. For detailed exposition of the von Zeipel method and its application to the lunar theory as well as to the artificial satellite problems, readers are referred to Brouwer and Clemence (1961).

In applying the von Zeipel method, S, F, and F' are all developed in series in accordance with the magnitude of the terms involved. Thus

$$S = S_0 + S_1 + S_2 + \dots ,$$

as is shown in (2.37), and

$$F = F_0 + F_1 + F_2 + \dots = F_0' + F_1' + F_2' + \dots , \tag{2.40}$$

where

$$F_0 = \frac{\mu_E^2}{2L^2} - \dot{\Omega}_M K_M - \dot{\alpha} K_S ,$$

and

$$F_0' = \frac{\mu_E^2}{2L'^2} - \dot{\Omega}_M K_M' - \dot{\alpha} K_S' .$$

In order to simplify our presentation, we express L, G, H, K_M, and K_S as p_1, p_2, p_3, p_4, and p_5, respectively, which may be considered as components of a vector $\vec{p}$. Similarly, ℓ, g, h, k_M, and k_S are replaced by q_1, q_2, q_3, q_4, and q_5, respectively and they are considered as components of a vector $\vec{q}$. By using the notation of summing over repeated subscripts, (2.37), (2.38), and (2.39) then become

$$S(\vec{p}', \vec{q}) = S_0 + S_1 + S_2 + \dots , \tag{2.41}$$

$$p_i = \frac{\partial S}{\partial q_i} , \quad q_i' = \frac{\partial S}{\partial p_i'} , \tag{2.42}$$

and

$$S_0 = p_i' q_i , \tag{2.43}$$

where i = 1, ..., 5.

Substituting (2.43) into (2.41), differentiating the resulting (2.41) with respect to p_i', q_i, and substituting the result in (2.42), we have

$$p_i = p_i' + \frac{\partial S_1}{\partial q_i} + \frac{\partial S_2}{\partial q_i} + \dots ,$$

and

$$q_i = q_i' - \frac{\partial S_1}{\partial p_i'} - \frac{\partial S_2}{\partial p_i'} + \dots . \tag{2.44}$$

Developing F_0, F_1, and F_2 in (2.40) in Taylor series of $(p_i - p_i')$ and $(q_i - q_i')$ and substituting for $(p_i - p_i')$ and $(q - q_i')$ from (2.44), we have

$$F_0' = F_0(\vec{p}') ,$$

$$F_1' = \frac{\partial F_0}{\partial p_i'} \frac{\partial S_1}{\partial q_i} + F_1(\vec{p}', \vec{q}) ,$$

and

$$F_2' + \frac{\partial F_1'}{\partial q_2'} \frac{\partial S_1}{\partial p_2'} + \frac{\partial F_1'}{\partial q_3'} \frac{\partial S_1}{\partial p_3'} = \frac{\partial F_0}{\partial p_i'} \frac{\partial S_2}{\partial q_i} + \frac{1}{2} \frac{\partial^2 F_0}{\partial p_i' \partial p_j'} \left(\frac{\partial S_1}{\partial q_i}\right)\left(\frac{\partial S_1}{\partial q_j}\right) + \frac{\partial F_1}{\partial p_i'} \frac{\partial S_1}{\partial q_i} + F_2(\vec{p}', \vec{q}) , \tag{2.45}$$

where the terms in $F_1(\vec{p}', \vec{q})$ may be separated into two groups; one, defined as $F_{1s}(\vec{p}', q_2, q_3)$, contains terms with arguments of $\vec{p}'$, q_2, and q_3, while the remaining terms are grouped together as $F_{1p}(\vec{p}', \vec{q})$. Expressions in (2.45) determine the terms in S, because F' contains only the variables of p_1', ..., p_5', q_2' and q_3', and the omitted variables q_1, q_4, and q_5 must be accounted for by S. In other words, the second equation in (2.45) can be resolved into two equations:

$$F_1' = [F_{1s}(\vec{p}', q_2, q_3)]_{q_2, q_3 \to q_2', q_3'} ,$$

and

$$\frac{\delta F_0}{\delta p_i'} \frac{\delta S_1}{\delta q_i} + F_{1p}(\vec{p}', \vec{q}) = 0 . \qquad (2.46)$$

The second equation in (2.46) can be integrated to obtain S_1. Similarly, S_2 can be obtained from the third equation in (2.45). Again it should be reminded that $F_2(\vec{p}', \vec{q})$ is resolved into $F_{2s}(\vec{p}', q_2, q_3)$ and $F_{2p}(\vec{p}', \vec{q})$, and $[F_{2s}(\vec{p}', q_2, q_3)]_{q_2, q_3 \to q_2', q_3'}$ will equal to F_2'. With known terms in S, p_i and q_i can be determined from (2.44).

The next step of the von Zeipel method employs a new force function F' defined as

$$F' = F_0' + F_1' + F_2' \qquad (2.47)$$

where F_0', F_1', and F_2' are previously obtained from (2.45). We shall repeat our process by introducing a determining function S^*, the omitted variable q_2 will be accounted for by S^* and the force functions, F_0'', F_1'', and F_2'' are obtained for use in the next step. This process will be continued until the entire problem is solved.

2.3.5. Recapitulation

We have completed our description of the canonical method to solve the geostationary satellites problem. Basically, the following steps have been taken:

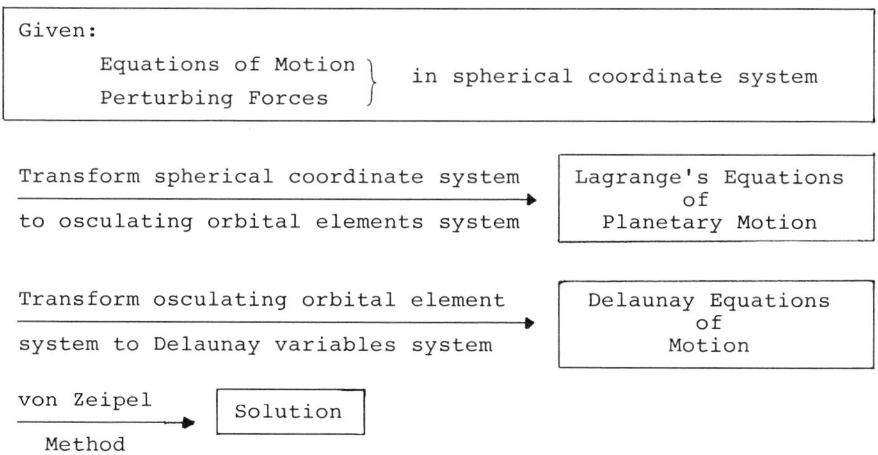

2.4. Other Standard Methods of Solution

There are many other methods or variations of the von Zeipel method to solve artificial satellite problems. We shall mention a few of them for readers interested in the methods of solution.

2.4.1. Brouwer's Modified von Zeipel Method

In his solution to the problem of artificial satellite theory without drag, Brouwer (1959) used the von Zeipel method described in Section 2.3.4. as his method of solution. When the non-period terms in F_2' (= F_{2s}') was sought in obtaining F_2'' for the solution of secular and long-period effects, Brouwer used the averaged value of the expressions in F_2' over a half-period of the mean anomaly.

When Hori (1973) described the von Zeipel method by using Brouwer's paper (1959) as his reference, he introduced the averaging technique to obtain F_1', S_1, F_2', S_2, etc.:

$$
\begin{aligned}
F_1' &= F_1'(L', G', H', -, g', h') \\
&= \langle F_1(L', G', H', \ell, g, h)\rangle \; g, h \to g', h' \\
S_1 &= -\left(\frac{dF_0}{dL'}\right)^{-1} \int (F_1 - \langle F_1\rangle)\, d\ell \\
F_2' &= \left\langle \frac{1}{2}\frac{d^2F_0}{dL'^2}\left(\frac{\partial S_1}{\partial \ell}\right)^2 + \frac{\partial F_1}{\partial L'}\frac{\partial S_1}{\partial \ell} + \frac{\partial F_1}{\partial G'}\frac{\partial S_1}{\partial g} + \frac{\partial F}{\partial H'}\frac{\partial S_1}{\partial h} - \right. \\
&\quad \left. - \frac{\partial F_1'}{\partial g}\frac{\partial S_1}{\partial G'} - \frac{\partial F_1'}{\partial h}\frac{\partial S_1}{\partial H'} + F_2 \right\rangle g, h \to g', h' \qquad (2.48) \\
S_2 &= -\left(\frac{dF_0}{dL'}\right)^{-1} \int \frac{1}{2}\frac{d^2F_0}{dL'^2}\left(\frac{\partial S_1}{\partial \ell}\right)^2 + \frac{\partial F_1}{\partial L'}\frac{\partial S_1}{\partial \ell} + \ldots + F_2 - \\
&\quad - \left\langle \frac{1}{2}\frac{d^2F_0}{dL'^2}\left(\frac{\partial S_1}{\partial \ell}\right)^2 + \frac{\partial F_1}{\partial L'}\frac{\partial S_1}{\partial \ell} + \ldots + F_2 \right\rangle d\ell
\end{aligned}
$$

where $\langle\ \rangle$ denotes the average with respect to ℓ.

From the above expressions, we can see that Brouwer's approach to obtain F_1', S_1, F_2', S_2, etc. is different from what Hori did. Hori used the averaging technique to obtain these functions, while Brouwer employed the averaging technique on F_2' over a half-period of ℓ for obtaining F_2'', and F_1' was derived by selecting terms independent on ℓ in F_1 as described in Section 2.3.4.

2.4.2. Kozai's Method

Kozai's method of solution (Kozai, 1959) is to solve the Langrange's equations of the general planetary motion, (2.21), directly. The disturbing function, R, is averaged over a 2π-period of the mean anomaly. The resulting disturbing function, $\bar{R}$, consists of secular terms (R_s) having no angular orbital parameters and long-period terms (R_L) containing angular orbital parameters, ω and Ω, while the short-period disturbing function (R_p) consists of the difference of the first-order terms containing the true anomaly, v, in R and those in R_s.

The first-order secular, long-period, and short-period solutions are then obtained by inserting the first-order terms in R_s, R_L, and R_p in (2.21) and solving them separately with the transformation of the independent variable from t to v by the differential equation

$$\frac{dv}{dM} = \frac{a^2}{r^2}(1 - e^2)^{1/2}, \qquad (2.49)$$

where the mean anomaly, M, is a linear function of the time, t.

The second-order solution is derived based on an expansion of the expressions on the right-hand side of (2.21) into a Taylor series with respect t to the mean orbital elements (i.e., the first-order secular solution) to the second order. In the process of solution, again the independent variable, v, is used and the averaged values of the functions over a 2π-period of M are employed.

Readers should notice that the disturbing function, R_d, shown in (2.22) derived by Kozai (1973) is a collection of the first-order terms in R_s and R_L in the sense that Kozai adopted in his method of solution.

2.4.3. Hori-Lie Method

This method (Hori, 1973) is an improvement of the von Zeipel method; it is considered to be the state-of-the-art in analytical artificial satellite methods of solution. In describing the method, we shall consider a set of canonical variables, x and y, to be transformed to another set of canonical variables, ξ and η, by means of a canonical transformation:

$$\begin{aligned} x &= \xi + \frac{\partial}{\partial y}\tilde{S}_1(\xi, y) + \frac{\partial}{\partial y}\tilde{S}_2(\xi, y) + \dots, \\ y &= \eta - \frac{\partial}{\partial \xi}\tilde{S}_1(\xi, y) - \frac{\partial}{\partial \xi}\tilde{S}_2(\xi, y) + \dots, \end{aligned} \qquad (2.50)$$

where

$$S = \Sigma\xi_i y_i + \tilde{S}_1(\xi, y) + \tilde{S}_2(\xi, y) + \dots$$

is the generating function of the canonical transformation, and subscripts in $\tilde{S}$ denote the small order of magnitude.

If $\tilde{S}_j(\xi, \eta)$ stands for $[\tilde{S}_j(\xi, y)]_{y\to\eta}$, and $\tilde{S}_{j\eta}$, $\tilde{S}_{j\xi}$, ... for $\delta\tilde{S}_j/\delta\eta$, $\delta\tilde{S}_j/\delta\xi$, ..., the explicit equations for x, y in terms of ξ, η are given by

$$\begin{aligned} x &= \xi + \tilde{S}_{1\eta} + \tilde{S}_{2\eta} - \Sigma\tilde{S}_{1\eta\eta_i}\tilde{S}_{1\xi_i} + \dots, \\ y &= \eta - \tilde{S}_{1\xi} - \tilde{S}_{2\xi} + \Sigma\tilde{S}_{1\xi\eta_i}\tilde{S}_{1\xi_i} + \dots. \end{aligned} \qquad (2.51)$$

Therefore, after obtaining $\tilde{S}_1(\xi, y) + \tilde{S}_2(\xi, y) + \dots$ by the von Zeipel method, if we introduce $\tilde{S}_1(\xi, \eta)$, $\tilde{S}_2(\xi, \eta)$, ... by

$$S_1 = \tilde{S}_1, \quad S_2 = \tilde{S}_2 - \frac{1}{2}\Sigma\tilde{S}_{1\xi_i}\tilde{S}_{1\eta_i} + \dots,$$

(2.51) becomes

$$\begin{aligned} x &= \xi + S_{1\eta} + S_{2\eta} + \frac{1}{2}\sum_i(S_{1\eta\xi_i}S_{1\eta_i} - S_{1\eta\eta_i}S_{\eta_i}) + \dots, \\ y &= \eta - S_{1\xi} - S_{2\xi} - \frac{1}{2}\sum_i(S_{1\xi\xi_i}S_{1\eta_i} - S_{1\xi\eta_i}S_{\eta_i}) + \dots. \end{aligned} \qquad (2.52)$$

If we introduce Poisson Brackets, { }, (Korn and Korn, 1968), which is

$$\{S_{1\eta}, S_1\} \equiv \sum_i\left(\frac{\partial S_{1\eta}}{\partial \xi_i}\frac{\partial S_1}{\partial \eta} - \frac{\partial S_{1\eta}}{\partial \eta_i}\frac{\partial S_1}{\partial \xi_i}\right),$$

then x, y in (2.52) have the form of

$$x = \xi + S_{1\eta} + S_{2\eta} + \frac{1}{2}\{S_{1\eta}, S_1\} + \dots ,$$
$$y = \eta - S_{1\xi} - S_{2\xi} - \frac{1}{2}\{S_{1\xi}, S_1\} - \dots , \tag{2.53}$$

and we have, for any function f(x, y) of x and y, the equation

$$f(x, y) = f(\xi, \eta) + \{f, S_1\} + \{f, S_2\} + + \frac{1}{2}\Big\{\{f, S_1\}, S_1\Big\} + \dots . \tag{2.54}$$

If the operator D_S is introduced by

$$D_S f(\xi, \eta) = \{f(\xi, \eta), S(\xi, \eta)\},$$
$$D_S^n f(\xi, \eta) = D_S\big(D_S^{n-1} f(\xi, \eta)\big) ,$$

then (2.52) becomes

$$x = \xi + \sum_{n=1}^{\infty} \frac{1}{n!} D_S^n \xi, \qquad y = \eta + \sum_{n=1}^{\infty} \frac{1}{n!} D_S^n \eta \tag{2.55}$$

and (2.54) may be written as

$$f(x, y) = f(\xi, \eta) + \sum_{n=1}^{\infty} \frac{1}{n!} D_S^n f(\xi, \eta). \tag{2.56}$$

The expressions in (2.55) and (2.56) are known as the Lie series.

As the expressions in (2.53), (2.54), (2.55), and (2.56) are applicable to any two sets of canonical variables, we shall use these expressions to solve the equations of motion (2.29) because the set of Delaunay variables in (2.29) is canonical.

The equations of motion shown in (2.29) can be written as

$$\frac{dx_i}{dt} = \frac{\partial F}{\partial y_i}, \qquad \frac{dy_i}{dt} = -\frac{\partial F}{\partial x_i}, \qquad i = 1, 2, 3 , \tag{2.57}$$

where L, G, and H are replaced by x_i's, and ℓ, g, and h by y_i's. The force function, F, (or Hamiltonian) in (2.30) may be expanded in powers of a small parameter as the orders of the smallness are denoted by the subscripts:

$$F = \frac{\mu_E^2}{2L^2} + R$$
$$= F_0(x, y) + F_1(x, y) + F_2(x, y) + \dots \tag{2.58}$$

with $F_0(x, y) \equiv \dfrac{\mu_E^2}{2x_1^2}$ and R being independent on t explicitly.

By a canonical transformation generated by (2.51), new equations of motion are obtained:

$$\frac{d\xi_i}{dt} = \frac{\partial F^*}{\partial \eta_i}, \qquad \frac{d\eta_i}{dt} = -\frac{\partial F^*}{\partial \xi_i}, \qquad i = 1, 2, 3 \tag{2.59}$$

with the new Hamiltonian

$$F^*(\xi, \eta) = F_0^*(\xi, \eta) + F_1^*(\xi, \eta) + F_2^*(\xi, \eta) + \dots = F(x, y).$$

Applying the expansion in (2.54) with $f(x, y) \equiv F(x, y)$ results in

$$\begin{aligned} &\text{Order 0} \quad F_0 = F_0^*, \\ &\text{Order 1} \quad \{F_0, S_1\} + F_1 = F_1^*, \\ &\text{Order 2} \quad \{F_0, S_2\} + \frac{1}{2}\Big\{\{F_0, S_1\}, S_1\Big\} \quad \{F_1, S_1\} + F_2 = F_2^*, \\ &\dots \quad . \end{aligned} \tag{2.60}$$

The expressions of order 2 may be simplified as

$$\{F_0, S_2\} + \frac{1}{2}\{F_1 + F_1^*, S_1\} + F_2 = F_2^*$$

with the aid of the equation of order 1.

We shall introduce a parameter τ by the following relationships:

$$\frac{d\xi_i}{d\tau} = \frac{\partial}{\partial \eta_i} F_0(\xi, \eta), \quad \frac{d\eta_i}{d\tau} = -\frac{\partial}{\partial \xi_i} F_0(\xi, \eta), \quad i = 1, 2, 3 \tag{2.61}$$

which are the equations of the unperturbed motion. Their solutions may be written as

$$\xi_i = \xi_i(\tau + C_1, C_2, \ldots, C_6), \quad \eta_i = \eta_i(\tau + C_1, C_2, \ldots, C_6),$$
$$i = 1, 2, 3, \tag{2.62}$$

where $C_1, \ldots, C_6$ are the integration constants with respect to τ.

The equation of order 1 in (2.60) now takes the form

$$-\frac{dS_1}{d\tau} + F_1\big(\xi(\tau + C_1, C_2, \ldots, C_6), \eta(\tau + C_1, C_2, \ldots, C_6)\big) = F_1^* . \tag{2.63}$$

Since F_1 is usually of multiple periodicity in τ, S_1 is so chosen that F_1^* is free from τ:

$$F_1^* = \langle F_1(C_2, \ldots, C_6)\rangle = F_1^*(\xi, \eta), \tag{2.64}$$

$$S_1 = \int (F_1 - \langle F_1\rangle)\, d\tau = S_1(\xi, \eta), \tag{2.65}$$

where $\langle\ \rangle$ denotes the average with respect to τ.

Similarly, the equation of order 2 in (2.60) gives

$$F_2^* = \left\langle \tfrac{1}{2}\{F_1 + F_1^*, S_1\} + F_2 \right\rangle = F_2^*(\xi, \eta), \tag{2.66}$$

$$S_2 = \int \left(\tfrac{1}{2}\{F_1 + F_1^*, S_1\} + F_2 - \left\langle \tfrac{1}{2}\{F_1 + F_1^*, S_1\} + F_2\right\rangle\right) d\tau, \tag{2.67}$$

and so on for equations of higher orders.

When a differentiation of composite functions is examined, the equations of motion shown in (2.57), (2.59), and (2.61) admit the following integrals:

$$F(x, y) = F^*(\xi, \eta) = \text{constant} \tag{2.68}$$

and

$$F_0^*(\xi, \eta) = F_0(x, y) = \text{constant}. \tag{2.69}$$

If the new equations of motion (2.59) do not admit an immediate solution, we then make a second canonical transformation $\xi, \eta \rightarrow p, q$ with a generating function

$$S^*(p, q) = S_1^* + S_2^* + \ldots \tag{2.70}$$

and the corresponding Hamiltonian is

$$F^*(\xi, \eta) = F_0^*(\xi, \eta) + F_1^*(\xi, \eta) + F_2^*(\xi, \eta) + \ldots$$
$$= F_0^{**}(p, q) + F_1^{**}(p, q) + F_2^{**}(p, q) + \ldots . \tag{2.71}$$

By utilizing (2.69), we have

$$F_0^*\big(\xi(p, q), \eta(p, q)\big) = F_0^{**}(p, q) = \text{constant},$$

and consequently, (2.71) results in

$$F_1^*(\xi, \eta) + F_2^*(\xi, \eta) + \ldots = F_1^{**}(p, q) + F_2^{**}(p, q) + \ldots . \tag{2.72}$$

Now we introduce a parameter τ^* by the following relationships:

$$\frac{dp_i}{d\tau^*} = \frac{\partial}{\partial q_i} F_1^*(p, q), \quad \frac{dq_i}{d\tau^*} = -\frac{\partial}{\partial p_i} F_1^*(p, q), \quad i = 1, 2, 3 \tag{2.73}$$

where

$$F_1^*(p, q) = \Big[F_1^*(\xi, \eta)\Big]_{\xi, \eta \rightarrow p, q} .$$

The corresponding expansion of $F^{**}(p, q)$ in the form of (2.60) is

$$\text{Order 1} \quad F_1^* = F_1^{**},$$
$$\text{Order 2} \quad \{F_1^*, S_1^*\} + F_2^* = F_2^{**}, \tag{2.74}$$
$$\ldots .$$

It can be seen that F_1^* plays the role of F_0 in the preceeding procedure, and consequently,

$$F_1^*(\xi, \eta) = F_1^{**}(p, q) = \text{constant} ,$$
$$F^{**}(p, q) = F^*(\xi, \eta) = F(x, y) = \text{constant} , \qquad (2.75)$$

and
$$-\frac{dS_1^*}{d\tau^*} + F_2^*(p, q) = F_2^{**} .$$

With S_1^* so chosen that F_2^{**} is free from τ^*,

$$F_2^{**}(p, q) = \langle F_2^* \rangle ,$$

and
$$S_1^*(p, q) = \int (F_2^* - \langle F_2^* \rangle)\, d\tau^* , \qquad (2.76)$$

where $\langle\ \rangle$ denotes the average with respect to τ^*.

We note that f(x, y) in (2.54) can be given directly in terms of p, q by

$$f(x, y) = f(p, q) + \{f, S_1 + S_1^*\} + \{f, S_2 + S_2^*\} + + \frac{1}{2}\left\{\{f, S_1 + S_1^*\}, S_1 + S_2\right\} + \frac{1}{2}\left\{f, \{S_1, S_1^*\}\right\} + \ldots , \qquad (2.77)$$

where S_i stands for $\left[S_i(\xi, \eta)\right]_{\xi, \eta \to p, q}$.

The inverse formula is

$$f(p, q) = f(x, y) + \{f, - S_1 - S_1^*\} + \{f, - S_2 - S_2^*\} + + \frac{1}{2}\left\{\{f, - S - S_1^*\}, - S_1 - S_1^*\right\} + + \frac{1}{2}\left\{f, \{- S_1 - S_1^*\}\right\} + \ldots , \qquad (2.78)$$

if S_i, S_i^* stand for $\left[S_i(\xi, \eta)\right]_{\xi, \eta \to x, y}$, $\left[S_i^*(p, q)\right]_{p, q \to x, y}$, respectively.

By f(x, y) $\equiv$ x or y, (2.77) and (2.78) can be used to obtain the desired expressions for the solution. Equation (2.78) is employed for obtaining the initial values of p and q from the known initial values of x and y, while (2.77) is employed for the expressions of x and y when the known expressions of p, q, S_1, S_1^*, etc. are obtained from the method of solution.

2.5. Proposed Method of Solution

We propose our approach to the solution of the geostationary satellites problem from the very basic equations of motion of the satellites in an Earth-centered spherical coordinate system (r, θ, ϕ) with time (t) as the independent variable, where θ is measured in the equatorial plane of the Earth. The spherical Earth forms the center force field, while the oblateness of the Earth, the Sun, the Moon, and solar radiation pressure, etc. constitute the perturbing forces. A change of the independent variable from t to θ transforms the equations of motion to their Laplace's form. It then follows that Lagrange's method of variation of parameters in the theory of differential equations is used to resolve the Laplace's form equations of motion into a system of six first-order differential equations, and the system of differential equations is solved by the asymptotic method in nonlinear mechanics. In the following sections, we shall discuss in detail the procedures of solution mentioned above.

2.5.1. Equations of Motion in Spherical Coordinates

Movements of all heavenly bodies are expressed in terms of orbital elements, which are closely associated with a spherical coordinate (r, θ, ϕ) system. This leads us to adopt our equations of motion in the Earth's center inertial spherical coordinate system with θ situated in the Earth equatorial plane and measured counter-clockwise from the vernal equinox as seen from the north pole, ϕ measured from the equatorial plane, and r being the radial distance from the center of the Earth. Since a geostationary satellite is basically attracted by the Earth, we shall designate μ_E/r^2 as the spherical Earth body force with μ_E being the universal constant of the Earth. This force results in a Keplerian orbit of the satellite. Because the body of the Earth can be expressed in terms of spherical harmonics, and the attractions of the Sun and Moon and the higher harmonics of the Earth on the satellite are rather small in comparison with the spherical body force, we shall be able to write the expressions of these attractions associated with a small parameter, ε, by designating εf_r, εf_θ, and εf_ϕ as the small attractive forces on a unit mass of the satellite in the positive directions of r, θ, and ϕ, respectively. With single and double dots on top of r, θ, and ϕ denoting the first and second derivatives, respectively, with respect to the time, t, our equations of motion are

$$\ddot{r} - r\dot{\theta}^2 \cos\phi - r\dot{\phi}^2 = -\frac{\mu_E}{r^2} + \varepsilon f_r, \tag{2.79}$$

$$\frac{1}{r\cos\phi}\frac{d}{dt}(r^2\dot{\theta}\cos^2\phi) = \varepsilon f_\theta, \tag{2.80}$$

and

$$\frac{1}{r}\frac{d}{dt}(r^2\dot{\phi}) + r\dot{\theta}^2\cos\phi\sin\phi = \varepsilon f_\phi . \tag{2.81}$$

2.5.2. Equations of Motion in Laplace's Form

We shall follow Laplace's approach (Brown, 1896) in obtaining his form of equations of motion by changing the independent variable from 't' to 'θ' and introducing the following dependent variables:

$$H = r^2\dot{\theta}\cos^2\phi , \tag{2.82}$$

$$Q = \frac{1}{r\cos\phi} , \tag{2.83}$$

and

$$S = \tan\phi . \tag{2.84}$$

After taking the first and second derivatives of Q and S with respect to θ, a substitution of the resulting expressions and (2.82)-(2.84) into the corresponding terms in (2.79)-(2.81) gives

$$\frac{d^2Q}{d\theta^2} + Q = \frac{\mu_E}{H^2(1+S^2)^{3/2}} + \varepsilon F_Q(Q, H, S, f_r, f_\theta, f_\phi), \tag{2.85}$$

$$\frac{dH}{d\theta} = \varepsilon F_H(Q, H, S, f_\theta), \tag{2.86}$$

and

$$\frac{d^2S}{d\theta^2} + S = \varepsilon F_S(Q,\ H,\ S,\ f_r,\ f_\theta,\ f_\phi)\ , \tag{2.87}$$

where F_Q, F_H, and F_S are the functions of parameters in the parentheses followed. In the following, we shall omit the parameters in writing the functions for simplicity. The readers will be able to see the expressions of εF_Q, εF_H, and εF_S in the following chapters as we proceed to study the geostationary satellites problem. It should be noted that in the process of obtaining (2.85), (2.87) has been used. The relation between time, t, and θ is obtained from (2.82) and (2.83) as

$$\frac{dt}{d\theta} = \frac{1}{HQ^2}\ . \tag{2.88}$$

It can be seen that once H, Q, and S are solved from (2.85)-(2.87) in terms of θ, t is computed directly by an integration of (2.88).

As a result of the above manipulation, (2.79)-(2.81) in the r-θ-ϕ-t system are replaced by (2.85)-(2.88) in our Q-S-H-θ-t system. The new system of equations certainly offers the advantages of determining the position of the satellite with knowing only the geometrical parameter θ, hence r and ϕ, while the old system requires the evaluation of the geometrical parameters based on the time, t. In addition, θ can be easily counted for a period of 2π in the analysis, whereas t is always related to an orbital period counting the time required for the movement of a satellite from perigee-to-perigee or node-to-node. It should be noted that the perigee and the node line usually revolve around the north-south pole line due to the oblateness of the Earth.

For readers familiar with the theory of mechanics, they already recognize that (2.85) and (2.87) are in a form of the general equation of free oscillations. These two equations include both nonlinear damping and restoring forces. They further notice that the terms on the right-hand side in (2.85)-(2.87) are associated with a small parameter, ε, except for the first term in (2.85). If ε becomes zero, the solutions to these equations will depict the well-known elliptical motion of a satellite around a spherical Earth. Since ε is small, we expect that the motion of the satellite in our problem will not deviate too much from the elliptical motion, and our solution to the problem is then based on a nominal ellipse with perturbed effects due to the ε-terms.

2.5.3. Equations of Motion - First-Order Differential Equations

In solving (2.85)-(2.87), we shall consider first the case of dropping the ε-terms in these equations; consequently, the resulting equations become a set of ordinary linear differential equations, that can be easily solved by the Lagrange's method of variation of parameters in the theory of differential equations (Saaty and Bram, 1964). The solutions to these equations are:

$$S = \tan\phi = X_1 \cos\theta + X_2 \sin\theta\ ,^{*} \tag{2.89}$$

$$Q = \frac{1}{r\cos\phi} = X_5 + X_3\cos\theta + X_4\sin\theta\ , \tag{2.90}$$

and

$$H = r^2\dot{\theta}\cos^2\phi = X_6, \tag{2.91}$$

where X's are the integration constants and

$$X_5 = \frac{\mu_E}{X_6^2}\frac{1}{1 + X_1^2 + X_2^2}[1 + (X_1\cos\theta + X_2\sin\theta)^2]^{1/2}\ .$$

The first derivatives of Q and S with respect to θ are

$$\frac{dQ}{d\theta} = \frac{\mu_E}{X_6^2}\frac{1}{1 + X_1^2 + X_2^2}\times\frac{(X_1\cos\theta + X_2\sin\theta)(-X_1\sin\theta + X_2\cos\theta)}{[1 + (X_1\cos\theta + X_2\sin\theta)^2]^{1/2}} - X_3\sin\theta + X_4\cos\theta\ , \tag{2.92}$$

and

$$\frac{dS}{d\theta} = -X_1\sin\theta + X_2\cos\theta\ . \tag{2.93}$$

The inclusion of the ε-terms in (2.85)-(2.87) requires that the X's in (2.89)-(2.91) no longer be considered as constants when the method of variation of parameters is carried out further. With the X's being considered as functions of θ, we differentiate (2.89)-(2.93) with respect to θ once, and equating the corresponding first derivatives to those in (2.86), (2.92), and (2.93), three equations are obtained. Two additional equations are derived by inserting the corresponding second derivatives together with (2.87)-(2.91) into (2.85) and (2.87). Consequently, we have

$$\frac{dX_1}{d\theta} = -\varepsilon\sin\theta\, F_S\ , \tag{2.94}$$

$$\frac{dX_2}{d\theta} = \varepsilon\cos\theta\, F_S\ , \tag{2.95}$$

$$\frac{dX_3}{d\theta} = \varepsilon\frac{2\mu_E}{X_6^3}\cos\theta\, F_H - \varepsilon\sin\theta\, F_Q\ , \tag{2.96}$$

$$\frac{dX_4}{d\theta} = -\varepsilon\frac{2\mu_E}{X_6^3}\sin\theta\, F_H + \varepsilon\cos\theta\, F_Q\ , \tag{2.97}$$

and

$$\frac{dX_6}{d\theta} = \varepsilon F_H\ , \tag{2.98}$$

where the terms containing $X_1(dX_1/d\theta)$, $X_1(dX_2/d\theta)$, etc. in (2.96) and (2.97) have been omitted because X_1, X_2 and their derivatives are very small as they are all associated with the small parameter, ε.

It should be mentioned that the parameters, $X_1, \ldots, X_6$, are here chosen particularly suitable for the study of low inclination and small eccentricity orbits**, which is the case of a geostationary satellite.

---

*To facilitate the attainment of a closed-form solution of the problem, a similar form of $S = X_7\sin(\theta + X_8)$ is used in Chapter 3. However, either form of S is applicable to our presentation of method of solution.

**The usual formulation of Lagrange's equations using six orbital elements as the dependent variables in Celestial Mechanics has a singularity when eccentricity or inclination vanishes; therefore a set of parameters, p-q-h-k, is used instead when orbits have low inclination and small eccentricity. Readers may refer to METHODS OF CELESTIAL MECHANICS by Brouwer and Clemence (1961) for detailed derivations.

If the expressions on the right-hand side of (2.94)-(2.98) are in terms of periodic functions of θ, the small parameter, ε, appearing on the right-hand side of these system of equations indicates that the X's remain practically constant within a 2π-period of θ. In other words, the X's can be represented by their averaged values during a period of θ. Consequently, the system of differential equations is quasi-linear. This fact suggests that (2.94)-(2.98) can be solved by the asymptotic method in nonlinear oscillations. The early version of the asymptotic method is the averaging method of Kryloff and Bogoliuboff (1943), which was widely used in the 1930's. As a refinement of the averaging method, the asymptotic method was developed, and the formulation of the method has been perfected since the late fifties due to Bogoliubov and Mitropolski (1961).

2.5.4. Asymptotic Method - Formal Solutions

As we examine the derivation of the asymptotic method in Bogoliubov and Mitropolski (1961), it is found that this method is basically a Taylor series expansion about a known solution to a differential equation. The order of the Taylor series expansion shows the accuracy of the solution. It is known that the higher order solution requires much more mathematical manipulations than the low order does. However, most of the problems in nonlinear mechanics require only the low order solution to reveal their essential characteristics. With this in mind, we shall present our formal solutions to the second order; however readers are referred to Bogoliubov and Mitropolski (1961) for the method of obtaining higher order solutions.

According to the asymptotic method, the general form of (2.94)-(2.98) can be represented as

$$\frac{dX_k}{d\theta} = Z_k(\theta,\ X), \quad k = 1,\ \ldots,\ 4,\ 6\ , \tag{2.99}$$

where X denotes $X_1, \ldots, X_4, X_6$, and $Z_k(\theta, X)$ are the functions of θ and X. The first-order solution of X_k is obtained from

$$\frac{d\bar{X}_k}{d\theta} = \varepsilon M_\theta[Z_k(\theta,\ \bar{X})]\ , \tag{2.100}$$

where

$$M_\theta[Z_k(\theta,\ \bar{X})] = \frac{1}{2\pi}\int_0^{2\pi} Z_k(\theta,\ X)\,d\theta \qquad \text{(X remains constant)}. \tag{2.101}$$

It can be seen that $\bar{X}_k$ is the averaged value of X_k over a 2π-period of θ. The improved first-order solution is

$$X_k = \bar{X}_k + \varepsilon F_k(\theta,\ \bar{X})\ , \tag{2.102}$$

where

$$F_k(\theta,\ \bar{X}) = \int\{Z_k(\theta,\ \bar{X}) - M_\theta[Z_k(\theta,\ \bar{X})]\}\,d\theta + f_k(\bar{X})\ , \tag{2.103}$$

and the integration is carried out by assuming that $\bar{X}$ remains constant, and $f_k(\bar{X})$ is determined by the requirement that $M_\theta[F_k(\theta,\ \bar{X})] = 0$.

The second-order solution of X_k is derived from

$$\frac{d\bar{X}_k}{d\theta} = \varepsilon M_\theta[Z_k(\theta,\ \bar{X})] + \varepsilon^2 \sum_{n=1\ (\neq 5)}^{6} M_\theta\left[F_n(\theta,\ \bar{X})\frac{\partial}{\partial \bar{X}_n} Z_k(\theta,\ \bar{X})\right]\ , \tag{2.104}$$

where $M_\theta[Z_k(\theta, \bar{X})]$ and $F_k(\theta, \bar{X})$ have been previously defined. The improved second-order solution has the form

$$X_k = \bar{X}_k + \varepsilon F_k(\theta, \bar{X}) + \varepsilon^2 G_k(\theta, \bar{X}) , \tag{2.105}$$

where

$$G_k(\theta, \bar{X}) = \sum_{n=1\ (\neq 5)}^{6} \int \left\{ F_n(\theta, \bar{X}) \frac{\partial}{\partial \bar{X}_n} Z_k(\theta, \bar{X}) - \right.$$
$$- M_\theta[F_n(\theta, \bar{X})] \frac{\partial}{\partial \bar{X}_n} Z_k(\theta, \bar{X}) -$$
$$\left. - \frac{\partial F_k}{\partial \bar{X}_n} M_\theta[Z_n(\theta, \bar{X})] \right\} d\theta + g_k(\bar{X}) . \tag{2.106}$$

The integrations are carried out by assuming that $\bar{X}$ remains constant and $g_k(\bar{X})$ is determined by the requirement that $M\,[G_k(\theta, \bar{X})] = 0$.

In (2.102) and (2.105), $\bar{X}_k$ depicts the secular part of X_k, and the remainder in these two equations gives the oscillatory aspect of X_k. A full derivation of (2.99)-(2.106) is presented in Bogoliubov and Mitropolski (1961).

2.5.5. Osculating Orbital Elements

Geometrically, the trajectory of a geostationary satellite under the influences of the Earth, Sun, and Moon is a space curve, which is described by (2.89)-(2.91). However, if the concept of an osculating orbit is introduced, the orbital elements can be expressed in terms of the X_k's (k = 1, ..., 6, ≠ 5) in (2.102) or (2.105). The orbit can be defined by (See Figure 2.1.)

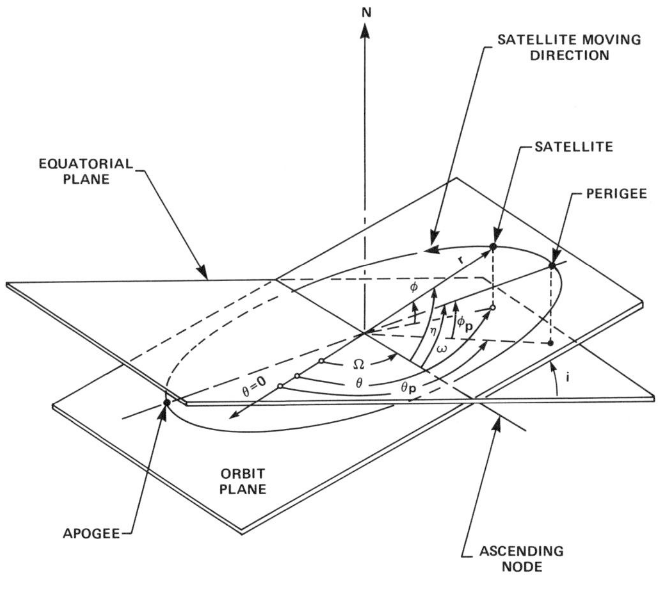

Fig. 2.1. Osculating orbital elements.

$$r = \frac{\ell}{1 + e\cos(\eta - \omega)} , \quad (2.107)$$

where ℓ is the semilatus rectum, e is the eccentricity, and η and ω are the arguments of latitude and perigee, respectively.

After combining (2.89), (2.90), and (2.91), and comparing the resulting expression with (2.107), we have

$$\frac{1}{r} = \frac{\mu_E}{X_6^2} \frac{1}{1 + X_1^2 + X_2^2} \left\{1 + \frac{X_6^2}{\mu_E} (1 + X_1^2 + X_2^2) \times \right.$$
$$\left. \times \frac{X_3 \cos\theta + X_4 \sin\theta}{[1 + (X_1 \cos\theta + X_2 \sin\theta)^2]^{1/2}} \right\} =$$
$$= \frac{1}{\ell} [1 + e\cos(\eta - \omega)] . \quad (2.108)$$

If we define $\delta X_k \equiv \varepsilon F_k(\theta, \bar{X})$ or $\delta X_k \equiv \varepsilon F_k(\theta, \bar{X}) + \varepsilon^2 G_k(\theta, \bar{X})$, which depict the oscillatory nature of X_k within a 2π-period of θ, then the expression, $\delta X_3 \cos\theta + \delta X_4 \sin\theta$, will have terms of constant, secular (containing only $\bar{X}$'s), long-period (containing combinations of $\bar{X}$'s, and trigonometric terms relating to the motions of the Sun and Moon), and short-period (containing combinations of $\bar{X}$'s and trigonometric terms relating to the motions of the Sun and Moon and the satellite (θ)) in nature. The short-period terms are basically associated with cos $n\theta$ or sin $n\theta$, wherein n is 1, 2, ..., etc. When the short-period terms containing either cos θ or sin θ are sorted out and lumped together as $\delta\bar{X}_3' \cos\theta + \delta\bar{X}_4' \sin\theta$, then $X_3 \cos\theta + X_4 \sin\theta$ in (2.108) results in

$$(\bar{X}_3 + \delta\bar{X}_3') \cos\theta + (\bar{X}_4 + \delta\bar{X}_4') \sin\theta + (\delta X_3'' \cos\theta + \delta X_4'' \sin\theta)$$
$$\equiv \bar{X}_3' \cos\theta + \bar{X}_4' \sin\theta + (\delta X_3'' \cos\theta + \delta X_4'' \sin\theta) , \quad (2.109)$$

where $(\delta X_3'' \cos\theta + \delta X_4'' \sin\theta)$ gives the constant, secular, long-period terms, and the short-period terms associated with cos $n\theta$ or sin $n\theta$ with $n \neq 1$.

A substitution of (2.109) into (2.108) results in

$$\frac{1}{\ell} = \frac{\mu_E}{X_6^2} \frac{1}{1 + X_1^2 + X_2^2} + \frac{(\delta X_3'' \cos\theta + \delta X_4'' \sin\theta)}{[1 + (X_1 \cos\theta + X_2 \sin\theta)^2]^{1/2}} , \quad (2.110)$$

and

$$e\cos(\eta - \omega) = \frac{\ell}{[1 + (X_1 \cos\theta + X_2 \sin\theta)^2]^{1/2}} \times$$
$$\times (\bar{X}_3' \cos\theta + \bar{X}_4' \sin\theta) . \quad (2.111)$$

From (2.110), it can be seen that $(\delta X_3'' \cos\theta + \delta X_4'' \sin\theta)$ constitutes a part of the semilatus rectum (ℓ). We shall derive various orbital elements in terms of X's.

2.5.5.1. Semilatus Rectum (ℓ)

The semilatus rectum is represented by (2.110), its averaged value over a 2π-period of θ may be approximated by

$$\frac{1}{\ell} = \frac{\mu_E}{\bar{X}_6^2} \frac{1}{1 + \bar{X}_1^2 + \bar{X}_2^2} + [\text{constant, secular and long-period terms in } (\delta X_3'' \cos\theta + \delta X_4'' \sin\theta)] , \quad (2.112)$$

where $[1 + (\bar{X}_1 \cos\theta + \bar{X}_2 \sin\theta)^2]^{1/2}$ is replaced by unity because this term is closely related to the inclination of orbit. The oscillatory characteristics of ℓ can be derived by applying a variational process to the first term on the right-hand side of (2.110) with known expressions of δX_1, δX_2, δX_6, $\bar{X}_1$, $\bar{X}_2$, and $\bar{X}_6$, and then by adding the short-period terms associated with cos nθ or sin nθ (with $n \neq 1$) in $(\delta X_3'' \cos\theta + \delta X_4'' \sin\theta)$ to the resulting expression of the first term.

2.5.5.2. Longitude of Ascending Node (Ω)

The node line of the osculating orbit can be determined by setting $\phi = 0$ in (2.89). This results in

$$\theta = \Omega = \tan^{-1}\left[-\frac{X_1}{X_2}\right], \tag{2.113}$$

which, in turn, gives

$$\Omega = \bar{\Omega} + \delta\Omega = \tan^{-1}\left(-\frac{\bar{X}_1}{\bar{X}_2}\right) + \frac{\bar{X}_1\delta X_1 - \bar{X}_2\delta X_2}{\bar{X}_1^2 + \bar{X}_2^2}. \tag{2.114}$$

2.5.5.3. Inclination of Orbital Plane (i)

The inclination of the osculating orbital plane may be determined by considering the geometry of any point in the orbit. As is seen in Figure 2.1.,

$$\cos\eta = \cos\phi\cos(\theta - \Omega), \tag{2.115}$$

and

$$\sin\phi = \sin i \sin\eta. \tag{2.116}$$

Manipulations between these two equations, (2.89), and (2.113), give

$$\tan i = (X_1^2 + X_2^2)^{1/2}, \tag{2.117}$$

and

$$i = \bar{i} + \delta i = \tan^{-1}(\bar{X}_1^2 + \bar{X}_2^2)^{1/2} + \frac{\bar{X}_1\delta X_1 + \bar{X}_2\delta X_2}{(\bar{X}_1^2 + \bar{X}_2^2)^{1/2}}. \tag{2.118}$$

2.5.5.4. Eccentricity (e) and Argument of Perigee (ω)

The position of the perigee of an osculating orbit can be designated by $\phi = \phi_p$ and $\theta = \theta_p$ as shown in Figure 2.1. Use of (2.89), (2.115), and (2.116) for the case of perigee gives

$$\cos\omega = \cos\Omega\cos\phi_p\cos\theta_p + \sin\Omega\cos\phi_p\sin\theta_p, \tag{2.119}$$

and

$$\sin\omega\sin i = X_1\cos\phi_p\cos\theta_p + X_2\cos\phi_p\sin\theta_p. \tag{2.120}$$

These two equations together with (2.111) for the cases of a satellite located at the perigee and at the ascending node result in

$$\tan\omega = \frac{X_1\bar{X}_3' + X_2\bar{X}_4'}{X_2\bar{X}_3' - X_1\bar{X}_4'} \times \frac{1}{(1 + X_1^2 + X_2^2)^{1/2}}, \tag{2.121}$$

and

$$e = \ell[\bar{X}_3'^2 + \bar{X}_4'^2 + (X_2\bar{X}_3' - X_1\bar{X}_4')^2]^{1/2}. \tag{2.122}$$

The averaged values of e and ω (i.e., $\bar{e}$ and $\bar{\omega}$) are obtained by inserting $\bar{X}_1$ and $\bar{X}_2$ into (2.121) and (2.122) with ℓ replaced by $\bar{\ell}$, and the oscillatory nature of e and ω over a 2π-period of θ can be brought out by applying the variational technique to (2.121) and (2.122) in terms of $\bar{X}_1$, $\bar{X}_2$, $\bar{X}_3'$, $\bar{X}_4'$, ℓ, δX_1, δX_2, and $\delta\ell$. It should be noted that $\bar{X}_3'$ and $\bar{X}_4'$ do not contain any oscillatory terms which vanish over a 2π-period of θ.

2.5.6. Recapitulation

Our procedures of solving the geostationary satellites problem can be summarized by the following steps:

Given:
Equations of Motion, Perturbing Forces } in spherical coordinate system

Change of the independent variable from t to θ → Equations of Motion in Laplace's Form

Lagrange's Method of variation of parameters → Equations of Motion - six first-order differential equations

Asymptotic Method → Solution

2.6. Comparison of Different Methods of Solution

After examining the different methods of solution described in the previous sections, the following statements can be made:

(1) Except for the original version of the von Zeipel method, all other methods involve the employment of the averaging technique in the method of solution.

(2) In the process of solution, all methods use an angular parameter (true anomoly, mean anomoly, spherical angle coordinate, etc.) as the independent variable.

(3) All methods utilize the Taylor series expansion technique to develop the terms of small order either for the force function (or Hamiltonian) or for the dependent variables (or orbital parameters). By virtue of the methods, the von Zeipel method and Hori-Lie method have their Taylor series version of the dependent variables obtained from a determination of the disturbing/generating function, while the other methods use a direct Taylor series expansion of the dependent variables about their mean values or zeroth-order solution.

(4) As the equations of motion and the perturbing forces are all basically given in the spherical coordinate system, the von Zeipel method and Hori-Lie method require two transformations (spherical coordinate system → osculating orbital element system → canonical variable system) to obtain a system of canonical equations, while the proposed approach attains the same goal by resolving the Laplace's form equations of motion with the standard Lagrange's method of variation of parameters into a system of six first-order differential equations. Kozai's approach requires the perturbing force functions to be transformed from the spherical coordinate system to

the osculating orbital element system. However, in the case of geostationary satellites, the near-zero eccentricity and near-zero inclination of the satellite orbit create a singularity problem in the Lagrange's equations of the general planetary motion, and further transformation is required as will be discussed in Section 4.11.1.

(5) The asymptotic method solves a system of differential equations associated with a small parameter. The method of solution is based on a Taylor series development around the solution of a system of differential equations when the small parameter becomes zero. In the present case, the zero-small-parameter solution to the geostationary satellites problem is an ellipse resulting from the central force field of the Earth. In other words, our proposed solution is a Taylor series development around a known ellipse solution. This is comparable to an ordinary Taylor series applied to a differential equation developed around a known point (the given initial condition).

(6) The application of the von Zeipel method, Hori-Lie method, and Kozai's method require a considerable background of celestial mechanics, a field most aerospace scientists and engineers are not familiar with.

2.7. Proposed Scheme of Solutions

In Section 2.3.2., we have examined the perturbing forces on a geostationary satellite; the perturbations of the Earth oblateness (J_2), Sun, Moon, and solar radiation pressure are considered to be the first-order of magnitude (J_2), while the perturbation of the Earth's equator ellipticity (J_{22}) is of the second-order (J_2^2). From the method of solution point of view, the solution to the geostationary satellite problem should be solved to the second-order; however, the complexity of the problem rules out the possibility of obtaining one integrated second-order analytical approximate solution encompassing all perturbing forces. From the mathematical point of view, the second-order J_{22}-term in the geostationary satellites problem should not have any effects on the first-order solution. If a second-order solution of the Earth oblateness and Earth's equator ellipticity problem is obtained, we shall be able to find those orbital elements which are basically dominated by the J_{22}-term, while the effects of the J_{22}-term on those orbital parameters, by definition, will not appear in the first-order solution. Therefore, our solution to the geostationary satellites problem can be represented by the J_{22}-term-dominated orbital parameters and those obtained from the first-order solution. In the following chapters, we shall attempt to obtain a second-order (J_2^2) solution to the Earth oblateness (J_2) and Earth's equator ellipticity problem of geostationary satellites first, and then solve the geostationary satellites problem of the combined perturbations of the Earth oblateness, Sun, Moon, and solar radiation pressure to the first-order. From these two solutions, we shall find that each perturbing force plays a dominant role only on certain orbital elements. If we are interested only in the first-order solution to the geostationary satellites problem, a combination of these two separate solutions will certainly satisfy our purpose of studying the behavior of geostationary satellites.

References

Abramowitz, M. and Stegun, I.A.; Handbook of Mathematical Functions with Formulas, Graphs, and Mathematical Tables, National Bureau of Standards, Applied Mathematics Series 55, U.S. Government Printing Office, Washington, D.C., 20402, 1964.

Bogoliubov, H. and Mitropolski, J.; Asymptotic Method in the Theory of Nonlinear Oscillations, Hindustan Publishing Corporation, Delhi-6, India, 1961.

Brouwer, D.; 'Solution of the Problem of Artificial Satellite Theory Without Drag', Astronomical Journal 64, 378-397 (1959).

Brouwer, D. and Clemence, G.M.; Method of Celestial Mechanics, Academic Press, New York and London, 1961.

Brown, E.W.; An Introductory Treatise on the Lunar Theory, Cambridge University Press, 1986 -- reprinted by Dover Publication, Inc., New York, 1960.

Gaposchkin, E.M.; 'Tesseral Harmonic Coefficients and Station Coordinates from Dynamic Method', Special Report 200, Smithsonian Astrophysical Observatory, Cambridge, Mass. (1966).

Gaposchkin, E.M. and Lambeck, K; '1969 Smithsonian Standard Earth (II)', Special Report 315, Smithsonian Astrophysical Observatory, Cambridge, Mass. (1970).

Gaposchkin, E.M.; 'Earth's Gravity Field to the Eighteenth Degree and Geocentric Coordinates for 104 Stations from Satellite and Terrestrial Data', J. Geophysical Research 79, 5377-5411 (1974).

Hori, G; 'Theory of General Perturbations', in B.D. Tapley and V. Szebehely (eds.), Recent Advances in Dynamical Astronomy, D. Reidel Publ. Co., Dordrecht, Holland (1973).

Kaula, W.M.; Theory of Satellite Geodesy, Blaisdell Publishing Company, Waltham, Mass., Toronto, London, 1966.

Korn, G.A. and Korn, T.M.; Mathematical Handbook for Scientists and Engineers, McGraw-Hill Book Company, New York, 1968.

Kozai, Y.; 'The Motion of a Close Earth Satellite', Astronomical Journal 64, 367-377 (1959).

Kozai, Y.; 'Revised Zonal Harmonics in the Geopotential', Special Report 295, Smithsonian Astrophysical Observatory, Cambridge, Mass. (1969).

Kozai, Y.; 'A New Method to Compute Lunisolar Perturbations in Satellite Motions', Special Report 349, Smithsonian Astrophysical Observatory, Cambridge, Mass. (1973).

Kryloff, N.M. and Bogoliuboff, N.N.; Introduction to Nonlinear Mathematics, translated by S. Lefschetz, Princeton University Press, Princeton, N.J. 1943.

Mello, S.F.; 'Analytical Studies of Earth's Shadowing Effects on Satellite Orbits', Celestial Mechanics **5**, 80-101 (1972).

Morando, M.B.; 'Libration d'un Satellite de 24h.', Compt. rend. **254**, 635-637 (1962a).

Morando, M.B.; 'Orbites de Résonance des Satellite de 24h.', Bulletin Astronomique **24**, 48-67 (1962b).

Saaty, T.L. and Bram, J.; Nonlinear Mathematics, McGraw-Hill Book Company, New York, 1964.

Seifert, H.; Space Technology, John Wiley & Sons, Inc., New York, 1959.

Smith, D.E., Lerch, F.J., Marsh, J.G., Wagner, C.A., Kelenkiewicz, R. and Klan, M.A.; 'Contribution to the National Geodetic Satellite Program by Goddard Space Flight Center', J. Geophysical Research **81**, 1006-1026 (1976).

Wagner, C.A.; 'Acceleration on 24-Hour Satellites and Low Order Longitude Terms in the Geopotential', X-552-69-508, Goddard Space Flight Center, NASA (1969).

Wills, A.P.; Vector Analysis with an Introduction to Tensor Analysis, Dover Publications, Inc., New York, 1958.

Chapter 3

THE EFFECT OF EARTH'S EQUATORIAL ELLIPTICITY

3.1. General

We shall study the effect of the Earth alone on a geostationary satellite. The equations of motion of a satellite solely under the influence of the Earth's gravitational field may be written in spherical coordinates (see Figure 3.1.) as

$$\ddot{r} - r\dot{\theta}^2 \cos^2\phi - r\dot{\phi}^2 = -\frac{\mu_E}{r^2} + F_r ,$$

$$\frac{1}{r\cos\phi}\frac{d}{dt}(r^2\dot{\theta}\cos^2\phi) = F_\theta , \tag{3.1}$$

and $$\frac{1}{r}\frac{d}{dt}(r^2\dot{\phi}) + r\dot{\theta}^2\cos\phi\sin\phi = F_\phi ,$$

where μ_E is Earth's gravitational constant, r is the radial distance from Earth's center, θ is the angle measured eastward (i.e., counter-clockwise looking toward the equatorial plane from the north pole), and ϕ is the geocentric latitude (north positive). The F's in the above equations are the Earth's gravitationary perturbing force components per unit mass of the satellite in the r-, θ-, and ϕ-directions as designated by the respective subscript. Single and double dots above a variable signify the first and the second time derivatives, respectively.

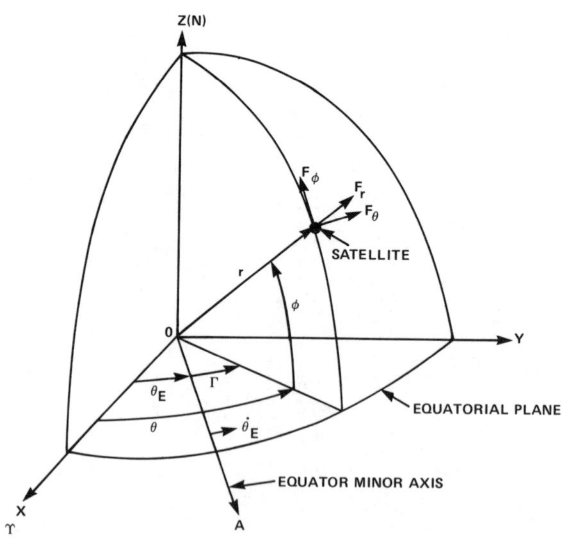

Fig. 3.1. Geocentric coordinate system.

The Earth's gravitational potential can be expressed in terms of spherical harmonics by means of the Legendre Polynomials as discussed in Chapter 2. The complicated nature of the expression of the polynomials prohibits any direct adoption of the expression in our approach to the problem. In order to obtain some meaningful results, only the dominant terms will be selected in our method of solution. We shall examine the fourth-order gravitational field of the Goddard Earth Model (Gem) 6 (Smith et al., 1976) by computing the perturbing forces on a geostationary satellite. These forces are estimated by setting $\sin\phi = 0$, $\cos\phi = 1$, and $r = r_c$ (the synchronous orbit radius) in the expressions of the radial (F_r), longitudinal (F_θ), and latitudinal (F_ϕ) forces shown in Appendix A.

A similar estimate can also be made by only taking the terms containing J_2 (oblateness coefficient) and J_{22} (equatorial ellipticity coefficient). As a result of $\sin\phi = 0$ and $\cos\phi = 1$, the J_2-effect only appears in the radial component of the gravitational acceleration, which is equal to $316.7 \times 10^{-7}\, g_s$ (g_s = the gravitational acceleration of the central force field of the Earth at the synchronous altitude). We shall plot these two estimates of the corresponding gravitational perturbing force components in Figures 3.2. to 3.4. as a means of comparing the fourth-order gravitational field and the triaxial (J_2 and J_{22} only) gravitational field in our case of interest.

Examination of these figures clearly shows that the triaxial field (J_2 and J_{22} only) plays a dominant role in determining the perturbing forces: The radial components in these two gravitational fields are in agreement with each other with a maximum deviation less than $0.5 \times 10^{-7}\, g_s$; the longitudinal accelerations of the triaxial gravitational field have a maximum deviation of $0.5 \times 10^{-7}\, g_s$ from that of the fourth-order one with an overall deviation of about ten percent on the average; and the latitudinal perturbation due to the triaxial field is zero, and the corresponding perturbing acceleration due to the fourth-order gravitational field is comparatively small with an order of the magnitude of $10^{-8}\, g_s$. Furthermore, in the fourth-order field, the minimum magnitude of the latitudinal perturbation occurs close to the dynamically unstable regions near the equator's major axis, and the maximum magnitude of the latitudinal perturbation occurs near the dynamically stable regions close to the minor axis. This coincidence tends to minimize the drift in the neighborhood of the major axis as any disturbance in the dynamically unstable region will cause drift although the latitudinal force is normal to the direction of drift. Based upon these observations, we are therefore justified to employ the triaxial field in our study of the behavior of near-equatorial geostationary satellites in the Earth's gravitational field. This simplification not only gives us a manageable approach to analyze the problem but also results in some meaningful expressions for physical interpretations and practical applications.

In realizing the role of the longitudinal perturbing force in our study dominated by J_{22}, we can modify the value of J_{22} to fit the

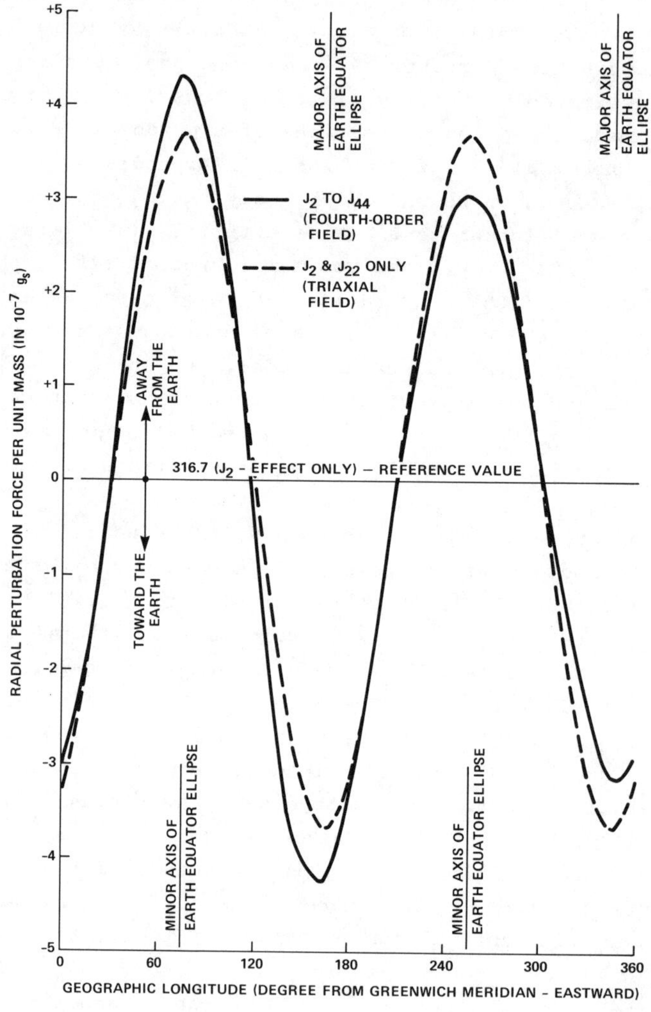

Fig. 3.2. Radial perturbation acceleration on a near-equatorial geostationary satellite.

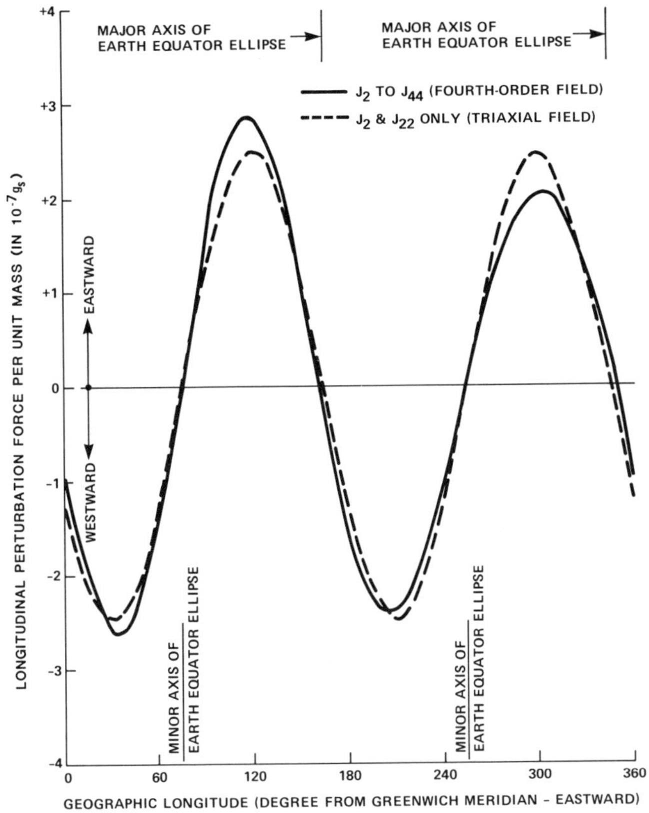

Fig. 3.3. Longitudinal perturbation acceleration on a near-equatorial geostationary satellite.

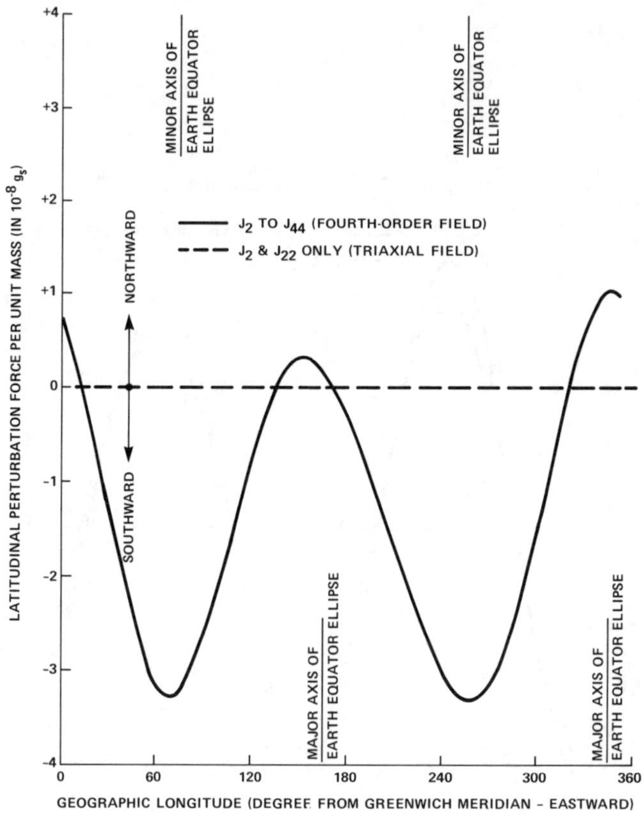

Fig. 3.4. Latitudinal perturbation acceleration on a near-equatorial geostationary satellite.

fourth-order gravitational field for geostationary satellites located at different longitudes. In other words, we adopt a particular value for J_{22} for a certain longitudinal region and, as soon as the satellite drifts out of that region, a new J_{22} value will be chosen. This procedure should reduce the errors introduced by truncating the higher order terms in the Earth's gravitational field.

3.2. Equatorial Ellipticity Constants of the Earth (λ_{22}, J_{22})

The shape of the Earth's equator can be best represented by an ellipse which, in turn, can be described by three constants: the mean radius (R_E), the eccentricity (related to J_{22}), and the geographic longitude of the major axis (λ_{22}) with respect to the Greenwich meridian. These constants appear in the Earth gravitational potential model, and their values are derived from actual observed orbit data. Many Earth potential models exist in open literature, among the better known models are the SAO-66-M1 model

(Gaposchkin, 1966), the SAO-69 (Gaposchkin, 1970; Kozai, 1969), Wagner's Model (1969), Gaposchkin's Model (1974), and Gem 6 (Smith et al., 1976). Their λ_{22}'s and J_{22}'s are given in Table II.

TABLE II. Earth equatorial ellipticity constants

Earth potential model	λ_{22}	$J_{22} \times 10^6$
SAO-66-M1	-14.796°	-1.766
SAO-69	-14.741°	-1.789
Wagner	-14.918°	-1.819
Gaposchkin	-14.924°	-1.771
Gem 6	-14.895°	-1.804

Generally, the model to be used in an application of the theory should be the one whose construction was based on data obtained from satellites similar to the satellite in the application. With this in mind, we choose the Wagner model for our study because this model has been derived from thirty-five synchronous orbits, nineteen of which have near-zero inclinations.

3.3. Equations of Motion

As we have previously stated, the triaxial Earth gravitational field will be selected as our Earth's gravitational potential model in our study of the behavior of near-equatorial geostationary satellites. Consequently, the Earth's gravitational potential field is in the form of (see Appendix A)

$$V_E = -\frac{\mu_E}{r}\left(\frac{R_E}{r}\right)^2[\tfrac{1}{2}J_2(3\sin^2\phi - 1) + 3J_{22}\cos^2\phi\cos 2(\lambda - \lambda_{22})] ,$$

and its gradients in the r-, θ-, and φ-directions are

$$F_r = \frac{\partial V_E}{\partial r} = \frac{\mu_E}{r^2}\left(\frac{R_E}{r}\right)^2[\tfrac{1}{2}J_2(3\sin^2\phi - 1) + 9J_{22}\cos^2\phi\cos 2(\lambda - \lambda_{22})],$$

$$F_\theta = \frac{1}{r\cos\phi}\frac{\partial V_E}{\partial\theta} = 6\frac{\mu_E}{r^2}\left(\frac{R_E}{r}\right)^2 J_{22}\cos^2\phi\sin 2(\lambda - \lambda_{22}) ,$$

and

$$F_\phi = \frac{\partial V_E}{\partial\phi} = \frac{\mu_E}{r^2}\left(\frac{R_E}{r}\right)^2[-3J^2\cos\phi\sin\phi + 6J_{22}\sin\phi\cos\phi \times \cos 2(\lambda - \lambda_{22})] ,$$

respectively.

In the above equations, λ_{22} denotes the geographic longitude of the major axis of Earth's equatorial ellipse and λ depicts a satellite's geographic longitude with respect to the Greenwich meridian (positive eastward). As indicated in Figure 3.1., we shall use the minor axis of Earth's equatorial ellipse as a reference line of the satellite angular position (Γ) with respect to a rotating Earth. This arrangement gives $\Gamma - 90° = \lambda - \lambda_{22}$, and consequently results in a change of sign of all J_{22}-terms in the above equations; therefore, the three equations in (3.1) take the form (see

Figure 3.1.):

$$\ddot{r} - r\dot{\theta}^2 \cos^2\phi - r\dot{\phi}^2 = -\frac{\mu_E}{r^2} + \frac{3}{2}J_2\frac{\mu_E R_E^2}{r^4}(3\sin^2\phi - 1) - 9J_{22}\frac{\mu_E R_E^2}{r^4}\cos^2\phi\cos 2\Gamma \,, \tag{3.2}$$

$$\frac{1}{r\cos\phi}\frac{d}{dt}(r^2\dot{\theta}\cos^2\phi) = -6J_{22}\frac{\mu_E R_E^2}{r^4}\cos\phi\sin 2\Gamma \,, \tag{3.3}$$

and

$$\frac{1}{r}\frac{d}{dt}(r^2\dot{\phi}) + r\dot{\theta}^2\cos\phi\sin\phi = -3J_2\frac{\mu_E R_E^2}{r^4}\sin\phi\cos\phi - 6J_{22}\frac{\mu_E R_E^2}{r^4}\sin\phi\cos\phi\cos 2\Gamma \,, \tag{3.4}$$

where

$$\Gamma = \theta - \theta_E = \theta - (\theta_{E0} + \dot{\theta}_E t) \,, \tag{3.5}$$

Γ is the drift angle, the satellite's angular position relative to the equator's minor axis (line 0A in Figure 3.1.), θ_E is the angular position of the minor axis relative to an inertial x-axis (♈, vernal equinox), θ_{E0} is the angular position of the minor axis at time zero, t is the time, and $\dot{\theta}_E$ is the Earth's angular rate of rotation.

In (3.2), if the angular velocity of a satellite along an equatorial circular orbit is $\dot{\theta}_E$, the resulting orbit is known as a geostationary orbit. This orbit has a radius of

$$r_c = \left[\frac{\mu_E}{\dot{\theta}_E^2}\left(1 + \frac{3}{2}J_2\frac{R_E^2}{r_c^2}\right)\right]^{1/3} ,$$

where the term containing J_{22} in (3.2) has been omitted, and r_c is known as the synchronous radius. In the usual form of a circular orbit, r_c may be written as

$$r_c = (\bar{\mu}_E/\dot{\theta}_E^2)^{1/3} = r_s\left(1 + \frac{1}{2}J_2\frac{R_E^2}{r_s^2}\right) ,$$

and

$$\bar{\mu}_E \equiv \left(1 + \frac{3}{2}J_2\frac{R_E^2}{r_s^2}\right)\mu_E \,,$$

where r_c in the J_2-term is approximated by r_s, and r_s is the synchronous radius for a spherical Earth of the same mass. The adoption of $\bar{\mu}_E$ herein is basically for the simplification of our presentation followed.

Following the Laplace's transformation of the parameters described in Chapter 2, we change the independent variable from 't' to 'θ' by introducing the following dependent parameters:

$$H = r^2\dot{\theta}\cos^2\phi \,, \tag{3.6a}$$

$$Q = \frac{1}{r\cos\phi} \,, \tag{3.6b}$$

and

$$S = \tan\phi \,. \tag{3.6c}$$

With (3.6a-c) and the first and second derivatives of Q and S with respect to θ, a direct substitution of the corresponding terms in (3.2)-(3.4) results in

$$\frac{d^2Q}{d\theta^2} + Q = \frac{\mu_E}{H^2(1 + S^2)^{3/2}} - \frac{1}{H}\frac{dQ}{d\theta}\frac{dH}{d\theta} + \frac{3}{2}J_2\mu_E R_E^2\frac{Q^2}{H^2} + 9J_{22}\mu_E R_E^2\frac{Q^2}{H^2}\cos 2\Gamma \,, \tag{3.7}$$

$$\frac{dH}{d\theta} = - 6J_{22}\mu_E R_E^2 \frac{Q}{H} \sin 2\Gamma \ , \tag{3.8}$$

and

$$\frac{d^2S}{d\theta^2} + S = - \frac{1}{H}\frac{dS}{d\theta}\frac{dH}{d\theta} - 3J_2\mu_E R_E^2 \frac{Q}{H^2} S - 6J_{22}\mu_E R_E^2 \frac{Q}{H^2} S \cos 2\Gamma \ , \tag{3.9}$$

where S^2 in the terms containing J_2 and J_{22} has been approximated by zero because orbits of geostationary satellites are near-equatorial.

Therefore, (3.2), (3.3), and (3.4) are replaced by (3.7), (3.8), and (3.9), respectively, in our H-Q-S parameters system. It should be noted that in the process of obtaining (3.7), (3.9) has been used.

It is reminded that when a comparison of (3.7)-(3.9) with (2.85)-(2.87) is made, the reader will be able to realize the actual complicated expressions of F_Q, F_H, and F_S in our discussion of the method of solution in Chapter 2.

3.4. Transformation of Equations of Motion - First-Order Differential Equations

For a reader familiar with the theory of mechanics, he already notices that (3.7) and (3.9) are in a form of the general equation of free oscillations. They include both nonlinear damping and restoring forces. He further notices that the terms on the right-hand side of all three equations are associated with small parameters, such as J_2 and J_{22}, except the first term in (3.7). If J_2 and J_{22} become zero, the solutions to these equations will depict the well-known elliptic motion of a satellite around a spherical Earth. Since J_2 and J_{22} are small, we expect the motion of a satellite in our problem will be in the neighborhood of the elliptical motion, and our solution to the problem is then based on an ellipse with perturbed effects due to J_2- and J_{22}-terms.

In solving (3.7)-(3.9), the Lagrange's method of variation of parameters in the theory of differential equations (Saaty and Bram, 1964) mentioned in Chapter 2 will be employed to obtain a system of first-order differential equations to replace (3.7)-(3.9). The parameters, $X_3, \ldots, X_8$, are introduced such that they are related to our Q-H-S parameters by the following expressions:

$$Q = X_5 + X_3 \cos\theta\ X_4 \sin\theta = \frac{\mu_E}{X_6^2(1 + X_7^2)}[1 + X_7^2 \sin^2(\theta + X_8)]^{1/2} + \\ + X_3 \cos\theta + X_4 \sin\theta \ , \tag{3.10}$$

$$H = X_6 \ , \tag{3.11}$$

and

$$S = X_7 \sin(\theta + X_8) \ . \tag{3.12}$$

In fact, (3.10), (3.11), and (3.12) are the solutions to (3.7), (3.8), and (3.9), respectively when terms containing J_2 and J_{22} are dropped, and $X_3, \ldots, X_8$ are then considered as integration constants in those solutions. As previously mentioned in Chapter 2, the parameters, $X_3, \ldots, X_8$, are here chosen particularly suitable for the study of a low eccentricity and low inclination orbit, which is the case of a geostationary satellite.

In terms of X_3, ..., X_8, the corresponding first derivatives of Q and S with respect to θ are

$$\frac{dQ}{d\theta} = \frac{\mu_E X_7^2 \sin(\theta + X_8) \cos(\theta + X_8)}{X_6^2(1 + X_7^2)[1 + X_7^2 \sin^2(\theta + X_8)]^{1/2}} - X_3 \sin\theta + X_4 \cos\theta , \tag{3.13}$$

and

$$\frac{dS}{d\theta} = X_7 \cos(\theta + X_8) . \tag{3.14}$$

As we further carry out the method of variation of parameters, X_3, ..., X_8 in (3.10)-(3.12) can no longer be constants; instead, they become functions of θ. Now differentiating (3.10)-(3.14) with respect to θ once, and then equating the corresponding first derivatives to those in (3.8), (3.13), and (3.14), three equations are obtained. Two additional equations are derived by inserting the corresponding second derivatives together with (3.10)-(3.12) into (3.7) and (3.9). After carrying out some algebraic manipulations among these resulting five equations, a system of five first-order differential equations and one algebraic equation are obtained, which replaces (3.7)-(3.9):

$$X_5 = \frac{\mu_E}{X_6^2(1 + X_7^2)}[1 + X_7^2 \sin^2(\theta + X_8)]^{1/2} , \tag{3.15}$$

$$\frac{dX_3}{d\theta} = -\frac{3}{2}J_2\mu_E R_E^2 \frac{Q^2}{X_6^2} \sin\theta - 9J_{22}\mu_E R_E^2 \frac{Q^2}{X_6^2} \sin\theta \cos 2\Gamma + + 2\mu_E \frac{1}{X_6^3}\frac{dX_6}{d\theta} \cos\theta + \frac{1}{X_6}\frac{dQ}{d\theta}\frac{dX_6}{d\theta} \sin\theta , \tag{3.16}$$

$$\frac{dX_4}{d\theta} = \frac{3}{2}J_2\mu_E R_E^2 \frac{Q^2}{X_6^2} \cos\theta + 9J_{22}\mu_E R_E^2 \frac{Q^2}{X_6^2} \cos\theta \cos 2\Gamma + + 2\mu_E \frac{1}{X_6^3}\frac{dX_6}{d\theta} \sin\theta - \frac{1}{X_6}\frac{dQ}{d\theta} \cos\theta , \tag{3.17}$$

$$\frac{dX_6}{d\theta} = -6J_{22}\mu_E R_E^2 \frac{Q}{X_6} \sin 2\Gamma , \tag{3.18}$$

$$\frac{dX_7}{d\theta} = -3J_2\mu_E R_E^2 \frac{Q}{X_6^2} X_7 \sin(\theta + X_8) \cos(\theta + X_8) - - 6J_{22}\mu_E R_E^2 \frac{Q}{X_6^2} X_7 \sin(\theta + X_8) \cos(\theta + X_8) \cos 2\Gamma - - \frac{1}{X_6}\frac{dX_6}{d\theta}\frac{dS}{d\theta} \cos(\theta + X_8) , \tag{3.19}$$

and

$$\frac{dX_8}{d\theta} = 3J_2\mu_E R_E^2 \frac{Q}{X_6^2} \sin^2(\theta + X_8) + 6J_{22}\mu_E R_E^2 \frac{Q}{X_6^2} \sin^2(\theta + X_8) \cos 2\Gamma + + \frac{1}{X_6}\frac{dX_6}{d\theta}\frac{dS}{d\theta} \sin(\theta + X_8) , \tag{3.20}$$

where X_7^2 has been approximated by zero. As will be seen later, X_7 is closely related to the inclination of the orbital plane.

In order to relate Γ to (3.15)-(3.20), (3.5) is differentiated with respect to θ as

$$\frac{d\Gamma}{d\theta} = 1 - \frac{d\theta_E}{d\theta} = 1 - \frac{d\theta_E}{dt}\frac{dt}{d\theta} = 1 - \dot{\theta}_E \frac{1}{HQ^2} << 1 , \tag{3.21}$$

where the expressions in (3.6) have been used.

Equations (3.15)-(3.21) constitute a system of first-order differential equations to be solved. These equations replace the original second-order differential equations (3.2)-(3.4) and the equation (3.5) for the longitudinal position of the satellite with respect to the Earth equator's minor axis.

If (3.16)-(3.21) could be integrated analytically with their known integration constants, the method of variation of parameters requires that the integrated expressions of the variation parameters be substituted into (3.10)-(3.12) to obtain the solutions to Q, H, and S. Unfortunately, complicated expressions of (3.16)-(3.21) rule out any possibility of integrating them analytically, so we shall rely on other means to proceed with our method of solution.

Since a near-equatorial geostationary satellite is nominally in a circular orbit, we have

$$Q \cong \frac{\mu_E}{X_6^2} + X_3 \cos\theta + X_4 \sin\theta , \tag{3.22}$$

and

$$\frac{dQ}{d\theta} \cong - \frac{2\mu_E}{X_6^3}\times\frac{dX_6}{d\theta} - X_3 \sin\theta + X_4 \cos\theta + \cos\theta \frac{dX_3}{d\theta} + \sin\theta \frac{dX_4}{d\theta} . \tag{3.23}$$

On observing (3.16)-(3.18), it is seen that in the above equations, $dX_6/d\theta \sim 0\ (J_2^2)$, and each of the remaining terms on the right-hand side is of $0(J_2)$; thus $dQ/d\theta \sim 0\ (J_2)$. With these observations, (3.16)-(3.20) can be represented in the form of

$$\frac{dX_k}{d\theta} = J_2 Z_k(\theta, X_3, X_4, X_6, X_7, X_8) + J_{22} Y_k(\theta, \Gamma, X_3, X_4, X_6, X_7, X_8) \quad (k = 3, 4, 6, 7, 8) , \tag{3.24}$$

where higher order terms such as

$$\frac{dQ}{d\theta}\times\frac{dX_6}{d\theta} \sim 0\ (J_2^3), \quad \frac{dX_6}{d\theta}\times\frac{dX_7}{d\theta} \sim 0\ (J_2^3), \quad \frac{dX_6}{d\theta}\times\frac{dX_8}{d\theta} \sim 0\ (J_2^3),$$

etc. are omitted. It is understood that Z_k and Y_k are functions of the arguments shown in the parentheses. In the following section, (3.24) will be solved by the asymptotic method in nonlinear mechanics (Bogoliubov and Mitropolski, 1961) described in Chapter 2.

3.5. Formal Solutions

Equation (3.24) can be written in the form of

$$\frac{dX_k}{d\theta} = \varepsilon Z_k(\theta, X) + \varepsilon^2 Y_k(\theta, \Gamma, X), \tag{3.25}$$

where ε is a small parameter, and X stands for all X_k's. This equation has a different form as is compared with (2.99) in Chapter 2; however, our method of solution remains the same. We shall carry out the formal solution in detail so that readers will fully grasp the procedure of solution as we proceed. The first-order solution of X_k is obtained from

$$\frac{d\bar{X}_k}{d\theta} = \varepsilon M_\theta[Z_k(\theta, \bar{X})] , \tag{3.26}$$

where

$$M_\theta[Z_k(\theta, \bar{X})] = \frac{1}{2\pi}\int_0^{2\pi} Z_k(\theta, X)\, d\theta \quad (X \text{ remains constant}) , \tag{3.27}$$

and $\bar{X}_k$ is the averaged value of X_k over a 2π-period of θ.

The improved first-order solution is

$$X_k = \bar{X}_k + \varepsilon F_k(\theta, \bar{X}) , \tag{3.28}$$

where

$$F_k(\theta, \bar{X}) = \int\left\{Z_k(\theta, \bar{X}) - M_\theta[Z_k(\theta, \bar{X})]\right\} d\theta + f_k(\bar{X}) , \tag{3.29}$$

and the integration is carried by assuming that $\bar{X}$ remains constant, and $f_k(\bar{X})$ is determined by the requirement that $M_\theta[F_k(\theta, \bar{X})] = 0$.

The second-order solution of X_k is derived from

$$\frac{d\bar{X}_k}{d\theta} = \varepsilon M_\theta[Z_k(\theta, \bar{X})] + \varepsilon^2\left\{M_\theta[Y_k(\theta, \Gamma, \bar{X})] + \right.$$
$$\left. + \sum_{n=\text{all } k\text{'s}} M_\theta[F_n(\theta, \bar{X}) \frac{\partial}{\partial \bar{X}_n} Z_k(\theta, \bar{X})] \right\}, \tag{3.30}$$

where $M_\theta[Z_k(\theta, \bar{X})]$ and $F_k(\theta, \bar{X})$ have been defined in (3.27) and (3.29), and

$$M_\theta[Y_k(\theta, \Gamma, \bar{X})] = \frac{1}{2\pi}\int_0^{2\pi} Y_k(\theta, \Gamma, X)\, d\theta \quad (X \text{ and } \Gamma \text{ remain constant}).$$

The improved second-order solution has the form

$$X_k = \bar{X}_k + \varepsilon F_k(\theta, \bar{X}) + \varepsilon^2[G_k(\theta, \bar{X}) + N_k(\theta, \Gamma, \bar{X})] , \tag{3.31}$$

where

$$G_k(\theta, \bar{X}) = \sum_{n=\text{all } k\text{'s}} \int\left\{F_n(\theta, \bar{X}) \frac{\partial}{\partial \bar{X}_n} Z_k(\theta, \bar{X}) - \right.$$
$$- M_\theta[F_n(\theta, \bar{X}) \frac{\partial}{\partial \bar{X}_n} Z_k(\theta, \bar{X})] -$$
$$\left. - \frac{\partial F_k(\theta, \bar{X})}{\partial \bar{X}_n} M_\theta[Z_n(\theta, \bar{X})]\right\} d\theta + g_k(\bar{X}) ,$$

and

$$N_k(\theta, \Gamma, \bar{X}) = \int\left\{Y_k(\theta, \Gamma, \bar{X}) - M_\theta[Y_k(\theta, \Gamma, \bar{X})]\right\} d\theta + n_k(\Gamma, \bar{X}) ,$$

and the integrations are again carried out by assuming that $\bar{X}$ and Γ remain constant. The expressions $g_k(\bar{X})$ and $n_k(\bar{X}, \Gamma)$ are determined by the requirements that $M_\theta[G_k(\theta, \bar{X})] = 0$ and $M_\theta[N_k(\theta, \Gamma, \bar{X})] = 0$, respectively.

It should be noted that, in (3.28) and (3.31), $\bar{X}_k$ denotes the secular part of X_k, and the remainder of the equation depict the oscillatory aspect of X_k. A full derivation of the equations in this section is presented in a book by Bogoliubov and Mitropolski (1961).

3.6. Approximate Solutions - Second-Order

If we consider J_2 as a small parameter (ε), then J_{22} ($\cong O(J_2^2)$) is of the second order (ε^2). As (3.24) has terms containing J_{22}, the approximate solution to (3.24) should be carried out to the second order. Therefore, with the aid of (3.30) and (3.31), (3.16)-(3.20) are ready to be solved.

Combining (3.18) with (3.21) and integrating the resulting expression yields

$$3J_{22}\mu_E^2R_E^2(\cos 2\Gamma - \cos 2\Gamma_0) = \frac{1}{4}(X_6^4 - X_{60}^4) - \frac{1}{7}\,\frac{\dot\theta_E}{\mu_E^2}(X_6^7 - X_{60}^7)\,, \quad (3.32)$$

where (3.22) has been used, and Γ_0 and X_{60} are the given initial conditions.

In order to find the relationship between X_6 and θ, (3.32) requires some modifications. From (3.2), (3.6a), and (3.11), a circular synchronous orbit which includes the effect of J_{22} can be written as

$$X_6^2 = \bar\mu_E r = \bar\mu_E(r_c + \Delta r_c) = \bar\mu_E r_c\left(1 + \frac{\Delta r_c}{r_c}\right)\,, \quad (3.33)$$

where $\ddot r$, ϕ, and $\dot\phi$ in (3.2) are approximated to be zero, and Δr_c accounts for the contribution of the J_{22}-term on r. As a result, X_6^4 and X_6^7 in (3.32) can be expanded in a series up to the term containing $(\Delta r_c/r_c)^2$ (which has a maximum value of $O(10^{-6})$ for the present problem), and (3.32) is then approximated by

$$\cos 2\Gamma_0 - \cos 2\Gamma = \frac{r_c^2}{8J_{22}R_E^2}\left[1 + 3J_2\frac{R_E^2}{r_s^2}\right]\left[\left(\frac{X_6^2}{\bar\mu_E r_c} - 1\right)^2 - \left(\frac{X_{60}^2}{\bar\mu_E r_c} - 1\right)^2\right]. \quad (3.34)$$

This last expression, in turn, can be written as

$$\cos 2\Gamma_0 - \cos 2\Gamma = \frac{r_c^2}{8J_{22}R_E^2}\left[1 + 3J_2\frac{R_E^2}{r_s^2}\right]\left[\left(\frac{\Delta r_c}{r_c}\right)^2 - \left(\frac{\Delta r_{c0}}{r_c}\right)^2\right] \quad (3.35)$$

with the aid of (3.33).

With (3.22) and the expression of sin 2Γ derived from (3.34), (3.18) can be integrated as

$$\theta - \theta_0 = \frac{1}{3}\left\{(2k_1)^{1/2}\,\mathrm{sd}^{-1}\left[\left(\frac{2k_1}{1 - k_2^2}\right)^{1/2} y \,\middle|\, \frac{1 - k_2}{2}\right] + \right.$$

$$\left. + \sin^{-1}(k_1y^2 + k_2)\right\}\Bigg|_{y_0}^{y}\,, \quad (3.36)$$

where

$$\left.\begin{aligned} y &= \frac{X_6^2}{\bar\mu_E r_c} - 1 \cong \frac{\Delta r_c}{r_c} \cong O(10^{-3})\,,\\ k_1 &= \frac{-r_c^2}{8J_{22}R_E^2}\left[1 + 3J_2\frac{R_E^2}{r_s^2}\right] = \text{constant} = O(10^6)\,,\\ k_2 &= \cos 2\Gamma_0 - k_1\left(\frac{X_{60}^2}{\bar\mu_E r_c} - 1\right)^2\,, \end{aligned}\right\} \quad (3.37)$$

and

$$\mathrm{sd}^{-1}\left[\left(\frac{2k_1}{1 - k_2^2}\right)^{1/2} y \,\middle|\, \frac{1 - k_2}{2}\right]$$

is an inverse Jacobian elliptic function, which can be determined from the table of elliptic integrals of the first kind (Abramowitz and Stugen, 1964). Equations (3.34) and (3.36) are the second-order approximate solutions to (3.18) and (3.21).

After achieving the form represented by (3.24), equations (3.16), (3.17), (3.19), and (3.20) may be expressed in the form of (3.30):

$$\frac{d\bar{X}_3}{d\theta} = -\frac{3}{2}\left(1 + \frac{3}{4}\mu_E^2 R_E^2 \frac{1}{X_6^4} J_2\right)\mu_E^2 R_E^2 \frac{1}{X_6^4} J_2 \bar{X}_4 \,, \tag{3.38}$$

$$\frac{d\bar{X}_4}{d\theta} = \frac{3}{2}\left(1 + \frac{3}{4}\mu_E^2 R_E^2 \frac{1}{X_6^4} J_2\right)\mu_E^2 R_E^2 \frac{1}{X_6^4} J_2 \bar{X}_3 \,, \tag{3.39}$$

$$\frac{d\bar{X}_7}{d\theta} = +\, 3J_{22}\mu_E^2 R_E^2 \frac{\bar{X}_7}{X_6^4} \sin 2\Gamma \,, \tag{3.40}$$

and

$$\frac{d\bar{X}_8}{d\theta} = \frac{3}{2}J_2\mu_E^2 R_E^2 \frac{1}{X_6^4} + \frac{9}{8}J_2^2\mu_E^4 R_E^4 \frac{1}{X_6^8} + 3J_{22}\mu_E^2 R_E^2 \frac{1}{X_6^4} \cos 2\Gamma \,, \tag{3.41}$$

where Q^2 is approximated by

$$\mu_E^2 \frac{1}{X_6^4}\left[1 + 2\,\frac{X_6^2}{\mu_E}(X_3 \cos\theta + X_4 \sin\theta)\right]$$

and terms containing $J_{22}X_3$ and $J_{22}X_4$ are omitted. Both approximations are permissible because X_3 and X_4 are related to the eccentricity of the synchronous orbit. The former is an approximation of a binominal expansion with a small parameter, while the latter may be considered as a third order (J_2^3) term, which can be ignored in our second-order solution.

Equations (3.38) and (3.39) have their solutions as

$$\bar{X}_3 = c_0 \cos[\omega_2(\theta - \theta_0) + \gamma_0] \,, \tag{3.42}$$

and

$$\bar{X}_4 = c_0 \sin[\omega_2(\theta - \theta_0) + \gamma_0] \,, \tag{3.43}$$

where X_6^2 is approximated by $\bar{\mu}_E r_c$ and

$$\omega_2 \equiv \frac{3}{2}J_2 \frac{\mu_E^2}{\bar{\mu}_E^2}\left(1 + \frac{3}{4}\frac{R_E^2}{r_c^2}\frac{\mu_E^2}{\bar{\mu}_E^2} J_2\right) = \text{constant} \,.$$

c_0 and γ_0 are integration constants to be determined from the given initial conditions.

(3.40) can be combined with (3.18) and upon integration results in

$$\bar{X}_7 = \bar{X}_{70}\left(\frac{X_{60}}{X_6}\right)^{1/2} \,, \tag{3.44}$$

where $\bar{X}_{70}$ is an integration constant.

A combination of (3.41), (3.18), (3.37), and the expression of sin 2Γ derived from (3.34) gives

$$\begin{aligned} d\bar{X}_8 = &-\frac{1}{8}\frac{J_2}{J_{22}}\frac{1}{1+y}\times\frac{1}{[1-(k_1y^2+k_2)^2]^{1/2}}\,dy - \\ &-\frac{3}{32}\frac{J_2^2}{J_{22}}\frac{R_E^2}{r_c^2}\frac{1}{(1+y)^3}\times\frac{1}{[1-(k_1y^2+k_2)^2]^{1/2}}\,dy - \\ &-\frac{1}{4}\frac{1}{1+y}\times\frac{k_1y^2+k_2}{[1-(k_1y^2+k_2)^2]^{1/2}}\,dy \,. \end{aligned} \tag{3.45}$$

As y is of $O(10^{-3})$, the approximations $(1 + y)^{-1} \cong 1 - y$, $(1 + y)^{-3} \cong 1 - 3y$ and $(k_1y^2 + k_2)/(1 + y) \cong k_2(1 - y)$ are permissible. Consequently, (3.45) can be integrated as

$$\bar{X}_8 - \bar{X}_{80} = \left\{ \frac{1}{16k_1} \left[\frac{J_2}{J_{22}} - \frac{9}{4} \frac{J_2^2}{J_{22}} \frac{R_E^2}{r_c^2} + 2k_2 \right] \sin^{-1}(k_1 y^2 + k_2) \right.$$
$$+ \frac{1}{8(2k_1)^{1/2}} \left[\frac{-J_2}{J_{22}} + \frac{3}{4} \frac{J_2^2}{J_{22}} \frac{R_E^2}{r_c^2} - 2k_2 \right] \times$$
$$\left. \times \; sd^{-1} \left[\left(\frac{2k_1}{1 - k_2^2} \right)^{1/2} y \left| \frac{1 - k_2}{2} \right. \right] \right\} \Bigg|_{y_0}^{y} . \tag{3.46}$$

Equations (3.42)-(3.44) and (3.46) are the second-order approximate solutions to X_3, X_4, X_7, and X_8 respectively.

After obtaining these solutions, the improved second-order approximate solutions may be found by employing the relationships (3.25)-(3.31). Consequently, the improved second-order approximate solutions to X_3, X_4, X_7, and X_8 are:

$$X_3 = \bar{X}_3 + \delta X_3$$
$$= \bar{X}_3 + \frac{3}{2} J_2 \mu_E^3 R_E^2 \frac{1}{X_6^6} \cos\theta + \frac{3}{4} J_2 \mu_E^2 R_E^2 \frac{1}{X_6^4} (\bar{X}_3 \cos 2\theta + \bar{X}_4 \sin 2\theta) +$$
$$+ \frac{9}{2} J_2^2 \mu_E^5 R_E^4 \frac{1}{X_6^{10}} \cos\theta +$$
$$+ 3J_{22} \mu_E^3 R_E^2 \frac{1}{X_6^6} (3 \cos 2\Gamma \cos\theta - 4 \sin 2\Gamma \sin\theta), \tag{3.47}$$

$$X_4 = \bar{X}_4 + \delta X_4$$
$$= \bar{X}_4 + \frac{3}{2} J_2 \mu_E^3 R_E^2 \frac{1}{X_6^6} \sin\theta + \frac{3}{4} J_2 \mu_E^2 R_E^2 \frac{1}{X_6^4} (\bar{X}_3 \sin 2\theta - \bar{X}_4 \cos 2\theta) +$$
$$+ \frac{9}{2} J_2^2 \mu_E^5 R_E^4 \frac{1}{X_6^{10}} \sin\theta +$$
$$+ 3J_{22} \mu_E^3 R_E^2 \frac{1}{X_6^6} (3 \cos 2\Gamma \sin\theta + 4 \sin 2\Gamma \cos\theta) , \tag{3.48}$$

$$X_7 = \bar{X}_7 + \delta X_7$$
$$= \bar{X}_7 + \frac{3}{4} J_2 \mu_E^2 R_E^2 \bar{X}_7 \frac{1}{X_6^4} \cos 2(\theta + \bar{X}_8) +$$
$$+ \frac{3}{4} J_2 \mu_E R_E^2 \frac{1}{X_6^2} \bar{X}_7 \{ \bar{X}_3 [\cos(2\bar{X}_8 + \theta) + \frac{1}{3} \cos(2\bar{X}_8 + 3\theta)] -$$
$$- \bar{X}_4 [\sin(2\bar{X}_8 + \theta) - \frac{1}{3} \sin(2\bar{X}_8 + 3\theta)] \} -$$
$$- \frac{9}{64} J_2^2 \mu_E^4 R_E^4 \bar{X}_7 \frac{1}{X_6^8} \cos 4(\bar{X}_8 + \theta) -$$
$$- \frac{3}{2} J_{22} \mu_E^2 R_E^2 \bar{X}_7 \frac{1}{X_6^4} \cos 2(\theta + \bar{X}_8 - \Gamma) , \tag{3.49}$$

and

$$X_8 = \bar{X}_8 + \delta X_8$$
$$= \bar{X}_8 - \frac{3}{4} J_2 \mu_E^2 R_E^2 \frac{1}{X_6^4} \sin 2(\theta + \bar{X}_8) +$$
$$+ \frac{3}{4} J_2 \mu_E R_E^2 \frac{1}{X_6^2} \{ \bar{X}_3 [2 \sin\theta - \sin(2\bar{X}_8 + \theta) - \frac{1}{3} \sin(2\bar{X}_8 + 3\theta)] -$$

$$
\begin{aligned}
&- \bar{X}_4[2\cos\theta + \cos(2\bar{X}_8 + \theta) - \tfrac{1}{3}\cos(2\bar{X}_8 + 3\theta)]\} + \\
&+ \tfrac{9}{32}J_2^2\mu_E^4R_E^4\,\frac{1}{X_6^8}[\sin 4(\theta + \bar{X}_8) + 12\sin 2(\theta + \bar{X}_8)] - \\
&- \tfrac{3}{2}J_{22}\mu_E^2R_E^2\,\frac{1}{X_6^4}\,\bar{X}_7\sin 2(\theta + \bar{X}_8 - \Gamma)\ ,
\end{aligned}
\tag{3.50}
$$

where again the terms containing $J_2^2X_3$, $J_2^2X_4$, $J_{22}X_3$, and $J_{22}X_4$ are omitted because they are of the third order (J_2^3) due to the small eccentricity of geostationary orbits.

The second-order approximate solution to X_5 and its improved version can be deduced from (3.15) after its constituents X_6, X_7, and X_8 are determined. As a result of the approximation made in (3.22), X_6 does not have any short-period oscillatory characteristics as may be seen from (3.18). Therefore, there is no improved second-order approximate solution for X_6.

3.7. Recapitulation

The derivation of the necessary equations is completed and a recapitulation of the process is in order. The following chart shows the steps that were taken:

	Basic Equations of Motion: (3.2), (3.3), (3.4), and satellite longitude with respect to the Minor Axis of Earth Equator (3.5).	→	H, Q, and S System: Equations (3.7), (3.8), (3.9), and (3.5).	→
→	First-Order Differential Equations (3.16)-(3.21).	→	Averaged First-Order Differential Equations (3.18), (3.38)-(3.41).	→
→	Second-Order Solutions to X_6, X_3, X_4, X_7, and X_8: (3.34), (3.36), (3.42)-(3.44), (3.46).	→	Improved Second-Order Solution to X_3, X_4, X_7, and X_8: (3.47)-(3.50).	.

From (3.6) and (3.10)-(3.12), we have

$$\frac{1}{r\cos\phi} = X_5 + X_3\cos\theta + X_4\sin\theta\ , \tag{3.51}$$

$$X_6 = r^2\dot{\theta}\cos^2\phi\ , \tag{3.52}$$

and

$$\tan\phi = X_7\sin(\theta + X_8)\ , \tag{3.53}$$

where θ and X_3, ..., X_8 are related by (3.36), (3.47)-(3.50), and the values of Γ, $\bar{X}_3$, $\bar{X}_4$, $\bar{X}_6$, $\bar{X}_7$, and $\bar{X}_8$ in the X's are given in (3.34), (3.42)-(3.44), and (3.46). (3.51)-(3.53) are the solutions of (3.2)-(3.5), which show the relations among the four basic geometric parameters, r, θ, ϕ, and Γ, and the angular momentum (H) about the polar axis. The time, t, is related to the other parameters by (3.5), where Γ and θ are connected by X_6 through (3.34) and (3.36).

The integration constants, c_0, γ_0, $\bar{X}_{70}$, and $\bar{X}_{80}$ in (3.42)-(3.44) and (3.46) are to be determined from the given initial conditions r_0, ϕ_0, θ_0, $\dot{r}_0$, $\dot{\phi}_0$, $\dot{\theta}_0$, and Γ_0. With these seven given values and (3.34) and (3.37), the values of X_{30}, X_{40}, X_{70}, and X_{80} can be obtained from (3.51)-(3.53) and their first derivatives with respect to θ evaluated at $\theta = \theta_0$. $\bar{X}_{30}$, $\bar{X}_{40}$, $\bar{X}_{70}$, and $\bar{X}_{80}$ can then be calculated from (3.47)-(3.50). With known $\bar{X}_{30}$ and $\bar{X}_{40}$, c_0 and γ_0 are determined from (3.42) and (3.43).

3.8. Osculating Orbital Elements

The trajectory of a near-equatorial geostationary satellite under the influences of Earth oblateness and the ellipticity of the equator is a space curve, which is described by (3.51)-(3.53). However, if the concept of an osculating orbit is introduced, the results derived from the present analysis can be expressed in terms of the orbital elements of the osculating orbit. The osculating orbit can be defined by (see Figure 2.1.)

$$r = \frac{\ell}{1 + e\cos(\eta - \omega)}, \tag{3.54}$$

where ℓ is the semilatus rectum, e the eccentricity, η the argument of latitude, and ω is the argument of perigee, both angles measured from the ascending node.

By substituting (3.47), (3.48), and (3.6b) into (3.10) and approximating $\cos\phi \cong 1$ and $1 + X_7^2 \cong 1$ in the high order terms, a comparison of the resulting expression with (3.54) gives

$$\ell = \frac{X_6^2}{\bar{\mu}_E}(1 + X_7^2)\left(1 - \frac{27}{4}J_2^2\frac{R_E^2}{r_s^2} - 9J_{22}\frac{R_E^2}{r_s^2}\cos 2\Gamma\right), \tag{3.55}$$

and

$$\frac{X_6^2}{\bar{\mu}_E}\left(1 + \frac{3}{4}J_2\frac{R_E^2}{r_s^2} - \frac{27}{4}J_2^2\frac{R_E^4}{r_s^4} - 9J_{22}\frac{R_E^2}{r_s^2}\cos 2\Gamma\right)(\bar{X}_3\cos\theta + \bar{X}_4\sin\theta)$$
$$= e\cos(\eta - \omega). \tag{3.56}$$

These two equations will be used in the following derivation of various orbital elements.

3.8.1. Drift Angle (Γ) and Drift Rate ($\dot{\Gamma}$)

Basically, (3.32) and (3.36) give the position of a geostationary satellite in relation to the equatorial minor axis of the Earth. However, when (3.32) is approximated by (3.34), (3.36) becomes

$$\theta - \theta_0 = \frac{1}{3}\left\{(2k_1)^{1/2}\mathrm{sd}^{-1}\left[\left(\frac{2(\cos 2\Gamma - \cos 2\Gamma_{max})}{1 - \cos^2 2\Gamma_{max}}\right)^{1/2}\left|\frac{1 - \cos 2\Gamma_{max}}{2}\right]\right. + \right.$$
$$\left. + 2(\Gamma_{max} - \Gamma)\right\}, \tag{3.57}$$

where the criterion $y_0 = 0$ and $\Gamma_0 = \Gamma_{max}$ is used. Since $k_1 \sim O(10^6)$ and $\Gamma_{max} - \Gamma \leqq \pi/2$, the second term in the braces on the right-hand side of (3.57) can be neglected. Since $\Gamma_{max} \leqq \pi/2$, (3.5) shows that when θ is large, θ is linearly dependent on time, (3.57) becomes the well-known libration equation of a geostationary satellite about the equatorial minor axis of the Earth (see Figure 3.5.). Further examination shows that if $y_0 (\cong \Delta r_c) = 0$, then $\Gamma_0 = \Gamma_{max} = \pi/2$ leads to an unbounded period of libration, whereas $\Gamma_0 = \Gamma_{max} = 0$ yields a zero period, implying that the satellite remains always on the minor axis as is viewed from the Earth's equator. The former condition is referred to as dynamic instability while the latter condition is considered dynamically stable.

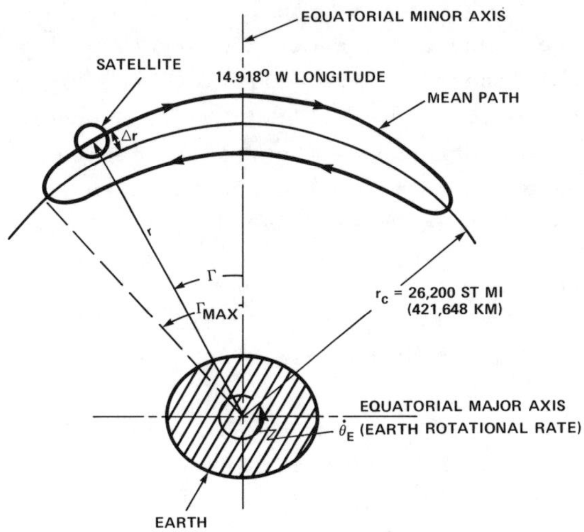

Fig. 3.5. Libration of geostationary satellites.

The drift rate of a geostationary satellite in relation to a fixed point on the Earth's equator can be examined by differentiating (3.5) with respect to the time, t:

$$\begin{aligned} \frac{d\Gamma}{dt} &= \frac{d\theta}{dt} - \dot{\theta}_E \\ &= \frac{H}{r^2 \cos^2 \phi} - \dot{\theta}_E \cong \frac{X_6}{r^2} - \dot{\theta}_E \\ &= \left(\frac{\mu_E}{(r_c + \Delta r_c)^3} \right)^{1/2} - \dot{\theta}_E = -\frac{3}{2} \frac{\Delta r_c}{r_c} \dot{\theta}_E \quad , \end{aligned} \tag{3.58}$$

where (3.6a), (3.33), and the definition of $\dot{\theta}_E = (\bar{\mu}_E / r_c^3)^{1/2}$ have been employed. In conjunction with (3.35), this expression shows that the drift rate is zero when $\Delta r_c = 0$ and $\Gamma = \Gamma_{max}$, and the drift rate reaches its maximums when $\Gamma = 0$ and $|\Delta r_c| =$ maximum. It can be seen that the drift rate may easily be computed when Δr_c is known. Figure 3.5. depicts the type of motion described herein (but exaggerated in the radial direction).

3.8.2. Semilatus Rectum (ℓ)

From (3.55), the semilatus rectum of an osculating orbit is

$$\ell = \bar{\ell} + \delta\ell$$

$$= \frac{X_6^2}{\bar{\mu}_E}\left(1 - \frac{27}{4}J_2^2 \frac{R_E^2}{r_s^2} - 9J_{22} \frac{R_E^2}{r_s^2} \cos 2\Gamma\right)[(1 + \bar{X}_7^2) + 2\bar{X}_7 \delta X_7] \ . \tag{3.59}$$

As will be seen later, X_7 is related to the inclination of the orbital plane; therefore, for a near equatorial satellite, $\bar{\ell}$ may be represented by $[1 - \frac{27}{4}J_2^2(R_E^2/r_s^2) - 9J_{22}(R_E^2/r_s^2)\cos 2\Gamma]X_6^2/\bar{\mu}_E$, which, when combined with (3.34), results in

$$\bar{\ell} = r_c\left(1 - \frac{27}{4}J_2^2 \frac{R_E^2}{r_s^2} - 9J_{22} \frac{R_E^2}{r_s^2} \cos 2\Gamma\right) \times$$

$$\times \left\{1 + \left[\frac{1}{k_1}(\cos 2\Gamma - \cos 2\Gamma_{max})\right]^{1/2}\right\} \ . \tag{3.60}$$

Equation (3.60) shows that the average semilatus rectum of a geostationary orbit has a small long-period variation related to the drift angle, Γ, as $(1/k_1)^{1/2} \sim O\ (10^{-3})$. Furthermore, short diurnal variations of ℓ are revealed by δX_7 in (3.59).

3.8.3. Longitude of Ascending Node (Ω)

The longitude of osculating orbit's ascending node can be determined by setting $\phi = 0$ in (3.53). The resulting expression is

$$\Omega = - X_8 . \tag{3.61}$$

This equation, in turn, yields

$$\Omega = \bar{\Omega} + \delta\Omega$$

$$= - \bar{X}_8 - \delta X_8 \ . \tag{3.62}$$

Since $J_2/J_{22} \sim O\ (10^3)$, $J_2^2/J_{22} \sim O\ (1)$, Equation (3.46) may be approximated by

$$\bar{X}_8 - \bar{X}_{80} = \frac{1}{8k_1}\frac{J_2}{J_{22}}(\Gamma_{max} - \Gamma) - \frac{1}{8(2k_1)^{1/2}}\frac{J_2}{J_{22}} \times$$

$$\times\ sd^{-1}\left[\left(\frac{2(\cos 2\Gamma - \cos 2\Gamma_{max})}{1 - \cos^2 2\Gamma_{max}}\right)^{1/2} \middle|\ \frac{1 - \cos 2\Gamma_{max}}{2}\right] \tag{3.63}$$

which, when combined with (3.57), results in

$$\bar{X}_8 - \bar{X}_{80} = \frac{3}{2}\frac{R_E^2}{r_c^2} J_2[(\theta - \theta_0) - \frac{4}{3}(\Gamma_{max} - \Gamma)] \ . \tag{3.64}$$

The first term in the brackets on the right-hand side of the equation provides the secular characteristics of Ω showing the effect of the Earth's oblateness and the second term gives the long-period variation of Ω arising from the rotating ellipsoidal potential field of the Earth. Since $\Gamma_{max} - \Gamma$ has a maximum value of $\pi/2$, and $\theta - \theta_0$ is a linear time-increasing function, it can be seen that the contribution of $\Gamma_{max} - \Gamma$ to $\bar{X}_8 - \bar{X}_{80}$ is of negligible importance. The short-period diurnal oscillation of Ω is brought out by δX_8.

3.8.4. Inclination of Orbital Plane (i)

The inclination of an osculating orbital plane may be obtained by manipulations of (2.84), (3.53), and (3.61) with $\theta - \Omega = \pi/2$:

$$\tan i = X_7. \tag{3.65}$$

For a near-equatorial satellite, $\tan i \cong i \cong X_7$, which gives

$$i = \bar{i} + \delta i$$
$$= \bar{X}_7 + \delta X_7; \tag{3.66}$$

consequently, (3.44) gives

$$\bar{i} = \bar{X}_{70}\left(\frac{X_{60}}{X_6}\right)^{1/2}$$
$$= \bar{X}_{70}\left\{1 + \left[\frac{1}{k_1}(\cos 2\Gamma - \cos 2\Gamma_{max})\right]^{1/2}\right\}^{-1/4}$$
$$\cong \bar{X}_{70}\left\{1 - \frac{1}{4}\left[\frac{-1}{k_1}(\cos 2\Gamma - \cos 2\Gamma_{max})\right]^{1/2}\right\}, \tag{3.67}$$

where (3.34) has been employed with the initial conditions $\Gamma_0 = \Gamma_{max}$ and $X_{60}^2 = \bar{\mu}_E r_c$. This equation shows that the inclination has a small long-period variation of O (10^{-3}) in addition to the short diurnal oscillation revealed by δX_7.

3.8.5. Eccentricity (e) and Argument of Perigee (ω)

The eccentricity and argument of perigee of the osculating orbit are derived simultaneously by applying (2.119) and (2.120) together with (3.56) to the orbit points at the perigee and at the node. The resulting expressions are

$$\tan \omega = \frac{1}{(1 + X_7^2)^{1/2}} \times \frac{\bar{X}_4 + \bar{X}_3 \tan X_8}{\bar{X}_3 - \bar{X}_4 \tan X_8}, \tag{3.68}$$

and

$$e = \frac{X_6^2}{\bar{\mu}_E}\left(1 + \frac{3}{4}J_2\frac{R_E^2}{r_s^2} - \frac{27}{4}J_2^2\frac{R_E^4}{r_s^4} - 9J_{22}\frac{R_E^2}{r_s^2}\cos 2\Gamma\right) \times$$
$$\times (1 + X_7^2)^{1/2}\left[\bar{X}_3^2 + \bar{X}_4^2 + \bar{X}_7^2(\bar{X}_4 \sin X_8 - \bar{X}_3 \cos X_8)^2\right]^{1/2}. \tag{3.69}$$

With

$$\bar{X}_3 = c_0 \cos[\omega_2(\theta - \theta_0) + \gamma_0]$$

and

$$\bar{X}_4 = c_0 \sin[\omega_2(\theta - \theta_0) + \gamma_0]$$

as the known solutions previously obtained in (3.42) and (3.43), Equation (3.68) can be simplified as

$$\tan \bar{\omega} = \frac{1}{(1 + \bar{X}_7)^{1/2}} \tan[\omega_2(\theta - \theta_0) + \bar{X}_8 + \gamma_0]. \tag{3.70}$$

For the case of a near-equatorial satellite, this equation may be approximated as

$$\bar{\omega} = \omega_2(\theta - \theta_0) + \bar{X}_8 + \gamma_0. \tag{3.71}$$

Similarly, (3.69) can be reduced to

$$\bar{e} = \frac{X_6^2}{\bar{\mu}_E}\left(1 + \frac{3}{4}J_2\frac{R_E^2}{r_s^2} - \frac{27}{4}J_2^2\frac{R_E^4}{r_s^4} - 9\frac{R_E^2}{r_s^2}J_{22}\cos 2\Gamma\right) \times$$
$$\times\ (1+\bar{X}_7^2)^{1/2}\left\{c_0^2 + \bar{X}_7^{\ 2}\cos^2[\omega_2(\theta-\theta_0) + \bar{X}_8 + \gamma_0]\right\} \qquad (3.72)$$

for the averaged value of eccentricity.

Equation (3.71), when combined with (3.64), shows that the line of apsides rotates in the same sense as the satellite. The rate of rotation is approximately twice that of the regression of the line of nodes, and both have the same long-period variation with magnitude of negligible importance..

On account of the smallness of $\bar{X}_7$, the value within the braces on the right-hand side of (3.72) remains practically constant. As may be seen from (3.60), $\bar{e}$ has the same characteristics as $\bar{\ell}$ does. The oscillatory natures of ω and e as θ ranges over a period of 2π can be brought out by applying the variational technique to (3.68) and (3.69) in terms of $\bar{X}$'s and δX_7 and δX_8. No δX_3 and δX_4 will appear in the oscillatory terms of ω and e, since X_3 and X_4 are already in the form of their averaged values as is seen in (3.56). The complete variational equations can be written out when a practical computation is performed.

3.9. True Equatorial Geostationary Orbit

Most literature dealing with the ellipticity effect of the Earth's equator on geostationary satellites are restricted to the ones in the equatorial plane. This restriction simplifies the result considerably; with i = 0 and e = 0, (3.10) becomes

$$\frac{1}{r} = \frac{\mu_E}{X_6^2} + \frac{3}{2}J_2\mu_E^3R_E^2\frac{1}{X_6^6} + \frac{9}{2}J_2^2\mu_E^5R_E^4\frac{1}{X_6^{10}} + 9J_{22}\mu_E^3R_E^2\frac{1}{X_6^6}\cos 2\Gamma\ , \qquad (3.73)$$

where (3.47) and (3.48) have been used. Equation (3.73) together with (3.5), (3.32), and (3.36) are required to describe the orbit of a true equatorial geostationary satellite.

3.10. Resonance and Libration

We have presented our approach to the solution of the problem without utilizing or referring to any previous studies or methodology on the same topic. Therefore our result is a mathematically independent and complete solution to the problem. We have also briefly mentioned the libration of geostationary satellites about the minor axis of the elliptical equator of the Earth as our mathematical solution indicates. In order to have a clear concept of the libration of geostationary satellites and to allow readers to compare our results with others' or with other approaches to the problem, we shall briefly describe the results of the problem from other approaches by other investigators.

3.10.1. Direct Solution from Lagrange's Equations (Allan, 1963; Kaula, 1966)

Since Table II shows $J_{22} < 0$, the disturbing function in (2.24) becomes

$$R_{22} = -3\frac{\mu_E}{a}J_{22}\left(\frac{R_E}{a}\right)^2 \cos 2(\omega + M + \Omega - \lambda_{22} - \theta_g), \quad (2.24a)$$

which has the mean anomaly M as one of its angular arguments. This fact results in a perturbation of the semimajor axis as is seen from the first equation of the Lagrange's equations of motion (2.21):

$$\frac{da}{dt} = \frac{2}{na}\frac{\partial R_{22}}{\partial M}$$

$$= +\frac{12}{na}\frac{\mu_E}{a}J_{22}\left(\frac{R_E}{a}\right)^2 \sin 2(\omega + M + \Omega - \lambda_{22} - \theta_g) . \quad (3.74)$$

This change in the semimajor axis causes an acceleration in the Earth-referred longitude of the satellite. Let

$$\lambda_A = \omega + M + \Omega - \theta_g , \quad (3.75)$$

which is the 'broken-legged' astronomical longitude, not the conventional geodetic longitude measured along the equator. However, for the present case of geostationary satellite orbits of near-zero inclination, λ_A can be approximated as the geodetic longitude of the satellite measured along the equator from the Greenwich meridian. Using Kepler's law of

$$n = \mu_E^{1/2}a^{-3/2} \quad (3.76)$$

and $\dot{M} = n$, we have

$$\ddot{\lambda}_A = \ddot{M} = \dot{n} = \frac{dn}{da}\frac{da}{dt}$$

$$= 6K_{22} \sin 2(\lambda_A - \lambda_{22}) , \quad (3.77)$$

where $K_{22} \equiv (3\mu_E/a^3)J_{22}(R_E/a)^2$ is introduced for the sake of simplicity.

Since $\dot{\lambda}_A = \dot{\omega} + \dot{M} + \dot{\Omega} - \dot{\theta}_g \cong n - \dot{\theta}_g$, both n and $\dot{\theta}_g$ are constants, and $n - \dot{\theta}_g$ is a very small number in the present case of geostationary satellites. Therefore the double integration of (3.77) would lead to a large $\Delta\lambda_A$ because of the small divisor $(n - \dot{\theta}_g)^2$. This fact indicates that R_{22} is a resonant disturbing function. In other words, the assumption that a valid first approximation of orbit is a secularly precessing ellipse in the method of solution breaks down in the vicinity of a resonance, and (3.77) must be examined directly.

Multiplying (3.77) by $2\dot{\lambda}_A$, we obtain a first integral:

$$(\dot{\lambda}_A)^2 = K - 6K_{22} \cos 2(\lambda_A - \lambda_{22}) , \quad (3.78)$$

where the constant K is defined as a combination of the initial longitude λ_{A0} and rate $\dot{\lambda}_{A0}$:

$$K \equiv (\dot{\lambda}_{A0})^2 + 6K_{22} \cos 2(\lambda_{A0} - \lambda_{22}) . \quad (3.79)$$

Since $(\dot{\lambda}_A)^2 \geq 0$, Equation (3.78) depicts

$$K \geq 6K_{22} \cos 2(\lambda_{A0} - \lambda_{22}) . \quad (3.80)$$

If the constant K is small enough, the criterion shown in (3.80) will prevent λ_A from going through a full cycle of 2π. The smallness of K requires that $\cos 2(\lambda_{A0} - \lambda_{22})$ has a value in the neighborhood of '-1' as is seen from (3.79); in other words, λ_{A0} should be situated in the vicinity of $\lambda_{22} \pm \pi/2$, which is the minor axis of the equator ellipse. On referring longitude to the minor axis by making the substitution

$$\Gamma = \lambda_A - \lambda_{22} - \frac{\pi}{2} , \tag{3.81}$$

(3.78) becomes

$$\begin{aligned}(\dot{\Gamma})^2 &= K + 6K_{22} \cos 2\Gamma \\ &= K + 6K_{22}(1 - 2 \sin^2 \Gamma) .\end{aligned} \tag{3.82}$$

The maximum departure Γ_{max} from the minor axis will thus correspond to zero $\dot{\Gamma}$, and

$$\sin \Gamma_{max} = \left(\frac{1}{2} + \frac{K}{12K_{22}} \right)^{1/2} . \tag{3.83}$$

From (3.82), we have

$$\begin{aligned}\frac{d\Gamma}{dt} &= (K + 6K_{22} - 12K_{22} \sin^2 \Gamma)^{1/2} \\ &= [12K_{22}(\sin^2 \Gamma_{max} - \sin^2 \Gamma)]^{1/2} .\end{aligned} \tag{3.84}$$

On defining

$$k^2 = 1 / \sin^2 \Gamma_{max} , \tag{3.85}$$

(3.84) becomes

$$\frac{d\Gamma}{dt} = \left[\frac{12K_{22}}{k^2} (1 - k^2 \sin^2 \Gamma)\right]^{1/2} , \tag{3.86}$$

which can be integrated as

$$\begin{aligned}t &= \frac{k}{(12K_{22})^{1/2}} \int_0^{\Gamma} \frac{d\Gamma}{(1 - k^2 \sin^2 \Gamma)^{1/2}} \\ &= \frac{k}{(12K_{22})^{1/2}} F(k, \Gamma) .\end{aligned} \tag{3.87}$$

F is the elliptic integral of the first kind. Equation (3.85) shows that if $k > 1$, $F(k, \Gamma)$ in (3.87) can be evaluated by the transformation

$$F(k, \Gamma) = \frac{1}{k} F[\frac{1}{k}, \sin^{-1}(k \sin \Gamma)] ; \tag{3.88}$$

therefore a complete cycle of the elliptic integral F in (3.87) is $(4/k)F(1/k, \pi/2)$, and the time T for a complete period will be

$$T = \frac{2}{(3K_{22})^{1/2}} F(1/k, \pi/2) . \tag{3.89}$$

If $k > 1$, and the initial longitude λ_{A0} is close to the stable point $\lambda_{22} \pm \pi/2$, $\sin \Gamma$ must vary between the limits $\pm(1/k)$; thus the satellite will librate about the longitude $\lambda_{22} \pm \pi/2$. If the initial longitude λ_{A0} is outside of the limits of libration, Γ may have any value; that is, the satellite will drift all the way forward or backward around the Earth. When the satellite's mean motion is greater than the Earth rotation rate $\dot{\theta}_E$, the satellite is located inside the zone of possible resonance, while the greater mean motion will mean that the satellite is situated outside the resonance zone. Figure 3.6. gives a schematic picture of the area of libration in an Earth-fixed reference frame.

Two extreme values of k should be mentioned, i.e., $k = 1$ and $k = \infty$; the former depicts $\Gamma_{max} = \pi/2$, and $T = \infty$, while the latter gives $\Gamma_{max} = 0$ and $T = 0$. Physically, the former case indicates that when the satellite is situated near the major axis of the Earth elliptical equator, the satellite will drift toward the minor axis and further to the other end of the major axis and then back to the starting point with an infinite period of time. The latter case shows that when a satellite is located along the minor axis of the Earth elliptical equator, the satellite will remain at its initial point and stabilize. For this reason, from the dynamic stability point of view, the former case is dynamically unstable and the latter case is dynamically stable.

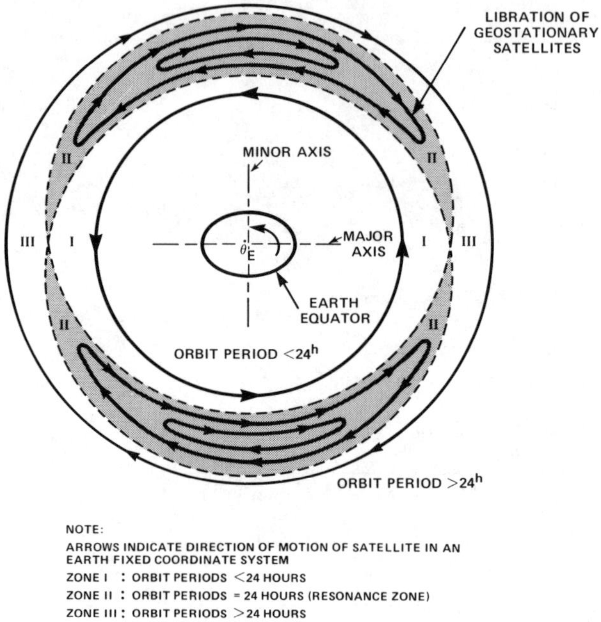

Fig. 3.6. Path of satellites in nearly circular, small-inclination orbits in an Earth-fixed reference coordinate system.

Equation (3.87) is equivalent to our findings in (3.5), (3.35), and (3.36) so far as the drift angle Γ is concerned; however, our solution reveals the complete information of the path of libration in terms of Γ, Δr_c, and time. In addition, we have also found the behavior of all orbital elements as a result of the perturbation of the ellipticity of the Earth's equator.

3.10.2. Solution by the von Zeipel Method

A direct application of the von Zeipel method to the geostationary satellites problem with the disturbing function R_{22} would lead to an infinite coefficient in the determining function S_2 (Morando, 1963). In order to avoid this awkward situation, Morando utilized Hori's method (Hori, 1960) to isolate the terms in the disturbing function that would lead to the infinite coefficient. Since θ_g appears in R_{22} and is explicitly dependent on time, the removal of θ_g in R_{22} can be made by defining

$$h = \Omega - \theta_g . \tag{3.90}$$

Then in order for the equations of motion (2.29) to apply, $\dot{\theta}_g H$ must be added to the force function F; thus, with the disturbing function R_{22}, we write the force functions in the order of magnitude of the small parameter in R_{22} as

$$F_0 = \frac{\mu_E^2}{2L^2} + \dot{\theta}_g H ,$$

$$F_1 = -6 \frac{\mu_E^4}{L^6} J_{22} R_E^2 \cos^2(g + \ell + h - \lambda_{22}) , \tag{3.91}$$

and

$$F_2 = 6 \frac{\mu_E^4}{L^6} J_{22} R_E^2 .$$

By employing the Taylor series development shown in (2.45), we make the following arrangement:

$$F_0 = F_0' ,$$

$$\frac{\partial F_0}{\partial L'} \frac{\partial S_1}{\partial \ell} + \frac{\partial F_0}{\partial H'} \frac{\partial S_1}{\partial h} + \frac{1}{2} \frac{\partial^2 F_0}{\partial L'^2}\left(\frac{\partial S_1}{\partial \ell}\right)^2 + F_1 = F_1' = 0 , \tag{3.92}$$

and

$$\frac{\partial F_0}{\partial L'} \frac{\partial S_2}{\partial \ell} + \frac{\partial F_0}{\partial H'} \frac{\partial S_2}{\partial h} + F_2 = F_2' .$$

It can be seen that F_1' equals zero since the corresponding part of F is entirely periodic, and F_1 must be absorbed by S_1.

It should be noted that the developments of (3.91) and (3.92) deviate slightly from the von Zeipel method described in Section 2.3.4. Since R_{22} is of the second order of magnitude of the small parameter, F_1 in (3.91) has a magnitude of the second order. In (3.92), $(\partial S_1/\partial \ell)^2$ is defined to have a magnitude of the second order, and should belong to F_2', yet in (3.92), it is arranged to be included in F_1'. If the term containing $(\partial S_1/\partial \ell)^2$ was dropped out from F_1' in (3.92) as the usual Taylor series arrangement of the order of magnitude, we have

$$\frac{\partial F_0}{\partial L'} \frac{\partial S_1}{\partial \ell} + \frac{\partial F_0}{\partial H'} \frac{\partial S_1}{\partial h} = 6 \frac{\mu_E^4}{L^6} J_{22} R_E^2 \cos^2(g + \ell + h - \lambda_{22}) . \tag{3.93}$$

From (3.91), we get $\partial F_0/\partial L' = -(\mu_E^2/L'^3)$ and $\partial F_0/\partial H' = \dot{\theta}_g$.
Since ℓ and h appear together as a sum $\ell + h$, $\partial S_1/\partial \ell$ is equal to $\partial S_1/\partial h$, and then (3.93) becomes

$$\left(\dot{\theta}_g - \frac{\mu_E^2}{L'^3}\right)\frac{\partial S_1}{\partial \ell} = 6 \frac{\mu_E^4}{L'^6} J_{22} R_E^2 \cos^2(g + \ell + h - \lambda_{22}) . \tag{3.94}$$

As (2.27) and (3.76) indicate that μ_E^2/L'^3 is the mean motion, then $\dot{\theta}_g - (\mu_E^2/L'^3)$ is a very small number. Because of the smallness of $\dot{\theta}_g - (\mu_E^2/L'^3)$, the phenomenon of resonance thus appear if (3.94) is integrated.

On substituting Γ as defined by (3.75) and (3.81), F_1' has the form of

$$F_1' = \left(\dot{\theta}_g - \frac{\mu_E^2}{L'^3}\right)\frac{\partial S_1}{\partial \Gamma} + \frac{3}{2} \frac{\mu_E^2}{L'^4}\left(\frac{\partial S_1}{\partial \Gamma}\right)^2 - 6 \frac{\mu_E^4}{L'^6} J_{22} R_E^2 \sin^2\Gamma = 0, \tag{3.95}$$

where $\partial S_1/\partial \ell = \partial S_1/\partial h = \partial S_1/\partial \Gamma$ is used because $\Gamma = g + \ell + h - \lambda_{22} - \pi/2$.
If we consider $\dot{\theta}_g - (\mu_E^2/L'^3)$ as having a magnitude of the first order, then all terms in (3.95) have the magnitude of the second order. Solving the

quadratic equation (3.95) for $\partial S_1/\partial\Gamma$ and integrating, we have

$$S_1 = -\frac{L'^4}{3\mu_E^2}\left(\dot\theta_g - \frac{\mu_E^2}{L'^3}\right)(\Gamma \pm E) , \tag{3.96}$$

where

$$E = E(k_1, \Gamma) = \int_0^\Gamma (1 - k_1^2 \sin^2 \Gamma)^{1/2} \, d\Gamma , \tag{3.97}$$

the elliptical integral of the second kind, in which the modulus is

$$k_1^2 \equiv \frac{-36J_{22}R_E^2}{\left(\dot\theta_g - \dfrac{\mu_E^2}{L'^3}\right)^2} \frac{\mu_E^6}{L'^{10}} . \tag{3.98}$$

For the first-order solution, because $F' = F_0' + F_1' = F_0'$, (2.34) gives

$$\ell' = -\frac{\partial F'}{\partial L'}(t - t_0) = -\frac{\partial F_0'}{\partial L'}(t - t_0) = \frac{\mu_E^2}{L'^3}(t - t_0) ,$$

$$g' = -\frac{\partial F'}{\partial G'}(t - t_0) = -\frac{\partial F_0'}{\partial G'}(t - t_0) = 0 , \tag{3.99}$$

and

$$h' = -\frac{\partial F'}{\partial H'}(t - t_0) = -\frac{\partial F_0'}{\partial H'}(t - t_0) = -\dot\theta_g(t - t_0) ,$$

and (2.44) shows

$$\ell = \ell' - \frac{\partial S_1}{\partial L'}, \quad g = g' - \frac{\partial S_1}{\partial G'}, \quad \text{and} \quad h = h' - \frac{\partial S_1}{\partial H'} . \tag{3.99}$$

Consequently,

$$\ell + g + h = \ell' + g' + h' - \left(\frac{\partial S_1}{\partial L'} + \frac{\partial S_1}{\partial G'} + \frac{\partial S_1}{\partial H'}\right)$$

$$= \left(\frac{\mu_E^2}{L'^3} - \dot\theta_g\right)(t - t_0) - \left(\frac{\partial S_1}{\partial L'} + \frac{\partial S_1}{\partial G'} + \frac{\partial S_1}{\partial H'}\right) . \tag{3.100}$$

As is shown in (3.96), S_1 contains L' only; therefore $\partial S_1/\partial G'$ and $\partial S_1/\partial H'$ are equal to zero. In $\partial S_1/\partial L'$, if we consider the terms containing $(\mu_E^2/L'^3 - \dot\theta_g)$ as of the higher order, then

$$\frac{\partial S_1}{\partial L'} \cong -(\Gamma \pm E) \pm \int_0^\Gamma \frac{k_1^2 \sin^2 \Gamma}{(1 - k_1^2 \sin^2 \Gamma)^{1/2}} \, d\Gamma$$

$$= -\Gamma \mp \int_0^\Gamma (1 - k_1^2 \sin^2 \Gamma)^{-1/2} \, d\Gamma . \tag{3.101}$$

Since $\Gamma = \ell + g + h - \lambda_{22} - \pi/2$, and we are interested only in the variation of Γ with respect to the minor axis of the elliptical equator of the Earth, (3.100) and (3.101) give

$$\left(\frac{\mu_E^2}{L'^3} - \dot\theta_g\right)(t - t_0) = \pm \int_0^\Gamma (1 - k_1^2 \sin^2 \Gamma)^{-1/2} \, d\Gamma =$$

$$= \pm F(k_1, \Gamma) , \tag{3.102}$$

which is similar to (3.87) and shows the libration of geostationary satellites.

3.11. Comparison of the Canonical and Proposed Solutions

As we have pointed out previously in Chapter 2, the canonical method of solving the geostationary satellites problem requires considerable background in celestial mechanics; readers will realize now that our approach to the problem is much simpler and direct. Basically, both the canonical and proposed solutions depict the main theme of interest on geostationary satellites - libration motion. However, our exploration of the subject is much more thorough and comprehensive. In addition to the libration motion, we have systematically studied all orbital elements as a result of the perturbation of the ellipticity of the Earth's equator, and the solution is carried out to the order of $(J_2)^2$, which is consistent with the set-up of the problem because J_{22} is of the order of $(J_2)^2$.

It should be mentioned that Vashkov'yak and Lidov (1973) also considered the Earth's equatorial ellipticity effect in their study of geostationary satellites, and the same conclusion of libration motion was reached. We shall compare their solution with our findings in Chapter 5.

3.12. Characteristics of Libration - Graphic Presentation

This section will present some numerical results obtained from the application of the foregoing theory. The characteristics of the libration of a true equatorial geostationary satellite about the minor axis of the Earth's elliptical equator are presented in Figures 3.7.-3.10. These figures contain the numerical results from (3.5), (3.35), and (3.36).

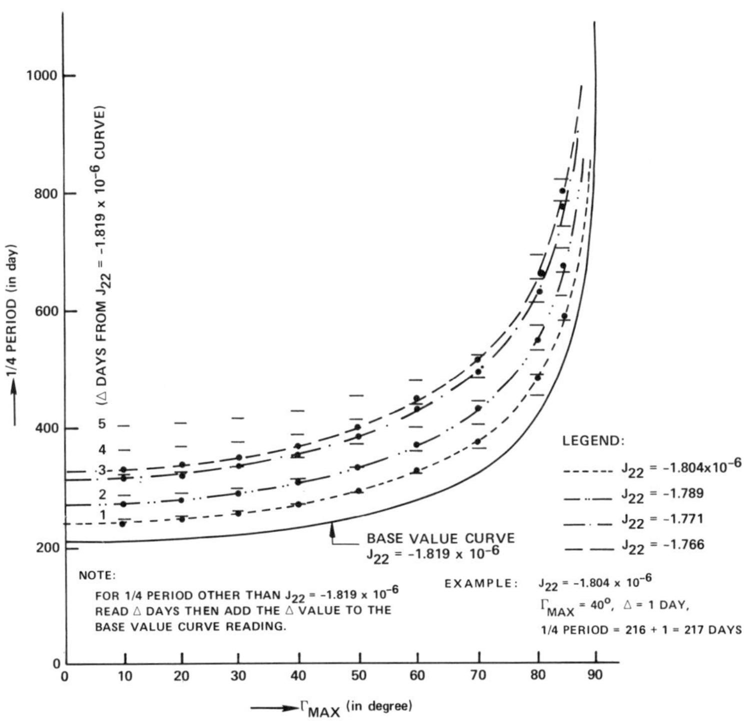

Fig. 3.7. Quarter-period of libration.

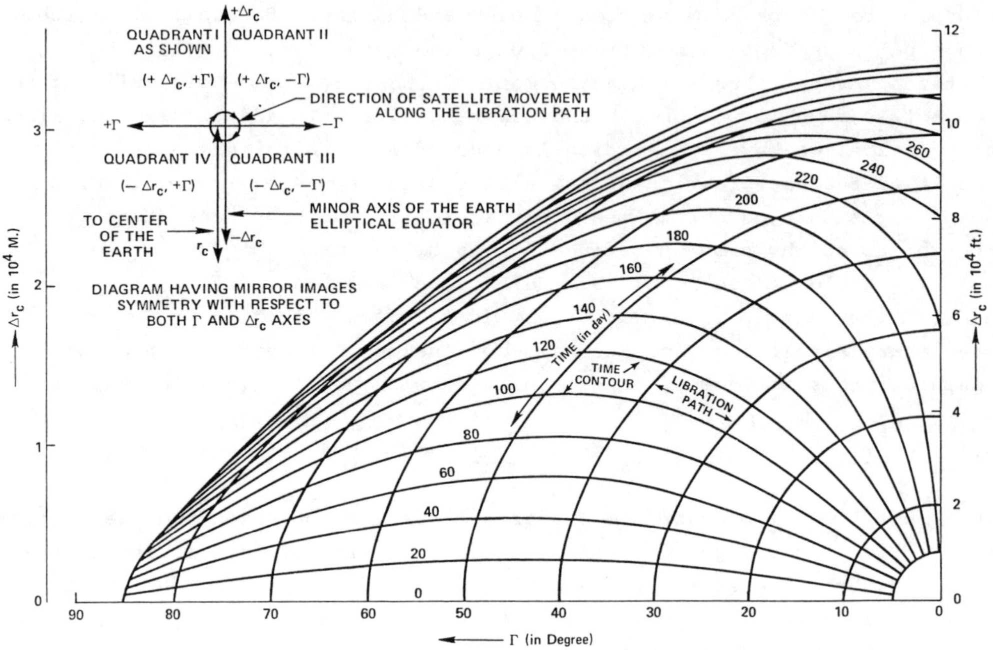

Fig. 3.8. 'Γ - Δr_c - time' relationships during libration (for J_{22} = $- 1.819 \times 10^{-6}$).

Figure 3.7. shows the quarter-period of the libration at different Γ_{max}'s for various J_{22}'s. Figure 3.8. presents a quarter of the libration paths in a rotating Earth coordinate system for various Γ_{max}'s shown on the Γ-axis with $J_{22} = - 1.819 \times 10^{-6}$. A complete libration cycle can be obtained by taking the mirror images of the quarter cycle with respect to both axes since (3.35) is an even function with respect to both axes. This figure can be used to trace the time parameter during the libration. With known initial values of Γ and Δr_c of a satellite, the libration path can be delineated on Figure 3.8 by following the plotted paths. The reader will be able to trace the satellite along the libration path to determine the time elapsed by counting the difference in time contours. Attention should be paid on the mirror image property of this figure, e.g., when the satellite is in the second quadrant, the libration path is traced from the '+ Δr_c' - axis to the '- Γ' - axis, the difference in time contours being a negative value. Consequently, the time spent in each quadrant should be counted separately, and the total time spent is the sum of the time spent in each quadrant.

Figure 3.9. is used to compute the impulse (ΔV) required for keeping a geostationary satellite within a longitudinal deviation limit (ΔΓ) for different libration paths (i.e., Γ_{max}'s, hence Δr_c). Figure 3.10. gives the time interval for the ΔV application in order to keep the geostationary satellite within the ΔΓ limit. These two figures are included here for their applications in station-keeping to be discussed in Chapter 6. They are simply a replotting of the data shown in Figure 3.8. The property of mirror image also applies to these two figures.

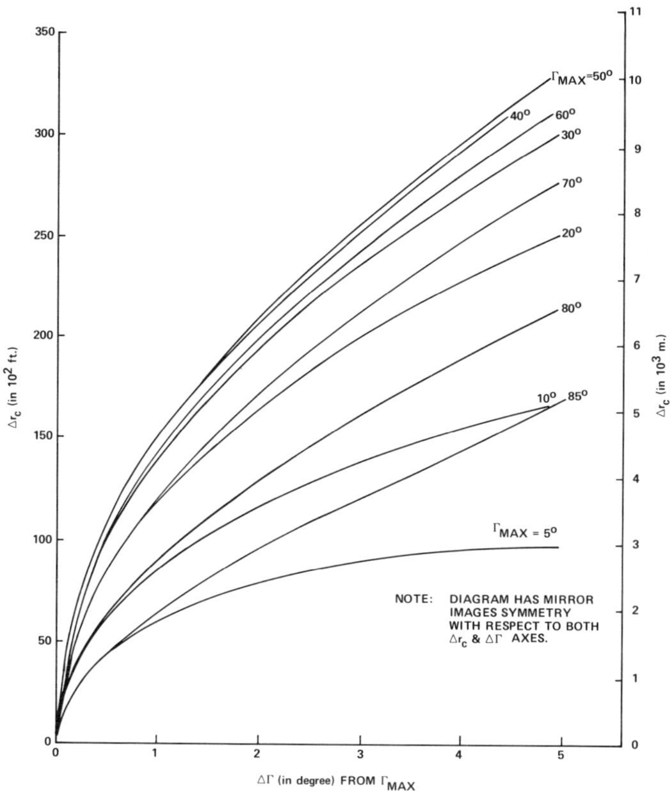

Fig. 3.9. 'Δr_c - $\Delta\Gamma_{\text{from } \Gamma_{max}}$' relationships for various Γ_{max}'s (for $J_{22} = -1.819 \times 10^{-6}$).

When (3.34) was derived, terms related to the orbital inclination (X_7) were omitted as was seen in (3.10) and (3.18). This omission does not create any serious error in X_6 because only X_7^2 is involved. Since the case we are dealing with is near-equatorial, X_7^2 is very small so that Figures 3.7.-3.10. will also be applicable to the near-equatorial geostationary satellites.

In addition to the 'Δr_c - $\Delta\Gamma_{\text{from } \Gamma_{max}}$' relation, the drift rate of a geostationary satellite at any point along the libration path can also be derived from Figure 3.8. if Δr_c is selected and (3.58) used.

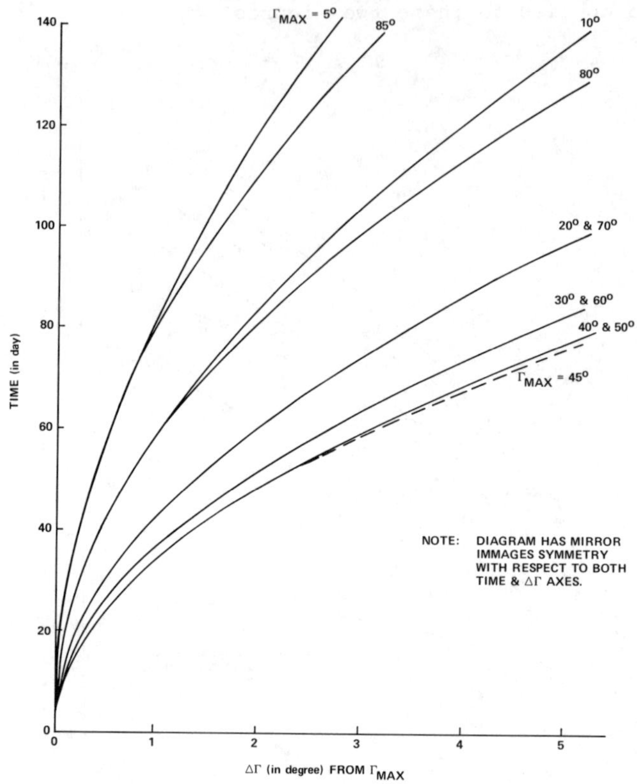

Fig. 3.10. 'Time - $\Delta\Gamma_{\text{from }\Gamma_{max}}$' relationships for various Γ_{max}'s (for $J_{22} = -1.819 \times 10^{-6}$).

3.13. Summary

The predominant effects of the triaxial field of the Earth on a geostationary satellite can be characterized by the libration of the satellite about the minor axis of the elliptical equator of the Earth. The libration contour (see Figure 3.5.) is expressed as

$$\cos 2\Gamma - \cos 2\Gamma_{max} = \frac{r_c^2}{8J_{22}R_E^2}\left(1 + 3J_2\frac{R_E^2}{r_s^2}\right)\left(\frac{\Delta r_c}{r_c}\right)^2 . \qquad (3.103)$$

The location of the satellite in an Earth-center inertial coordinate system is related to the time by

$$\Gamma = \theta - \theta_E = \theta - (\theta_{E0} + \dot{\theta}_E t) , \qquad (3.104)$$

and

$$\theta - \theta_0 = \frac{1}{3}\left\{2(\Gamma_{max} - \Gamma) + (2k_1)^{1/2} \times \right. \qquad (3.105)$$

$$\left. \times \ \mathrm{sd}^{-1}\left[\left(\frac{2(\cos 2\Gamma - \cos 2\Gamma_{max})}{2}\right)^{1/2} \middle| \ \frac{1 - \cos 2\Gamma_{max}}{2}\right]\right\}$$

The drift rate of the satellite with respect to the minor axis of the equator of the Earth is

$$\frac{d\Gamma}{dt} = -\frac{3}{2}\frac{\Delta r_c}{r_c}\dot{\theta}_E . \tag{3.106}$$

In addition to the libration, various orbital elements are experiencing both secular (due to oblateness) and long-term (due to equator ellipticity) variations as:

$$\bar{\Omega}_0 - \bar{\Omega} = \frac{3}{2}\frac{R_E^2}{r_c^2} J_2[(\theta - \theta_0) - \frac{4}{3}(\Gamma_{max} - \Gamma)] , \tag{3.107}$$

$$\bar{i} = \bar{i}_0\left\{1 - \frac{1}{4}\left[\frac{1}{k_1}(\cos 2\Gamma - \cos 2\Gamma_{max})\right]^{1/2}\right\} , \tag{3.108}$$

$$\tan\bar{\omega} = \sec\bar{i}\tan[\omega_2(\theta - \theta_0) - \bar{\Omega} + \gamma_0] , \tag{3.109}$$

$$\bar{e} = \frac{\bar{X}_6^2}{\bar{\mu}_E}(1 - \frac{27}{8}J_2^2\frac{R_E^4}{r_s^4} - 9\frac{R_E^2}{r_s^2}\cos 2\Gamma)\sec\bar{i}\times$$

$$\times\left\{c_0^2 + \tan^2\bar{i}\cos^2[\omega_2(\theta - \theta_0) - \bar{\Omega} + \gamma_0]\right\} , \tag{3.110}$$

and

$$\bar{X}_6^2 = \bar{\mu}_E r_c\left(1 + \frac{\Delta r_c}{r_c}\right) ,$$

where $\bar{i}_0$, $\bar{\Omega}_0$, c_0, and γ_0 are constants derived from given initial conditions.

Bibliography

Early interests in the effects of Earth's equatorial ellipticity on synchronous satellites involved the determination of the constants (J_{mn}, λ_{mn}) of tesseral harmonics (Allan, 1967; Blitzer, 1968). After obtaining these constants, four symmetric equilibrium positions in an equatorial synchronous orbit are exhibited mainly due to the J_{22} tesseral harmonics associated with the equatorial ellipticity. When the entire spectrum of tesseral harmonics is included, the dominance of the J_{22} harmonic still limits the number of equilibrium positions to four; however, the symmetry of the positions is destroyed, and the longitudinal deviations can be 3° or more (Blitzer, 1965; Morando, 1962a).

Existing treatments of the effect of the Earth's equatorial ellipticity on a synchronous satellite fall into two categories:

The first category is more or less limited to the planar linearized version. Frick and Garber (1962, 1963) and Perkins (1963) solved the linearized equations of motion in a polar coordinate system. Blitzer, etc. (1962a, 1963, 1966) solved a set of linearized equations of motion in an Earth-fixed coordinate system. As the presence of tesseral harmonics results in a tangential force (longitudinal direction) acting on a satellite, Allan (1963) utilized this fact to derive expressions for the long-period change in energy, and consequently , also in the semimajor axis and in mean motion. For the case of the Earth's equatorial ellipticity, an equation of libration about the minor axis of the elliptical equator was obtained.

Musen (1962a) derived a set of variational equations resulting from the ellipticity effects on a satellite orbiting around an oblated spherical Earth. His findings were claimed as a semi-analytical method for treating the problem since further numerical integration was required to obtain the final solution. Wagner (1964) gave a linearized solution to the drift of a 24-hour equatorial satellite due to an Earth potential field including up through the 4th-order, which accounts for the effect of the equatorial ellipticity. Kamel et al. (1973) set up the differential equations governing the variations of the semimajor axis and the longitude due to the Earth triaxiality as well as the lunisolar perturbations. The equations were then linearized around a given station longitude and a certain semimajor axis whose value was determined such that the constant part in the drift equation, due to zonal harmonics and some of the lunisolar effects, is zero. Closed-form solutions were obtained, and the solutions were checked reasonably well with the numerically integrated results.

The second category characterizes the studies of the resonant effects on satellites brought about by the longitude dependent terms of the Earth's potential. Morando (1962a, b, 1963) employed the von Zeipel method, as modified by Hori, to separately obtain the stable positions and the period of libration resulting from each of the tesseral harmonics. Musen and Bailie (1962b) solved the problem by using canonical elements and the Hamilton-Jacobi partial differential equations. The small divisors in the solution were removed by adjusting the constant of integration. Allan (1965, 1967b, c) derived the expressions of the time rate of all orbital elements due to the tesseral harmonics. Garfinkel (1966) used his 'Small Divisors' solution to investigate the resonant effect. Gedeon (1969) made a comprehensive study of resonance effects on satellite orbit due to tesseral harmonics; the special case of geostationary satellites was studied by comparing the librations due to one critical tesseral (J_{22}) and three critical tesserals ($J_{22} + J_{31} + J_{33}$). He concluded that J_{22} is the dominant tesseral, and this is also true for the case of other resonant circular orbits.

Vagners (1968) combined the von Zeipel method with the perturbation technique of Hamiltonian mechanics to investigate the long-period behavior of satellites due to the tesseral harmonics. Unfortunately, his solution has a singularity for satellites in the equatorial plane. Sih and Eckstein (1968) employed Zee's formulation (Zee, 1965) of the equations of motion for a satellite around an oblate Earth and solved them by an application of the two-variable expansion for the resonance problem. Yet the most important case for practical application - a 24-hr geostationary satellite - was discussed only briefly and not investigated in detail. Voshkov'yak and Lidov (1973) included the Earth's equatorial ellipticity effect in their studies of the perturbations of the Sun, Moon, and Earth on geostationary satellites. All these investigations concluded that the relationship between time and the mean longitude of the satellite (measured from the Earth's equatorial minor axis) is analogous to that of the motion of a simple pendulum.

As cross-track sustaining schemes for a geostationary satellite (Billick, 1967) require all inputs of orbital plane movements, the satellite may deviate from the equatorial plane because of the Sun and Moon perturbations (Frick, 1962; Zee, 1968) ; the ellipticity effect on orbital inclination and regression of the line of nodes should also be investigated. The effect on these two orbital elements has not been revealed explicitly in previous investigations and cannot be brought out from the studies of the purely planar case. Furthermore, since J_{22} is of the order of 10^{-6} and J_2 is of the order of 10^{-3}, the solution to the effect of the Earth's equatorial ellipticity problem should be obtained to the order of J_2^2 ($\sim 10^{-6}$). Except for Musen (1962a) and Shi (1968), all previous investigations were carried out to the first order of J_2 and J_{22}, with the second order of J_2 omitted. In 1971, Zee presented the solution to the problem to the order of J_2^2, and the long-period variations of all orbital elements were revealed (Zee, 1971).

References

Abramowitz, M. and Stegun, I.A.; Handbook of Mathematical Functions with Formulas, Graphs, and Mathematical Tables, Applied Mathematics Series-55, National Bureau of Standards, U.S. Dept. of Commerce, June, 1964.

Allan, R.R.; 'Perturbations of Geostationary Satellite by the Longitude-Dependent Terms in the Earth Gravitational Field', Planet. Space Sci. **11**, 1325-1334 (1963).

Allan, R.R.; 'On the Motion of Nearly Synchronous Satellites', Royal Society Proceedings **A 288**, 60-68 (1965).

Allan, R.R. and Piggot, B.A.M.; 'Longitude Dependence of the Geopotential Deduced from Synchronous Satellites', Phil. Trans. of Royal Society of London **A 262**, 137-147 (1976a).

Allan, R.R.; 'Resonance Effects Due to the Longitude Dependence of the Gravitational Field of a Rotating Primary', Planet. Space Sci. **15**, 53-76 (1967b).

Allan, R.R.; 'Satellite Resonance with Longitude-Dependent Gravity-II, Effect Involving the Eccentricity', Planet. Space Sci. **15**, 1829-1845 (1967c).

Billick, R.H.; 'Cross-Track Sustaining Requirements for a 24-Hour Satellite', J. Spacecraft and Rockets **4**, 297-301 (1967).

Blitzer, L; 'Synchronous and Resonant Orbits Associated with Equatorial Ellipticity', ARS J. **32**, 1016-1019 (1962a).

Blitzer, L., Boughton, E.M., Kang, G., and Page, R.M.; 'Effect of Ellipticity of the Equator on the 24-Hour Nearly Circular Satellite Orbits', J. Geophysical Research **67**, 329-335 (1962b).

Blitzer, L., Kang, G., and McGuire, J.B.; 'The Perturbed Motion of 24-Hour Satellite due to Equatorial Ellipticity', J. Geophysical Research **68**, 950-952 (1963).

Blitzer, L.; 'Equilibrium Positions and Stability of 24-hour Satellite Orbits', J. Geophysical Research 70, 3987-3992 (1965).

Blitzer, L.; 'Satellite Resonances and Libration Associated with Tesseral Harmonics of the Geopotential', J. Geophysical Research 71, 3557-3565 (1966).

Blitzer, L., Davis, D.R., and DeSulima, T.H.; 'Equilibrium Longitude of Geostationary Satellites', AIAA J. 6, 1199-1200 (1965).

Bogoliubov, H. and Mitropolsky, J.; Asymptotic Method in the Theory of Nonlinear Oscillations, Hindustan Publishing Corporation, Delhi-6, India, 1961.

Brouwer, D. and Clemence, G.M.; Methods of Celestial Mechanics, Academic Press, New York and London, 1961.

Brown, E.W.; An Introductory Treatise on the Lunar Theory, Cambridge University Press, 1896 - reprinted by Dover Publication, Inc., 1960.

Frick, R.H. and Garber, T.B.; 'Perturbation of a Synchronous Satellite', R-399-NASA, RAND Corporation, Santa Monica, California (1962).

Frick, R.H.; 'Perturbation of Equatorial Satellites Due to Equatorial Ellipticity', RM-3861-NASA, RAND Corporation, Santa Monica, California (1963).

Gaposchkin, E.M.; 'Tesseral Harmonic Coefficients and Station Coordinates from the Dynamic Method', Smithsonian Astrophysical Observatory Special Report 200 (1966).

Gaposchkin, E.M. and Lambeck, K.; '1969 Smithsonian Standard Earth (II)', Smithsonian Astrophysical Observatory Special Report 315 (1970).

Gaposchkin, E.M.; 'Earth's Gravity Field to the Eighteenth Degree and Geocentric Coordinates for 104 Stations from Satellite and Terrestrial Data', J. Geophysical Research 79, 5377-5411 (1974).

Garfinkel, B.; 'Formal Solution in the Problem of Small Divisors', Astron. J. 77, 657-699 (1966).

Gedeon, G.S.; 'Tesseral Resonance Effects on Satellite Orbits', Celestial Mechanics 1, 167-189 (1969).

Hori, G.I.; 'The Motion of an Artificial Satellite in the Vicinity of the Critical Inclination', Astron. J. 65, 291-300 (1960).

Kamel, A., Ekman, B., and Tibbitts, R.; 'East-West Stationkeeping Requirements of Nearly Synchronous Satellites Due to Earth's Triaxiality and Luni-Solar Effects', Celestial Mechanics 8, 129-148 (1973).

Kaula, W.M.; Theory of Satellite Geodesy, Blaisdell Publishing Company, Waltham, Mass., Toronto, London (1966).

Kozai, Y.; 'Revised Zonal Harmonics in the Geopotential', Smithsonian Astrophysical Observatory Special Report 295 (1969).

Morando, M.B.; 'Libration d'un Satellite de 24 h.', Compt. rend. 254, 635-637 (1962a).

Morando, M.B.; 'Orbites de Résonance des Satellite de 24 h.', Bulletin Astronomique 24, 48-67 (1962b).

Morando, M.B.; 'Recherches sur les Orbites de Résonance', in *Proceedings of the First International Symposium on the Use of Artificial Satellites for Geodesy*, Washington, D.C., 42-51, North Holland Publ. Co., Amsterdam (1963).

Musen, P.; 'On the Long-Period Effects in the motion of an Artificial Satellite Caused by the Ellipticity of the Equator of the Earth', *J. Geophysical Research* **67**, 313-319 (1962a).

Musen, P. and Bailie, A.E.; 'On the Motion of a 24-Hour Satellite', *J. Geophysical Research* **67**, 1123-1132 (1962b).

Perkins, F.M.; 'Flight Mechanics of the 24-Hour Satellite', *AIAA J.* **1**, 848-851 (1963).

Saaty, T.L. and Bram, J.; *Nonlinear Mathematics*, McGraw-Hill Book Company, New York, 1964.

Shi, Y.Y. and Eckstein, M.C.; 'Application of Singular Perturbation Methods to Resonance Problems', *Astron. J.* **71**, 275-289 (1968).

Smith, D.E., Lerch, F.J., Marsh, J.G., Wagner, C.A., Kolenkiewicz, R., and Khan, M.A.; 'Contribution to the National Geodetic Satellite Program by Goddard Space Flight Center', *J. Geophysical Research* **81**, 1006-1026 (1976).

Vagners, J.; 'Modified Long-Period Behavior Due to Tesseral Harmonics', *AIAA J.* **6**, 1229-1234 (1968).

Voshkov'yak, M.A. and Lidov, M.L.; 'An Approximate Description of the Evolution of the Orbit of a Stationary Artificial Earth Satellite', *Cosmic Research* **11**, 305-316 (1973).

Wagner, C.A.; 'The Drift of a 24-Hour Equatorial Satellite Due to an Earth Gravity Field Through 4th Order', TN D-2103, Goddard Space Flight Center, NASA, Washington, D.C. (1964).

Wagner, C.A.; 'Acceleration on 24-Hour Satellites and Low Order Longitude Terms in the Geopotential', X-552-69-508, Goddard Space Flight Center, NASA, Washington, D.C. (1969).

Zee, C.H.; 'Trajectories of Satellites Under the Influence of Earth Oblateness', Report No. ADR 06-06-64.2, Grumman Aircraft Engineering Corporation, Bethpage, N.Y. (1965).

Zee, C.H.; 'Effects of the Sun and the Moon on a Synchronous Satellite', Report No. FSR-AD6-06-68.2, Grumman Aircraft Engineering Corporation, Bethpage, N.Y. (1968).

Zee, C.H.; 'Effect of Earth Oblateness and Equator Ellipticity on a Synchronous Satellite', *Astronautica Acta* **16**, 143-153 (1971).

Chapter 4

SUN AND MOON EFFECT

4.1. General

In the Sun-Earth-Moon system, the Earth, moving around the Sun in the ecliptic plane, travels in an orbit with an eccentricity of 0.017. The Earth's equatorial plane has a mean inclination of 23°27' from the ecliptic plane. The Moon moves around the Earth in its own orbit, which has a mean inclination of 5°09' from the ecliptic plane and an orbit of eccentricity 0.055. The Moon's and the Earth's orbits are not contained in the same plane, the line of nodes of the Moon's orbit regresses in the ecliptic plane with a period of 18.6 yr, while the line of apsides progresses in its own orbital plane with a period of 8.8 yr. Consequently, the angle between the equatorial plane of the Earth and the orbital plane of the Moon varies from 28°36' to 18°18'.

In studying the effect of the lunisolar perturbation on satellite motion by means of a disturbing function (Kozai, 1959; Allan and Cook, 1963), the squared values of the eccentricity of the Earth's and Moon's orbits appear in the function; the former has an order of magnitude of 10^{-4}, while the latter is of 10^{-3}. The assumption then that both the Earth's and Moon's orbits are circular would yield good approximate results. Since synchronous orbits are relatively small in comparison with that of the Earth and the Moon, only the lowest-order (second harmonic) term in the expansion of the Earth's disturbing function needs to be retained to obtain a good representation of the satellite's motion. The parallactic term (third harmonic) in the lunar disturbing function will change the eccentricity of a circular orbit. A small initial eccentricity, such as may be built up by the parallactic term, may increase rapidly if the orbit is sufficiently large (Lidov, 1963). As the parallactic term becomes effective when satellites are at a distance of about 10 Earth radii, its effect on synchronous satellites (at a distance of 6.61 Earth radii) may not need to be considered. In their study of the long-period motion of the orbital plane of a satellite, Allan and Cook (1963) averaged the disturbing function analytically over the mean anomalies of both the satellite and the disturbing bodies (i.e., the Sun and Moon). This procedure will yield a good result only when the period of the predicated long-period motion of the satellite orbital plane is reasonably long compared with the periods of the disturbing bodies. This criterion requires that satellites are at distances greater than 3 Earth radii. We shall use both of the above arguments as our criterion for judging the accuracy of the results in the method of solution.

In the following sections, we shall present our solution to the problem by taking the terms to the first-order in the expansion of the Sun's attraction function and to the second-order in the expansion of the Moon's attraction function (i.e. the parallactic term is included), while the eccentricity of both the Earth's and Moon's orbits are taken into consideration. Furthermore, an oblate Earth (J_2 effect) is taken as the body of the center of attraction.

Owing to the complicated nature of the problem, we shall present only the first-order approximate solution and the improved first-order solution in a sense of the asymptotic method described in Chapter 2. The former shows the secular effect and the latter both the secular and short-period effects.

4.2. Force Components

The Sun-Earth-Moon system together with a satellite constitute a four-body problem in celestial mechanics. In order to study the satellite motion around the Earth, various forces acting on the satellite must be analyzed. Since a geostationary satellite has an angular travelling rate equal to the rotation rate of the Earth about its polar axis, we choose an intertial geocentric spherical coordinate (r, θ, ϕ) system as the basic reference frame (as shown in Figure 4.1.); r is the radial distance from the center of the Earth to the satellite, θ is the satellite longitude relative to the vernal equinox, ϕ is the satellite latitude, and the corresponding unit vectors are $\vec{n}_r$, $\vec{n}_\theta$, and $\vec{n}_\phi$, respectively. Because the motions of the Sun and Moon are involved in the analysis, another coordinate system which is Earth-centered, inertial rectangular is also used, where x is in the direction of the vernal equinox, y is in the ecliptic plane, z is normal to the ecliptic plane, and their respective unit vectors are designated by $\vec{x}$, $\vec{y}$, and $\vec{z}$.

The relationships between the unit vectors $\vec{n}_r$, $\vec{n}_\theta$, $\vec{n}_\phi$ and $\vec{x}$, $\vec{y}$, $\vec{z}$ can be expressed as

$$\vec{n}_r = a_x\vec{x} + a_y\vec{y} + a_z\vec{z} ,$$

$$\vec{n}_\theta = b_x\vec{x} + b_y\vec{y} + b_z\vec{z} , \tag{4.1}$$

and

$$\vec{n}_\phi = c_x\vec{x} + c_y\vec{y} + c_z\vec{z} ,$$

where

$$\left.\begin{aligned}
a_x &= \cos\theta\cos\phi , \\
a_y &= \sin\theta\cos\phi\cos\lambda - \sin\lambda\sin\phi , \\
a_z &= \sin\theta\cos\phi\sin\lambda + \cos\lambda\sin\phi , \\
b_x &= -\sin\theta , \\
b_y &= \cos\theta\cos\lambda , \\
b_z &= \cos\theta\sin\lambda , \\
c_x &= -\cos\theta\sin\phi , \\
c_y &= -\sin\theta\sin\phi\cos\lambda - \cos\phi\sin\lambda , \\
c_z &= -\sin\theta\sin\phi\sin\lambda + \cos\phi\cos\lambda ,
\end{aligned}\right\} \tag{4.2}$$

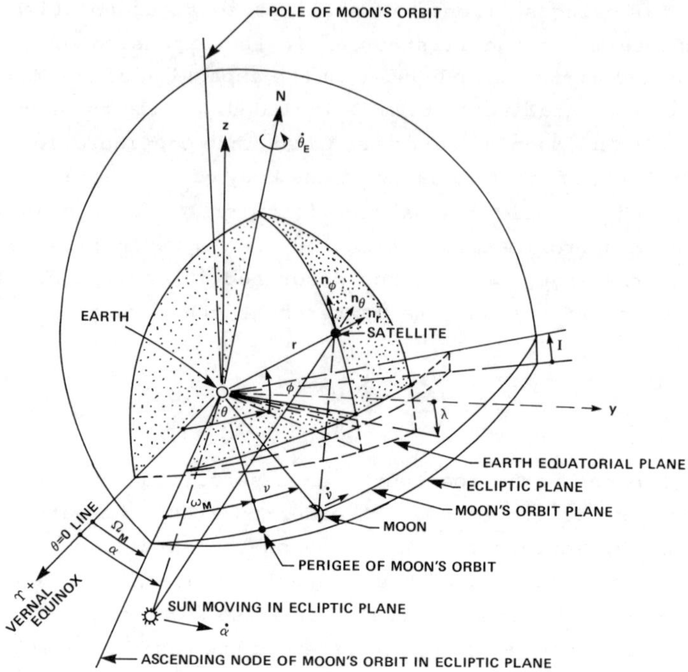

Fig. 4.1. Geocentric coordinate system for motions of Sun, Moon, and satellite.

and λ is the angle between the Earth's equatorial plane and the plane of the ecliptic ($\lambda = -23°27'$).

As in the classical n-body problem, the acceleration of the satellite relative to the Earth ($\ddot{\vec{r}}$) results not only from the attractions of the Sun, Moon, and Earth, but also from the gravitational forces of the Sun and Moon on the Earth. Thus,

$$\ddot{\vec{r}} = \frac{1}{M}(\vec{F}_E + \vec{F}_S + \vec{F}_M) - \frac{1}{M_E}(\vec{F}_{SE} + \vec{F}_{ME}) \ , \tag{4.3}$$

where M and M_E are the masses of the satellite and the Earth, respectively. Also $\vec{F}_E$, $\vec{F}_S$, and $\vec{F}_M$ are the gravitational attractions of the Earth, Sun, and Moon on the satellite, while $\vec{F}_{SE}$ and $\vec{F}_{ME}$ are the gravitational attractions of the Sun and Moon on the Earth. The acceleration components, a_r, a_θ, and a_ϕ along the directions $\vec{n}_r$, $\vec{n}_\theta$, and $\vec{n}_\phi$, respectively, are then

$$a_r = \ddot{\vec{r}}.\vec{n}_r = \ddot{r} - r\dot{\theta}^2 \cos^2 \phi - r\dot{\phi}^2 \ , \tag{4.4}$$

$$a_\theta = \ddot{\vec{r}}.\vec{n}_\theta = \frac{1}{r \cos \phi} \frac{d}{dt}(r^2\dot{\theta} \cos^2 \phi) \ , \tag{4.5}$$

and

$$a_\phi = \ddot{\vec{r}}.\vec{n}_\phi = \frac{1}{r} \frac{d}{dt}(r^2\dot{\phi}) + r\dot{\theta}^2 \cos \phi \sin \phi \ . \tag{4.6}$$

Equations (4.4)-(4.6) are clearly recognized as the equations of motion in the spherical coordinate system.

The forces which appear in (4.3) will be evaluated. The magnitude of the forces is either obtained by taking the gradient of the Earth's potential field or by the simple inverse square distance law between two bodies of mutual attractions. The components of the forces in the r-, θ-, and φ-directions are then computed by the Earth's potential field gradient in the r-, θ-, and φ-directions and by taking the dot-product of the unit vectors $\vec{n}_r$, $\vec{n}_\theta$, and $\vec{n}_\phi$ with the attractive forces.

(1) $\vec{F}_E$, the gravitational attraction of the Earth on the satellite, can be expressed by means of the Earth's potential function, as we are limiting ourselves to take the oblate Earth as the only attraction from the Earth:

$$V_E = \frac{\mu_E}{r}\left[1 - \tfrac{1}{2}J_2 \frac{R_E^2}{r^2}(3 \sin^2 \phi - 1)\right] , \tag{4.7}$$

where V_E is the Earth's gravitational potential, μ_E is the gravitational constant of the Earth, J_2 is the Earth oblateness coefficient, and R_E is the Earth's mean equatorial radius.

It will be seen in Section 4.3. that the perturbation of the J_2-term at the synchronous radius is comparable to that of the Sun and Moon.

The three components of $\vec{F}_E$ in the $\vec{n}_r$, $\vec{n}_\theta$, and $\vec{n}_\phi$ directions are given by the relations

$$\frac{\vec{F}_E\cdot\vec{n}_r}{M} = \frac{\partial V_E}{\partial r} = -\frac{\mu_E}{r^2} + \tfrac{3}{2}J_2\mu_E \frac{R_E^2}{r^4}(3 \sin^2 \phi - 1) , \tag{4.8}$$

$$\frac{\vec{F}_\theta\cdot\vec{n}_\theta}{M} = \frac{1}{r \cos \phi}\frac{\partial V_E}{\partial \theta} = 0 , \tag{4.9}$$

and

$$\frac{\vec{F}_\phi\cdot\vec{n}_\phi}{M} = \frac{1}{r}\frac{\partial V_E}{\partial \phi} = -3J_2\mu_E \frac{R_E^2}{r^4} \cos \phi \sin \phi . \tag{4.10}$$

(2) F_S, the gravitational attraction of the Sun on the satellite, is given by

$$\frac{\vec{F}_S}{M} = -\frac{\mu_S}{\rho_S^3}\vec{\rho}_S , \tag{4.11}$$

where $\vec{\rho}_S$ is the vector from the Sun to the satellite, and μ_S is the gravitational constant of the Sun. The vector $\vec{\rho}_S$ can be expressed as

$$\vec{\rho}_S = \vec{r}_{SE} + \vec{r} , \tag{4.12}$$

where $\vec{r}_{SE}$ is the vector from the Sun to the Earth. Combining (4.11) and (4.12) results in

$$\frac{\vec{F}_S}{M} = -\frac{\mu_S}{r_{SE}^3}\left[1 - \frac{3(\vec{r}_{SE}\cdot\vec{r})}{r_{SE}^2}\right](\vec{r}_{SE} + \vec{r}) \tag{4.13}$$

if high order terms such as $(r/r_{SE})^2$ are neglected, because r/r_{SE} is of the order of 10^{-4} for the case of geostationary satellites. The components of $\vec{F}_S$ in the $\vec{n}_r$, $\vec{n}_\theta$, and $\vec{n}_\phi$ directions are given by

$$\frac{\vec{F}_S\cdot\vec{n}_r}{M} = -\frac{\mu_S}{r_{SE}^3}\left[1 - \frac{3(\vec{r}_{SE}\cdot\vec{r})}{r_{SE}^2}\right](\vec{r}_{SE}\cdot\vec{n}_r) , \tag{4.14}$$

$$\frac{\vec{F}_S\cdot\vec{n}_\theta}{M} = \frac{\mu_S}{r_{SE}^3}\left[1 - \frac{3(\vec{r}_{SE}\cdot\vec{r})}{r_{SE}^2}\right](\vec{r}_{SE}\cdot\vec{n}_\theta), \tag{4.15}$$

and

$$\frac{\vec{F}_S \cdot \vec{n}_\phi}{M} = -\frac{\mu_S}{r_{SE}^3}\left[1 - \frac{3(\vec{r}_{SE}\cdot\vec{r})}{r_{SE}^2}\right](\vec{r}_{SE}\cdot\vec{n}_\phi) , \tag{4.16}$$

respectively, where $\vec{r}.\vec{n}_r$ in the parentheses of (4.14) has been omitted because of its smallness when compared with $\vec{r}_{SE}.\vec{n}_r$.

(3) F_M, the gravitational attraction of the Moon on the satellite, is given by the expression

$$\frac{\vec{F}_M}{M} = -\frac{\mu_M}{\rho_M^3}\vec{\rho}_M , \tag{4.17}$$

where $\vec{\rho}_M$ is the vector from the Moon to the satellite, and μ_M is the gravitational constant of the Moon. The vector $\vec{\rho}_M$ can be expressed as

$$\vec{\rho}_M = \vec{r}_{ME} + \vec{r} , \tag{4.18}$$

where $\vec{r}_{ME}$ is the vector from the Moon to the Earth. Combining (4.17) and (4.18) gives

$$\frac{\vec{F}_M}{M} = -\frac{\mu_M}{r_{ME}^3}\left[1 - \frac{3(\vec{r}_{ME}\cdot\vec{r})}{r_{ME}^2} - \frac{3}{2}\frac{r^2}{r_{ME}^2} + \frac{15}{2}\left(\frac{\vec{r}_{ME}\cdot\vec{r}}{r_{ME}^2}\right)^2\right](\vec{r}_{ME} + \vec{r}) , \tag{4.19}$$

where $(r/r_{ME})^3$ and the high order terms are neglected, because at the synchronous radius, r/r_{ME} is of the order of 10^{-1}.

The three components along $\vec{n}_r$, $\vec{n}_\theta$, and $\vec{n}_\phi$ are

$$\frac{\vec{F}_M\cdot\vec{n}_r}{M} = -\frac{\mu_M}{r_{ME}^3}\left[1 - \frac{3(\vec{r}_{ME}\cdot\vec{r})}{r_{ME}^2} - \frac{3}{2}\frac{r^2}{r_{ME}^2} + \frac{15}{2}\left(\frac{\vec{r}_{ME}\cdot\vec{r}}{r_{ME}^2}\right)^2\right][r + (\vec{r}_{ME}\cdot\vec{n}_r)] , \tag{4.20}$$

$$\frac{\vec{F}_M\cdot\vec{n}_\theta}{M} = -\frac{\mu_M}{r_{ME}^3}\left[1 - \frac{3(\vec{r}_{ME}\cdot\vec{r})}{r_{ME}^2} - \frac{3}{2}\frac{r^2}{r_{ME}^2} + \frac{15}{2}\left(\frac{\vec{r}_{ME}\cdot\vec{r}}{r_{ME}^2}\right)^2\right](\vec{r}_{ME}\cdot\vec{n}_\theta) , \tag{4.21}$$

and

$$\frac{\vec{F}_M\cdot\vec{n}_\phi}{M} = -\frac{\mu_M}{r_{ME}^3}\left[1 - \frac{3(\vec{r}_{ME}\cdot\vec{r})}{r_{ME}^2} - \frac{3}{2}\frac{r^2}{r_{ME}^2} + \frac{15}{2}\left(\frac{\vec{r}_{ME}\cdot\vec{r}}{r_{ME}^2}\right)^2\right](\vec{r}_{ME}\cdot\vec{n}_\phi) , \tag{4.22}$$

respectively.

(4) $\vec{F}_{SE}$, the gravitational attraction of the Sun on the Earth, is given by the expression

$$\frac{\vec{F}_{SE}}{M_E} = \frac{-\mu_S}{r_{SE}^3}\vec{r}_{SE} , \tag{4.23}$$

where M_E is the mass of the Earth, and the three components in the $\vec{n}_r$, $\vec{n}_\theta$, and $\vec{n}_\phi$ directions are

$$\frac{\vec{F}_{SE}\cdot\vec{n}_r}{M_E} = -\frac{\mu_S}{r_{SE}^3}(\vec{r}_{SE}\cdot\vec{n}_r) , \tag{4.24}$$

$$\frac{\vec{F}_{SE}\cdot\vec{n}_\theta}{M_E} = -\frac{\mu_S}{r_{SE}^3}(\vec{r}_{SE}\cdot\vec{n}_\theta) , \tag{4.25}$$

and

$$\frac{\vec{F}_{SE}\cdot\vec{n}_\phi}{M_E} = -\frac{\mu_S}{r_{SE}^3}(\vec{r}_{SE}\cdot\vec{n}_\phi) , \tag{4.26}$$

respectively.

(5) $\vec{F}_{ME}$, the gravitational attraction of the Moon on the Earth, is

given by the expression

$$\frac{\vec{F}_{ME}}{M_E} = - \frac{\mu_M}{r_{ME}^3} \vec{r}_{ME} \tag{4.27}$$

from which the components in the $\vec{n}_r$, $\vec{n}_\theta$, and $\vec{n}_\phi$ directions are

$$\frac{\vec{F}_{ME}\cdot\vec{n}_r}{M_E} = - \frac{\mu_M}{r_{ME}^3} (\vec{r}_{ME}\cdot\vec{n}_r) , \tag{4.28}$$

$$\frac{\vec{F}_{ME}\cdot\vec{n}_\theta}{M_E} = - \frac{\mu_M}{r_{ME}^3} (\vec{r}_{ME}\cdot\vec{n}_\theta) , \tag{4.29}$$

and

$$\frac{\vec{F}_{ME}\cdot\vec{n}_\phi}{M_E} = - \frac{\mu_M}{r_{ME}^3} (\vec{r}_{ME}\cdot\vec{n}_\phi) , \tag{4.30}$$

respectively.

In order to evaluate the components of the forces in (4.11)-(4.30), the vectors $\vec{r}_{SE}$ and $\vec{r}_{ME}$ are to be determined in terms of the unit vectors $\vec{x}$, $\vec{y}$, and $\vec{z}$ by simple geometric relationships. They are (see Figure 4.1.)

$$\vec{r}_{SE} = r_{SE}(\vec{x} \cos \alpha + \vec{y} \sin \alpha) , \tag{4.31}$$

where α is the angle between $\vec{x}$ and $(-\vec{r}_{SE})$, and

$$\vec{r}_{ME} = - r_{ME}(M_x\vec{x} + M_y\vec{y} + M_z\vec{z}) , \tag{4.32}$$

where

$M_x = \cos(\omega_M + \nu) \cos \Omega_M - \cos I \sin \Omega_M \sin(\omega_M + \nu)$,

$M_y = \cos(\omega_M + \nu) \sin \Omega_M + \cos I \cos \Omega_M \sin(\omega_M + \nu)$,

$M_z = \sin I \sin(\omega_M + \nu)$,

ν = the true anomaly of the Moon,

ω_M = the argument of perigee of the Moon's orbit,

I = the mean inclination of the Moon's orbit with respect to the ecliptic plane (5°09'),

and

Ω_M = the longitude of ascending node of the Moon's orbit in the ecliptic plane.

Since both the Sun and Moon are moving in elliptical orbits with respect to the Earth, their mean motion can be derived from the well-known Kepler's third law in orbit mechanics:

$$\mu_S = a_S^3 n_S^2$$

and

$$\mu_E + \mu_M = a_M^3 n_M^2 \equiv \sigma\mu_M ,$$

where a_S and a_M are the semimajor axes of the Sun's and Moon's orbits, respectively, and n_S and n_M are the corresponding mean motion, and σ is defined as

$$\frac{\text{mass of Earth + mass of Moon}}{\text{mass of Moon}} = \frac{M_E + M_M}{M_M} .$$

Furthermore, if the relationship between the true anomaly, ν, and the mean anomaly, M_M,

$$\frac{d\nu}{dM_M} = \frac{a_M^2}{r_{ME}^2} (1 - e_M^2)^{1/2}$$

for the Moon's orbit (Escobal, 1965) and the identity

$$d\theta = \frac{dt}{dM_M} dM_M \frac{d\theta}{dt}$$

are used, we have

$$d\theta = \frac{\dot{\theta}}{n_M}\left(\frac{r_{ME}}{a_M}\right)^2 \frac{1}{(1 - e_M^2)^{1/2}} d\nu ,$$

where e_M is the eccentricity of the Moon's orbit. The characteristics of small eccentricity of both the synchronous and Moon's orbits allow us to replace $(\dot{\theta}/n_M)\, d\nu (\cong (\dot{\theta}_E/n_M)\, d\nu)$ by $\overline{d\theta}$, where $\overline{d\theta}$ is the averaged value of $d\theta$ over a period of the Moon's orbit. Thus

$$\int \frac{\mu_M f_1(\nu)}{r_{ME}^3} d\theta = \int \frac{n_M^2 f_1(\nu)}{\sigma(1 - e_M^2)^{3/2}} (1 + e_M \cos \nu)\, \overline{d\theta} \tag{4.34a}$$

and

$$\int \frac{\mu_M f_2^2(\nu)}{r_{ME}^4} d\theta = \int \frac{n_M^2 f_2(\nu)}{\sigma a_M (1 - e_M^2)^{5/2}} (1 + e_M \cos \nu)^2\, \overline{d\theta}$$

$$\cong \frac{n_M^2 f_2(\nu)}{\sigma a_M} (1 + 3e_M^2)(1 + 2e_M \cos \nu)\, \overline{d\theta} , \tag{4.34b}$$

where $f_1(\nu)$ and $f_2(\nu)$ are the functions associated with $1/r_{ME}^3$ and $1/r_{ME}^4$ in (4.20)-(4.22). Equation (4.34) shows that μ_M/r_{ME}^3 and μ_M/r_{ME}^4 may be replaced by $n_M^2(1 + e_M \cos \nu)/\sigma(1 - e_M^2)^{3/2}$ and $n_M^2(1 + e_M \cos \nu)^2/\sigma a_M(1 - e_M^2)^{5/2}$, respectively, if we replace ν by $(n_M/\dot{\theta}_E)\theta$. The use of the averaged value, $\overline{d\theta}$, as well as the replacement of ν by $(n_M/\dot{\theta}_E)\theta$, correspond to the Allan's procedure of averaging the disturbing function over the mean anomalies of the disturbing bodies as previously mentioned in Section 4.1.

Similarly, we have

$$\int \frac{\mu_S f_3(\alpha)}{r_{SE}^3} d\theta = \int \frac{n_S^2 f_3(\alpha)}{(1 - e_S^2)^{3/2}} [1 + e_S \cos(\alpha - \omega_S)]\, \overline{d\theta} \tag{4.35}$$

for the Sun's orbit, where $f_3(\alpha)$ is the function associated with $1/r_{SE}^3$ in (4.24)-(4.26), e_S is the eccentricity, and ω_S is the argument of perigee of the Sun's orbit measured from the vernal equinox. Again (4.35) shows that μ_S/r_{SE}^3 may be replaced by $n_S^2[1 + e_S \cos(\alpha - \omega_S)]/(1 - e_S^2)^{3/2}$ and α by $(n_S/\dot{\theta}_E)\theta$.

With (4.31), (4.32), (4.34), and (4.35), the forces $\vec{F}_E$, $\vec{F}_S$, $\vec{F}_M$, $\vec{F}_{SE}$, and $\vec{F}_{ME}$ in (4.3) in the directions of $\vec{n}_r$, $\vec{n}_\theta$, and $\vec{n}_\phi$ are now evaluated by (4.8)-(4.10), (4.14)-(4.16), (4.20)-(4.22), (4.24)-(4.26), and (4.28)-(4.30), respectively. The resulting expressions are then substituted into (4.4)-(4.6) to yield a set of equations of motion, which will be presented in the next section.

4.3. Equations of Motion

With the preparatory work presented in the last section, we may write the equations of motion of a satellite under the influence of the potential field, V_E, in (4.7) and the perturbations due to the gravitational attractions of the Sun and Moon as

$$\ddot{r} - r\dot{\theta}^2 \cos^2 \phi - r\dot{\phi}^2$$

$$= - \frac{\mu_E}{r^2} + \frac{3}{2}J_2\mu_E \frac{R_E^2}{r^4}(3 \sin^2 \phi - 1) +$$

$$+ r \frac{n_S^2[1 + e_S \cos(\alpha - \omega_S)]}{(1 - e_S^2)^{3/2}}[3(a_x \cos \alpha + a_y \sin \alpha)^2 - 1] +$$

$$+ \frac{1}{\sigma} r \frac{n_M^2(1 + e_M \cos \nu)}{(1 - e_M^2)^{3/2}}[3(a_xM_x + a_yM_y + a_zM_z)^2 - 1 -$$

$$- \frac{9}{2} \frac{r}{r_{ME}}(a_xM_x + a_yM_y + a_zM_z) +$$

$$+ \frac{15}{2} \frac{r}{r_{ME}}(a_xM_x + a_yM_y + a_zM_z)^3] , \tag{4.36}$$

$$\frac{1}{r \cos \phi} \frac{d}{dt}(r^2\dot{\theta} \cos^2 \phi)$$

$$= 3r \frac{n_S^2[1 + e_S \cos(\alpha - \omega_S)]}{(1 - e_S^2)^{3/2}}(a_x \cos \alpha + a_y \sin \alpha) \times$$

$$\times (b_x \cos \alpha + b_y \sin \alpha) + 3r \frac{1}{\sigma} \frac{n_M^2(1 + e_M \cos \nu)}{(1 - e_M^2)^{3/2}} \times$$

$$\times (b_xM_x + b_yM_y + b_zM_z)[(a_xM_x + a_yM_y + a_zM_z) -$$

$$- \frac{1}{2} \frac{r}{r_{ME}} + \frac{5}{2} \frac{r}{r_{ME}}(a_xM_x + a_yM_y + a_zM_z)^2] , \tag{4.37}$$

and

$$\frac{1}{r} \frac{d}{dt}(r^2\dot{\phi}) + r\dot{\theta}^2 \cos \phi \sin \phi$$

$$= - 3J_2\mu_E \frac{R_E^2}{r^4} \cos \phi \sin \phi +$$

$$+ 3r \frac{n_S^2[1 + e_S \cos(\alpha - \omega_S)]}{(1 - e_S^2)^{3/2}}(a_x \cos \alpha + a_y \sin \alpha) \times$$

$$\times (c_x \cos \alpha + c_y \sin \alpha) + 3r \frac{1}{\sigma} \frac{n_M^2(1 + e_M \cos \nu)}{(1 - e_M^2)^{3/2}} \times$$

$$\times (c_xM_x + c_yM_y + c_zM_z)[(a_xM_x + a_yM_y + a_zM_z) -$$

$$- \frac{1}{2} \frac{r}{r_{ME}} + \frac{5}{2} \frac{r}{r_{ME}}(a_xM_x + a_yM_y + a_zM_z)^2] , \tag{4.38}$$

where the perturbing forces of the Sun and Moon are carried out to the order of r/r_{SE} and $(r/r_{ME})^2$, respectively, as we have previously stated.

For a satellite located at a distance of the synchronous altitude ($r = 6.61\ R_E$, $r/r_{ME} = 0.11$) from the center of the Earth, the perturbing forces are estimated as:

$$J_2\mu_E \frac{R_E^2}{r^4}\text{(Earth oblateness)} \sim 5.49 \times 10^{-6}\ \text{m/sec.}^2\ (1.8 \times 10^{-5}\ \text{ft/sec.}^2),$$

$$rn_S^2\text{(Sun)} \sim 1.67 \times 10^{-6}\ \text{m/sec.}^2\ (0.55 \times 10^{-5}\ \text{ft/sec.}^2),$$

and

$$\frac{1}{\sigma}\, rn_M^2\text{(Moon)} \sim 3.62 \times 10^{-6}\ \text{m/sec.}^2\ (1.19 \times 10^{-5}\ \text{ft/sec.}^2).$$

These three forces are comparable in their order of magnitude. Our first-order solution is therefore to the accuracy of 10^{-6} m/sec.^{2} (10^{-5} ft/sec.^{2}) so far as the perturbing forces are concerned. It should be noted that the terms containing r/r_{ME} in the expression of the lunar perturbing force refine the accuracy of the lunar attractions to 10^{-7} m/sec.^{2}. In the following, a treatment of terms containing J_2, n_S^2, and n_M^2/σ of the same order of magnitude will be utilized in the process of obtaining a solution to the equations of motion.

We shall follow the procedures described in Chapter 2 by changing the independent variable from 't' to 'θ' and introducing the dependent parameters H, Q, and S:

$$H = r^2\dot{\theta}\cos^2\phi, \tag{4.39a}$$

$$Q = \frac{1}{r\cos\phi}, \tag{4.39b}$$

and

$$S = \tan\phi . \tag{4.39c}$$

For a near-equatorial geostationary satellite, approximations of $\sin\phi \cong \phi$ and $\cos\phi \cong 1$ in (4.2) can be made because $rn_S^2 \sin\phi$ and $(1/\sigma)rn_M^2 \sin\phi$ are rather small. With (4.39) and when the approximations $S = \tan\phi = \phi$ are also made, (4.36)-(4.38) become

$$\begin{aligned}
\frac{d^2Q}{d\theta^2} + Q = {} & \frac{\mu_E}{H^2(1+S^2)^{3/2}} + \frac{3}{2}\mu_E \frac{R_E^2Q^2}{H^2} J_2 - \\
& - \frac{1}{H^2Q^3}\frac{n_S^2[1 + e_S\cos(\alpha - \omega_S)]}{(1 - e_S^2)^{3/2}} \times \\
& \times [3(\cos\theta\cos\alpha + \sin\theta\cos\lambda\sin\alpha)^2 - 3S\sin\lambda\sin\alpha \times \\
& \times (\cos\theta\cos\alpha + \sin\theta\cos\lambda\sin\alpha) - 1] - \frac{1}{H^2Q^3}\frac{n_M^2}{\sigma} \times \\
& \times \frac{(1 + e_M\cos\nu)}{(1 - e_M^2)^{3/2}}\{3(\cos\theta\, M_x + \sin\theta\cos\lambda\, M_y + \\
& + \sin\theta\sin\lambda\, M_z)^2[1 + \frac{5}{2Qr_{ME}}(\cos\theta\, M_x + \sin\theta\cos\lambda\, M_y + \\
& + \sin\theta\sin\lambda\, M_z)] - 3[S(\sin\lambda\, M_x - \cos\lambda\, M_z) + \frac{3}{2}\frac{1}{Qr_{ME}}] \times \\
& \times (\cos\theta\, M_x + \sin\theta\cos\lambda\, M_y + \sin\theta\sin\lambda\, M_z) - 1\} - \\
& - \frac{1}{H}\frac{dH}{d\theta}\frac{dQ}{d\theta},
\end{aligned} \tag{4.40}$$

$$\begin{aligned}
\frac{dH}{d\theta} = \frac{3}{HQ^4}\Bigg\{ & \frac{n_S^2[1 + e_S\cos(\alpha - \omega_S)]}{(1 - e_S^2)^{3/2}}(\cos\theta\cos\alpha + \sin\theta\cos\lambda\sin\alpha - \\
& - S\sin\lambda\sin\alpha)(-\sin\theta\cos\alpha + \cos\theta\cos\lambda\sin\alpha) +
\end{aligned}$$

$$+ \frac{n_M^2(1 + e_M \cos \nu)}{\sigma(1 - e_M^2)^{3/2}}[\cos \theta M_x + (\sin \theta \cos \lambda - S \sin \lambda)M_y +$$

$$+ (\sin \theta \sin \lambda + S \cos \lambda)M_z - \frac{1}{2}\frac{1}{Qr_{ME}} + \frac{5}{2Qr_{ME}}(\cos \theta M_x +$$

$$+ \sin \theta \cos \lambda M_y + \sin \theta \sin \lambda M_z)^2] \times$$

$$\left. \times (- \sin \theta M_x + \cos \theta \cos \lambda M_y + \cos \theta \sin \lambda M_z) \right\} , \quad (4.41)$$

and

$$\frac{d^2S}{d\theta^2} + S = - 3J_2\mu_E R_E^2 \frac{SQ}{H^2} - 3 \frac{1}{H^2Q^4} \frac{n_S^2[1 + e_S \cos(\alpha - \omega_S)]}{(1 - e_S^2)^{3/2}} \times$$

$$\times [(\cos \theta \cos \alpha + \sin \theta \cos \lambda \sin \alpha)\sin \lambda \sin \alpha +$$

$$+ S(\cos \theta \cos \alpha + \sin \theta \cos \lambda \sin \alpha)^2 -$$

$$- S \sin^2 \lambda \sin^2 \alpha] - 3 \frac{1}{H^2Q^4} \frac{n_M^2}{\sigma} \frac{1 + e_M \cos \nu}{(1 - e_M^2)^{3/2}} \times$$

$$\times \left\{ [\cos \theta M_x + \sin \theta \cos \lambda M_y + \sin \theta \sin \lambda M_z - \right.$$

$$- \frac{1}{2}\frac{1}{Qr_{ME}} + \frac{5}{2Qr_{ME}}(\cos \theta M_x + \sin \theta \cos \lambda M_y +$$

$$+ \sin \theta \sin \lambda M_z)^2](\sin \lambda M_y - \cos \lambda M_z) +$$

$$+ S(\cos \theta M_x + \sin \theta \cos \lambda M_y + \sin \theta \sin \lambda M_z)^2 -$$

$$\left. - S(\sin \lambda M_y - \cos \lambda M_z)^2 \right\} - \frac{1}{H}\frac{dH}{d\theta}\frac{dS}{d\theta} , \quad (4.42)$$

where S^2 and S/Qr_{ME} in the terms containing J_2, n_S^2, and n_M^2 have been neglected because $1/Qr_{ME} \cong 0.11$ and S is closely associated with the inclination of the orbital plane, and the latter is known to be small for a geostationary orbit.

Additional equations are introduced to describe the motion of the Sun and Moon around the Earth (Escobal, 1965) and the Sun's perturbation on the motion of the Moon:

(a) Sun's Orbital Movement

$$\frac{d\alpha}{dt} = n_S(1 - e_S^2)^{-3/2}[1 + e_S \cos(\alpha - \omega_S)]^2$$

$$\cong n_S(1 + 2e_S^2)[1 + 2e_S \cos(\alpha - \omega_S)] , \quad (4.43)$$

(b) Moon's Orbital Movement

$$\frac{d\nu}{dt} = c\bar{n}_M(1 - e_M^2)^{-3/2}(1 + e_M \cos \nu)^2$$

$$\cong c\bar{n}_M(1 + 2e_M^2)(1 + 2e_M \cos \nu) , \quad (4.44)$$

where the mean motion of the Moon, n_M, is replaced by $c\bar{n}_M$ as a result of the Sun's perturbation (see (c) below).

The approximations of the above two equations are obtained by omitting the higher order terms, $\frac{1}{2}e_S^2 \cos 2(\alpha - \omega_S)$ and $\frac{1}{2}e_M^2 \cos 2\nu$, etc. in the binomial expansions of $[1 + e_S \cos(\alpha - \omega_S)]^2$ and $(1 + e_M \cos \nu)^2$, respectively.

(c) Sun's Perturbation on the Moon's Motion

Some of the most significant terms in the Moon's motion can be illustrated by limiting the disturbing function (R_{SM}) of the Sun to the following selected terms (Brouwer and Clemence, 1961):

$$R_{SM} = n_S^2 a_M^2 [\tfrac{1}{4} + \tfrac{3}{4}\cos(2\lambda_M - 2\lambda_S) - \tfrac{9}{4}e_M \cos(\lambda_M - 2\lambda_S + \tilde{\omega}_M) + \\ + \tfrac{3}{4}e_M \cos(3\lambda_M - 2\lambda_S - \tilde{\omega}_M) + \tfrac{3}{4}e_S \cos(\lambda_S - \omega_S) + \tfrac{3}{8}e_M^2 + \\ + \tfrac{15}{8}e_M^2 \cos(2\lambda_S - 2\tilde{\omega}_M) + \tfrac{3}{8}e_S^2 - \tfrac{3}{8}I^2 + \tfrac{3}{8}I^2 \cos(2\lambda_M - 2\Omega_M)] , \tag{4.45}$$

where

$\lambda_M = M_M + \omega_M + \Omega_M$ = mean longitude of the Moon,

$\lambda_S = M_S + \omega_S$ = mean longitude of the Sun,

$\tilde{\omega}_M = \omega_M + \Omega_M$,

M_M = mean anomaly of the Moon,

and

M_S = mean anomaly of the Sun.

The method of the variation of an arbitrary constant is convenient for obtaining the solar perturbations in the motion of the Moon for which the ratio $n_S/\bar{n}_M$ ($\bar{n}_M$ = observed mean value of Moon's mean motion; 27.321611/365.25637 = 0.0748013 ≡ m) is very small compared with its value for the Earth-Moon system. In such a case, the first-order approximation is sufficient for obtaining the perturbations with all the accuracy needed for a comparison with observations. With the disturbing function, R_{SM}, in (4.45), the first-order effects on the various Moon's orbital elements are:

$$\Omega_M = (1 - g)\bar{n}_M t + \Omega_{M0} + \tfrac{3}{8}m \sin(2\lambda_S - 2\Omega_M)$$

$$= (1 - g)\bar{n}_M t + \Omega_{M0} + \delta_1\Omega_M , \tag{4.46a}$$

$$\omega_M = (g - c)\bar{n}_M t + \omega_{M0} + \\ + \frac{1}{4}\frac{m^2}{e_M}[\sin(3\lambda_M - 2\lambda_S - \tilde{\omega}_M) - 9\sin(\lambda_M - 2\lambda_S + \tilde{\omega}_M)] + \\ + \tfrac{15}{8}m \sin(2\lambda_S - 2\tilde{\omega}_M) - \tfrac{3}{8}m \sin(2\lambda_S - 2\Omega_M)$$

$$= (g - c)\bar{n}_M t + \omega_{M0} + \delta_1\omega_M , \tag{4.46b}$$

$$I = I_0 + \tfrac{3}{8}mI_0 \cos(2\lambda_S - 2\Omega_M)$$

$$= I_0 + \delta_1 I , \tag{4.46c}$$

and

$$\delta_1\psi = \delta_1(\nu + \Omega_M + \omega_M)$$

= solar perturbation on longitude of the Moon

$$= \tfrac{11}{8}m^2 \sin(2\lambda_M - 2\lambda_S) + \tfrac{15}{4}me_M \sin(\lambda_M - 2\lambda_S + \tilde{\omega}_M) - \\ - 3me_S \sin(\lambda_M - \omega_S) , \tag{4.46d}$$

where c = 0.9914825, g = 1.003999175, and δ_1( ) denotes the long-period term (or terms) of the solar perturbation on an orbital element ( ). In deriving the expressions in (4.46a)-(4.46d), the perturbation on other orbital elements, such as a_M and e_M, are implicitly included.

Readers are referred to the chapter on lunar theory in Methods of

Celestial Mechanics by Brouwer and Clemence (1961) for detailed derivations of the expressions in (4.46) wherein the effects of evection, variation, etc. are represented by various terms in R_{SM}. However, we summarize herein only the resulting expressions for our use in the following analyses.

With the secular terms of Ω_M and ω_M in (4.46a) and (4.46b), the corresponding mean longitude of the Moon is

$$\bar{\lambda}_M = M_M + (1 - c)\bar{n}_M t + \Omega_{M0} + \omega_{M0} = \bar{n}_M t + \bar{\lambda}_0 \; . \tag{4.47}$$

Consequently,

$$M_M = c\bar{n}_M t + M_{M0}, \quad \text{and} \quad \bar{\lambda}_{M0} = M_{M0} + \Omega_{M0} + \omega_{M0} \; . \tag{4.48}$$

The above relationship justifies the replacement of n_M by $c\bar{n}_M$ in (4.44), as it was pointed out, for the reason of solar perturbation.

For the sake of convenience, we shall proceed our method of solution by replacing the secular parts of Ω_M and ω_M in (4.46a) and (4.46b) with the following differential equations:

$$\frac{d\Omega_M}{dt} = \dot{\Omega}_M = (1 - g)\bar{n}_M = \text{constant} \; , \tag{4.49a}$$

$$\frac{d\omega_M}{dt} = \dot{\omega}_M = (g - c)\bar{n}_M = \text{constant} \; , \tag{4.49b}$$

and the long-period terms $\delta_1(\;)$'s in (4.46) will be considered later.

With the aid of (3.39a) and (3.39b), the four equations, (4.43), (4.44), (4.49a), and (4.49b), can be transformed into the H, Q, and S system:

$$\frac{d\alpha}{d\theta} = \frac{n_S}{HQ^2}(1 + 2e_S^2)[1 + 2e_S \cos(\alpha - \omega_S)] \; , \tag{4.50}$$

$$\frac{d\nu}{d\theta} = \frac{c\bar{n}_M}{HQ^2}(1 + 2e_M^2)(1 + 2e_M \cos \nu) \; , \tag{4.51}$$

$$\frac{d\Omega_M}{d\theta} = \frac{\dot{\Omega}_M}{HQ^2} \; , \tag{4.52a}$$

and

$$\frac{d\omega_M}{d\theta} = \frac{\dot{\omega}_M}{HQ^2} \; . \tag{4.52b}$$

It can be seen that (4.40)-(4.42) and (4.50)-(4.52) constitute a new basic set of differential equations for the present problem with H, Q, S, α, ν, Ω_M, and ω_M as the dependent variables and θ as the independent variable.

Readers are reminded that we shall proceed with our method of solution by solving (4.40)-(4.42) and (4.50)-(4.52) without considering $\delta_1(\;)$'s in (4.46). However, after obtaining the solution, the $\delta_1(\;)$'s will be taken into consideration by examining the effects of $\delta_1(\;)$'s on the dominate terms in the solution with a perturbed solution technique since $\delta_1(\;)$'s contain a small parameter, m. The incorporation of $\delta_1(\;)$'s into our solution and the perturbed solution technique are described in Section 4.6. and Appendix C, respectively.

4.4. Transformation of Equations of Motion - First-Order Differential Equations

We shall follow the same procedures described in Section 2.5.3. to resolve (4.40)-(4.42) into a system of first-order differential equations. The solutions to (4.40)-(4.42) when the terms containing the small parameters, J_2, n_S^2, and n_M^2/σ, are dropped out are:

$$S = X_1 \cos\theta + X_2 \sin\theta , \tag{4.53}$$

$$\begin{aligned} Q &= X_5 + X_3 \cos\theta + X_4 \sin\theta , \\ &= \frac{\mu_E}{X_6^2} \frac{1}{1 + X_1^2 + X_2^2} [1 + (X_1 \cos\theta + X_2 \sin\theta)^2]^{1/2} + \\ &\quad + X_3 \cos\theta \, X_4 \sin\theta , \end{aligned} \tag{4.54}$$

and

$$H = X_6 , \tag{4.55}$$

where $X_1, \ldots, X_4, X_6$ are conceived as integration constants as an intermediate step in carrying out the method of variation of parameters. The reader may already notice that the expressions of S, Q, and H in (4.53), (4.54), and (4.55) have been previously described as (2.89), (2.90), and (2.91), respectively, in Chapter 2.

Similarly, the corresponding first derivatives of S and Q with respect to θ in terms of $X_1, \ldots, X_4, X_6$ are

$$\frac{dS}{d\theta} = - X_1 \sin\theta + X_2 \cos\theta , \tag{4.56}$$

and

$$\begin{aligned} \frac{dQ}{d\theta} &= \frac{\mu_E}{X_6^2} \frac{1}{1 + X_1^2 + X_2^2} \frac{(X_1 \cos\theta + X_2 \sin\theta)(- X_1 \sin\theta + X_2 \cos\theta)}{[1 + (X_1 \cos\theta + X_2 \sin\theta)^2]^{1/2}} - \\ &\quad - X_3 \sin\theta + X_4 \cos\theta , \end{aligned} \tag{4.57}$$

respectively, as they are previously written as (2.92) and (2.93).

As a further step in the method of variation of parameters, $X_1, \ldots, X_4$, X_6 in (4.53)-(4.55) can no longer be constants; instead they become functions of θ. Differentiating (4.53)-(4.57) once with respect to θ and then equating the corresponding first derivatives to those in (4.41), (4.56), and (4.57), three equations are obtained. Two additional equations are derived by inserting the corresponding second derivatives together with (4.53) and (4.54) into (4.40) and (4.42). After performing the necessary algebraic manipulations among these resulting five equations, a system of five first-order differential equations and one algebraic equation are obtained to replace the equations (4.40)-(4.42):

$$\begin{aligned} \frac{dX_1}{d\theta} &= \frac{1}{X_6} \frac{dX_6}{d\theta} \frac{dS}{d\theta} \sin\theta + 3J_2\mu_E R_E^2 S \frac{Q}{X_6^2} \sin\theta + \\ &\quad + 3 \frac{\sin\theta}{X_6^2 Q^4} \frac{n_S^2[1 + e_S \cos(\alpha - \omega_S)]}{(1 - e_S^2)^{3/2}} [(\cos\theta \cos\alpha + \\ &\quad + \sin\theta \cos\lambda \sin\alpha)\sin\lambda \sin\alpha + S(\cos\theta \cos\alpha + \\ &\quad + \sin\theta \cos\lambda \sin\alpha)^2 - S \sin^2\lambda \sin^2\alpha] + 3 \frac{\sin\theta}{X_6^2 Q^4} \frac{n_M^2}{\sigma} \times \\ &\quad \times \frac{(1 + e_M \cos\nu)}{(1 - e_M^2)^{3/2}} \Big\{[\cos\theta \, M_x + \sin\theta \cos\lambda \, M_y + \end{aligned}$$

$$+ \sin\theta \sin\lambda\, M_z - \frac{1}{2}\frac{1}{Qr_{ME}} + \frac{5}{2Qr_{ME}}(\cos\theta\, M_x +$$

$$+ \sin\theta\cos\lambda\, M_y + \sin\theta\sin\lambda\, M_z)^2](\sin\lambda\, M_y - \cos\lambda\, M_z) +$$

$$+ S(\cos\theta\, M_x + \sin\theta\cos\lambda\, M_y + \sin\theta\sin\lambda\, M_z)^2 -$$

$$- S(\sin\lambda\, M_y - \cos\lambda\, M_z)^2\Big\}\ , \tag{4.58}$$

$$\frac{dX_2}{d\theta} = -\frac{1}{X_6}\frac{dX_6}{d\theta}\frac{dS}{d\theta}\cos\theta - 3J_2\mu_E R_E^2 S\frac{Q}{X_6^2}\cos\theta - 3\frac{\cos\theta}{X_6^2Q^4} \times$$

$$\times \frac{n_S^2[1 + e_S\cos(\alpha - \omega_S)]}{(1 - e_S^2)^{3/2}}[(\cos\theta\cos\alpha + \sin\theta\cos\lambda\sin\alpha) \times$$

$$\times \sin\lambda\sin\alpha + S(\cos\theta\cos\alpha + \sin\theta\cos\lambda\sin\alpha)^2 -$$

$$- S\sin^2\lambda\sin^2\alpha] - 3\frac{\cos\theta}{X_6^2Q^4}\frac{n_M^2}{\sigma}\frac{(1 + e_M\cos\nu)}{(1 - e_M^2)^{3/2}} \times$$

$$\times \Big\{[\cos\theta\, M_x + \sin\theta\cos\lambda\, M_y + \sin\theta\sin\lambda\, M_z - \frac{1}{2}\frac{1}{Qr_{ME}} +$$

$$+ \frac{5}{2Qr_{ME}}(\cos\theta\, M_x + \sin\theta\cos\lambda\, M_y + \sin\theta\sin\lambda\, M_z)^2] \times$$

$$\times (\sin\lambda\, M_y - \cos\lambda\, M_z) + S(\cos\theta\, M_x + \sin\theta\cos\lambda\, M_y +$$

$$+ \sin\theta\sin\lambda\, M_z)^2 - S(\sin\lambda\, M_y - \cos\lambda\, M_z)^2\Big\}\ , \tag{4.59}$$

$$\frac{dX_3}{d\theta} = \frac{2\mu_E}{X_6^3}\cos\theta\frac{dX_6}{d\theta} - \frac{3}{2}J_2\mu_E R_E^2\frac{Q^2}{X_6^2}\sin\theta + \frac{\sin\theta}{X_6^2Q^3} \times$$

$$\times \frac{n_S^2[1 + e_S\cos(\alpha - \omega_S)]}{(1 - e_S^2)^{3/2}}[3(\cos\theta\cos\alpha +$$

$$+ \sin\theta\cos\lambda\sin\alpha)^2 - 3S\sin\lambda\sin\alpha(\cos\theta\cos\alpha +$$

$$+ \sin\theta\cos\lambda\sin\alpha) - 1] + \frac{\sin\theta}{X_6^2Q^3}\frac{n_M^2}{\sigma}\frac{(1 + e_M\cos\nu)}{(1 - e_M^2)^{3/2}} \times$$

$$\times \Big\{3(\cos\theta\, M_x + \sin\theta\cos\lambda\, M_y + \sin\theta\sin\lambda\, M_z)^2 \times$$

$$\times \left[1 + \frac{5}{2Qr_{ME}}(\cos\theta\, M_x + \sin\theta\cos\lambda\, M_y + \sin\theta\sin\lambda\, M_z)\right] -$$

$$- 3\left[S(\sin\lambda\, M_y - \cos\lambda\, M_z) + \frac{3}{2}\frac{1}{Qr_{ME}}\right](\cos\theta\, M_x +$$

$$+ \sin\theta\cos\lambda\, M_y + \sin\theta\sin\lambda\, M_z) - 1\Big\} + \frac{1}{X_6}\frac{dX_6}{d\theta}\frac{dQ}{d\theta}\sin\theta\ , \tag{4.60}$$

$$\frac{dX_4}{d\theta} = \frac{2\mu_E}{X_6^3}\sin\theta\frac{dX_6}{d\theta} + \frac{3}{2}J_2\mu_E R_E^2\frac{Q^2}{X_6^2}\cos\theta - \frac{\cos\theta}{X_6^2Q^3} \times$$

$$\times \frac{n_S^2[1 + e_S\cos(\alpha - \omega_S)]}{(1 - e_S^2)^{3/2}}[3(\cos\theta\cos\alpha +$$

$$+ \sin\theta\cos\lambda\sin\alpha)^2 - S\sin\lambda\sin\alpha(\cos\theta\cos\alpha +$$

$$+ \sin\theta\cos\lambda\sin\alpha) - 1] - \frac{\cos\theta}{X_6^2Q^3}\frac{n_M^2}{\sigma}\frac{(1 + e_M\cos\nu)}{(1 - e_M^2)^{3/2}} \times$$

$$\times \left\{ 3(\cos\theta\, M_x + \sin\theta\cos\lambda\, M_y + \sin\theta\sin\lambda\, M_z)^2 \times \right.$$

$$\times \left[1 + \frac{5}{2Qr_{ME}}(\cos\theta\, M_x + \sin\theta\cos\lambda\, M_y + \sin\theta\sin\lambda\, M_z) \right] -$$

$$-3\left[S(\sin\lambda\, M_y - \cos\lambda\, M_z) + \frac{3}{2}\frac{1}{Qr_{ME}} \right](\cos\theta\, M_x +$$

$$\left. + \sin\theta\cos\lambda\, M_y + \sin\theta\sin\lambda\, M_z) - 1 \right\} - \frac{dX_6}{d\theta}\frac{dQ}{d\theta}\cos\theta , \qquad (4.61)$$

$$\frac{dX_6}{d\theta} = \frac{3}{X_6 Q^4}\left\{ \frac{n_S^2[1 + e_S\cos(\alpha - \omega_S)]}{(1 - e_S^2)^{3/2}}(\cos\theta\cos\alpha + \sin\theta\cos\lambda\sin\alpha - \right.$$

$$- S\sin\lambda\sin\alpha)(-\sin\theta\cos\alpha + \cos\theta\cos\lambda\sin\alpha) +$$

$$+ \frac{n_M^2(1 + e_M\cos\nu)}{\sigma(1 - e_M^2)^{3/2}}[\cos\theta\, M_x + (\sin\theta\cos\lambda - S\sin\lambda)M_y +$$

$$+ (\sin\theta\sin\lambda + S\cos\lambda)M_z - \frac{1}{2}\frac{1}{Qr_{ME}} + \frac{5}{2Qr_{ME}}(\cos\theta\, M_x +$$

$$+ \sin\theta\cos\lambda\, M_y + \sin\theta\sin\lambda\, M_z)^2](-\sin\theta\, M_x +$$

$$\left. + \cos\lambda\cos\theta\, M_y + \cos\theta\sin\lambda\, M_z) \right\} , \qquad (4.62)$$

and

$$X_5 = \frac{\mu_E}{X_6^2}\frac{1}{1 + X_1^2 + X_2^2}\left[1 + (X_1\cos\theta + X_2\sin\theta)^2\right]^{1/2} . \qquad (4.63)$$

S and Q and their first derivatives are defined in (4.53), (4.54), (4.56), and (4.57), respectively, with $X_1, \ldots, X_4, X_6$ being considered as functions of θ. The terms containing $X_1(dX_1/d\theta)$, $X_1(X_2/d\theta)$, etc. in (4.60) and (4.61) have been omitted because X_1, X_2, and their derivatives are very small. The quantities α, ν, Ω_M, and ω_M in the above equations can be obtained from (4.50)-(4.52), which, in terms of X's are:

$$\frac{d\alpha}{d\theta} = \frac{n_S}{X_5^2 X_6}(1 + 2e_S^2)\left[1 - \frac{2}{X_5}(X_3\cos\theta + X_4\sin\theta)\right] \times$$

$$\times [1 + 2e_S\cos(\alpha - \omega_S)] , \qquad (4.64)$$

$$\frac{d\nu}{d\theta} = \frac{c\bar{n}_M}{X_5^2 X_6}(1 + 2e_M^2)\left[1 - \frac{2}{X_5}(X_3\cos\theta + X_4\sin\theta)\right] \times$$

$$\times (1 + 2e_M\cos\nu) , \qquad (4.65)$$

$$\frac{d\Omega_M}{d\theta} = \frac{\dot{\Omega}_M}{X_5^2 X_6}\left[1 - \frac{2}{X_5}(X_3\cos\theta + X_4\sin\theta)\right] , \qquad (4.66a)$$

and

$$\frac{d\omega_M}{d\theta} = \frac{\dot{\omega}_M}{X_5^2 X_6}\left[1 - \frac{2}{X_5}(X_3\cos\theta + X_4\sin\theta)\right] . \qquad (4.66b)$$

In obtaining (4.64)-(4.65), $1/Q^2$ on the right-hand side of (4.50)-(4.52) has been expanded into a series, and only two terms are retained. This is permissible because the combined expression of (4.39b) and (4.54),

$$\frac{1}{r} = X_5 \cos \phi \left[1 + \frac{1}{X_5}(X_3 \cos \theta + X_4 \sin \theta)\right] , \tag{4.67}$$

may be considered as the equation of an ellipse, and this ellipse has a small eccentricity for the present problem.

In order to carry out our method of solution, (4.58)-(4.62) can be further simplified based on the following known physical characteristics of our problem:

(a) Since a near-equatorial geostationary satellite is nominally in a circular orbit with near-zero inclination, the Q^n's in these equations may be approximated by

$$Q^n \cong \left(\frac{\mu_E}{X_6^2}\right)^n \left[1 + n \frac{X_6^2}{\mu_E}(X_3 \cos \theta + X_4 \sin \theta)\right], \quad n = 1, 2, -3, -4 . \tag{4.68}$$

(b) On account of the smallness of I, the approximations, $\sin^2 I = 0$ and $\cos I = 1$, can be made in these equations.

(c) Furthermore, the terms containing the expressions X_1^2, X_2^2, and also the products of X_1, X_2, X_3, and X_4, are of the higher order and may be neglected because all these parameters are themselves small quantities. The parameters X_1 and X_2 are related to S (and therefore ϕ) by (4.53). $(\bar{X}_6^2/\mu_E)X_3$ and $(\bar{X}_6^2/\mu_E)X_4$ are associated with the eccentricity of the synchronous orbit. Both S and the eccentricity are known to be small for the present problem.

The above approximations and simplifications will be utilized in the following section when we carry out the averaging of (4.58)-(4.62) in the asymptotic method.

4.5. Averaged First-Order Differential Equations of Motion

Because of the smallness of n_S^2, n_M^2/σ, J_2, n_M, n_S, $\dot{\Omega}_M$, and $\dot{\omega}_M$ on the right-hand side of (4.58)-(4.62) and (4.64)-(4.66), the asymptotic method (Bogoliubov and Mitropolsky, 1961) in nonlinear mechanics can be employed in solving this system of differential equations as previously described in Chapter 2.

As a result of the approximations and simplifications described in the last section, the application of the asymptotic method to (4.58)-(4.62) and (4.64)-(4.66) yields

$$\begin{aligned}\frac{\overline{dX_1}}{d\theta} = \bar{X}_2\left[\frac{3}{2}J_2\mu_E^2R_E^2 \frac{1}{\bar{X}_6^4} + \frac{3}{4}\frac{\bar{X}_6^6}{\mu_E^4} \cos 2\lambda \left(n_S'^2 + \frac{n_M'^2}{\sigma}\right)\right] + \frac{3}{4}\frac{\bar{X}_6^6}{\mu_E^4} n_S'^2 \times \\ \times \left\{\frac{1}{2} \sin 2\lambda (1 - \cos 2\bar{\alpha})[1 + e_S \cos(\bar{\alpha} - \omega_S)] + \right. \\ \left. + \bar{X}_1 \cos \lambda \sin 2\bar{\alpha} - \bar{X}_2 \cos 2\lambda \cos 2\bar{\alpha}\right\} + \frac{3}{4}\frac{\bar{X}_6^6}{\mu_E^4}\frac{n_M'^2}{\sigma} \times \\ \times \left\{[\frac{1}{2} \sin 2\lambda \left(1 - \cos 2(\bar{\omega}_M + \bar{\nu} + \bar{\Omega}_M)\right) - \cos 2\lambda \sin I \times \right. \\ \times \left(\cos \bar{\Omega}_M - \cos(2\bar{\omega}_M + 2\bar{\nu} + \bar{\Omega}_M)\right)] (1 + e_M \cos \bar{\nu}) + \\ + \bar{X}_1[\cos \lambda \sin 2(\bar{\omega}_M + \bar{\nu} + \bar{\Omega}_M) - \sin \lambda \sin I \times\end{aligned}$$

$$\begin{aligned}
&\times \left(\sin\bar\Omega_M - \sin(2\bar\omega_M + 2\bar\nu + \bar\Omega_M)\right)] + \bar X_2[-\cos 2\lambda \times\\
&\times \cos 2(\bar\omega_M + \bar\nu + \bar\Omega_M) + 2\sin 2\lambda \sin I\left(\cos\bar\Omega_M -\right.\\
&\left.-\cos(2\bar\omega_M + 2\bar\nu + \bar\Omega_M)\right)]\Big\}, \qquad (4.69)
\end{aligned}$$

$$\begin{aligned}
\frac{\overline{dX_2}}{d\theta} = &-\bar X_1\left[\frac{3}{2}J_2\mu_E^2R_E^2\frac{1}{\bar X_6^4} + \frac{3}{4}\frac{\bar X_6^6}{\mu_E^4}\cos^2\lambda\left(n_S'^2 + \frac{n_M'^2}{\sigma}\right)\right] - \frac{3}{4}\frac{\bar X_6^6}{\mu_E^4}n_S'^2 \times\\
&\times\Big\{\sin\lambda\sin 2\bar\alpha[1 + e_S\cos(\bar\alpha - \omega_S)] + \bar X_1(1 + \sin^2\lambda)\times\\
&\times\cos 2\bar\alpha + \bar X_2\cos\lambda\sin 2\bar\alpha\Big\} - \frac{3}{4}\frac{\bar X^6}{\mu_E^4}\frac{n_M'^2}{\sigma}\\
&\times\Big\{[\sin\lambda\sin 2(\bar\omega_M + \bar\nu + \bar\Omega_M) + \cos\lambda\sin I\left(\sin\bar\Omega_M -\right.\\
&\left.-\sin(2\bar\omega_M + 2\bar\nu + \bar\Omega_M)\right)](1 + e_M\cos\bar\nu) + \bar X_1[(1 + \sin^2\lambda)\times\\
&\times\cos 2(\bar\omega_M + \bar\nu + \bar\Omega_M) + \sin 2\lambda\sin I\left(\cos\bar\Omega_M -\right.\\
&\left.-\cos(2\bar\omega_M + 2\bar\nu + \bar\Omega_M)\right)] + \bar X_2[\cos\lambda\sin 2(\bar\omega_M + \bar\nu + \bar\Omega_M) +\\
&+\sin\lambda\sin I\left(-\sin\bar\Omega_M + \sin(2\bar\omega_M + 2\bar\nu + \bar\Omega_M)\right)]\Big\}, \qquad (4.70)
\end{aligned}$$

$$\begin{aligned}
\frac{\overline{dX_3}}{d\theta} = &-\bar X_4\left[\frac{3}{2}J_2\mu_E^2R_E^2\frac{1}{\bar X_6^4} + \frac{3}{4}\frac{\bar X_6^6}{\mu_E^4}(-4\sin^2\lambda + 1)\left(n_S'^2 + \frac{n_M'^2}{\sigma}\right)\right] -\\
&-\frac{3}{4}\frac{\bar X_6^6}{\mu_E^4}n_S'^2[5\bar X_3\cos\lambda\sin 2\bar\alpha - \bar X_4(1 + 4\cos^2\lambda)\cos 2\bar\alpha] -\\
&-\frac{3}{4}\frac{\bar X_6^6}{\mu_E^4}\frac{n_M'^2}{\sigma}\Big\{5\bar X_3[\cos\lambda\sin 2(\bar\omega_M + \bar\nu + \bar\Omega_M) - \sin\lambda\sin I\times\\
&\times\left(\sin\bar\Omega_M - \sin(2\bar\omega_M + 2\bar\nu + \bar\Omega_M)\right)] - \bar X_4[(1 + 4\cos^2\lambda)\times\\
&\times\cos 2(\bar\omega_M + \bar\nu + \bar\Omega_M) - 4\sin 2\lambda\sin I\left(\cos\bar\Omega_M -\right.\\
&\left.-\cos(2\bar\omega_M + 2\bar\nu + \bar\Omega_M)\right)]\Big\} + \frac{15}{16}\frac{\bar X_6^6}{\mu_E^4}\frac{n_M'^2}{\sigma}(1 + \tfrac{3}{2}e_M^2)\frac{1}{a_M}\times\\
&\times(1 + 2e_M\cos\bar\nu)\Big\{\cos\lambda[(1 - \tfrac{15}{4}\sin^2\lambda)\sin(\bar\omega_M + \bar\nu + \bar\Omega_M) +\\
&+\tfrac{5}{4}\sin^2\lambda\sin 3(\bar\omega_M + \bar\nu + \bar\Omega_M)] + \sin\lambda\sin I\times\\
&\times[(8 - \tfrac{15}{2}\sin^2\lambda)\sin(\bar\omega_M + \bar\nu) + (-\tfrac{6}{4} + \tfrac{15}{4}\sin^2\lambda)\times\\
&\times\left(\sin(3\bar\omega_M + 3\bar\nu + 2\bar\Omega_M) - \sin(\bar\omega_M + \bar\nu + 2\bar\Omega_M)\right)]\Big\}, \qquad (4.71)
\end{aligned}$$

$$\begin{aligned}
\frac{\overline{dX_4}}{d\theta} = &\bar X_3[\tfrac{3}{2}J_2\mu_E^2R_E^2\frac{1}{\bar X_6^4} + \frac{3}{4}\frac{\bar X_6^6}{\mu_E^4}(1 + \sin^2\lambda)\left(n_S'^2 + \frac{n_M'^2}{\sigma}\right)] + \frac{3}{4}\frac{\bar X_6^6}{\mu_E^4}n_S'^2\times\\
&\times[\bar X_3(4 + \cos^2\lambda)\cos 2\bar\alpha + 5\bar X_4\cos\lambda\sin 2\bar\alpha] + \frac{3}{4}\frac{\bar X_6^6}{\mu_E^4}\frac{n_M'^2}{\sigma}\times\\
&\times\Big\{\bar X_3[(4 + \cos^2\lambda)\cos 2(\bar\omega_M + \bar\nu + \bar\Omega_M) - \sin 2\lambda\sin I\times\\
&\times\left(\cos\bar\Omega_M - \cos(2\bar\omega_M + 2\bar\nu + \bar\Omega_M)\right)] +\\
&+5\bar X_4[\cos\lambda\sin 2(\bar\omega_M + \bar\nu + \bar\Omega_M) + \sin\lambda\sin I\left(-\sin\bar\Omega_M +\right.\\
&\left.+\sin(2\bar\omega_M + 2\bar\nu + \bar\Omega_M)\right)] +
\end{aligned}$$

$$+ \frac{15}{16}\frac{\bar{X}_6^6}{\mu_E^4}\frac{n_M'^2}{\sigma}(1 + \tfrac{3}{2}e_M^2)\frac{1}{a_M}(1 + 2e_M\cos\bar{\nu})\Big\{(\tfrac{5}{4}\sin^2\lambda - 1)\times$$
$$\times\cos(\bar{\omega}_M + \bar{\nu} + \bar{\Omega}_M) - \tfrac{3}{4}\sin^2\lambda\cos 3(\bar{\omega}_M + \bar{\nu} + \bar{\Omega}_M) +$$
$$+ \tfrac{5}{4}\sin 2\lambda\sin I[\cos(3\bar{\omega}_M + 3\bar{\nu} + 2\bar{\Omega}_M) -$$
$$- \cos(\bar{\omega}_M + \bar{\nu} + 2\bar{\Omega}_M)]\Big\}, \tag{4.72}$$

$$\overline{\frac{dX_6}{d\theta}} = \frac{3}{4}\frac{\bar{X}_6^7}{\mu_E^4}n_S'^2[\bar{X}_2\sin\lambda\sin 2\bar{\alpha} - \bar{X}_1\sin\lambda\cos\lambda(1 - \cos 2\bar{\alpha})] +$$
$$+ \frac{3}{4}\frac{\bar{X}_6^7}{\mu_E^4}\frac{n_M'^2}{\sigma}\Big\{\bar{X}_1\Big(\frac{-\sin 2\lambda}{2}[1 - \cos 2(\bar{\omega}_M + \bar{\nu} + \bar{\Omega}_M)] -$$
$$- \cos 2\lambda\sin I\big(\cos\bar{\Omega}_M - \cos(2\bar{\omega}_M + 2\bar{\nu} + \bar{\Omega}_M)\big)\Big) +$$
$$+ \bar{X}_2[\sin\lambda\sin 2(\bar{\omega}_M + \bar{\nu} + \bar{\Omega}_M) - \cos\lambda\sin I\big(-\sin\bar{\Omega}_M +$$
$$+ \sin(2\bar{\omega}_M + 2\bar{\nu} + \bar{\Omega}_M)\big)]\Big\}, \tag{4.73}$$

$$\overline{\frac{d\alpha}{d\theta}} = \frac{n_S\bar{X}_6^3}{\mu_E^2}(1 + 2e_S^2)[1 + 2e_S\cos(\bar{\alpha} - \omega_S)], \tag{4.74}$$

$$\overline{\frac{d\nu}{d\theta}} = \frac{c\bar{n}_M\bar{X}_6^3}{\mu_E^2}(1 + 2e_M^2)(1 + 2e_M\cos\bar{\nu}), \tag{4.75}$$

$$\overline{\frac{d\Omega_M}{d\theta}} = \frac{\dot{\Omega}_M\bar{X}_6^3}{\mu_E^2}, \tag{4.76a}$$

and

$$\overline{\frac{d\omega_M}{d\theta}} = \frac{\dot{\omega}_M\bar{X}_6^3}{\mu_E^2}. \tag{4.76b}$$

In the above equations, the terms containing products of e_S and e_M with the $\bar{X}$'s have been omitted, and furthermore, $n_S^2(1 - e_S^2)^{3/2}$ and $n_M^2(1 - e_M^2)^{3/2}$ are written as $n_S'^2$ and $n_M'^2$, respectively, for brevity. In obtaining (4.69)-(4.73), Equations (4.34) and (4.35) have been used.

It should be noted that the 'bar'($^-$) appearing on top of a parameter designates the averaged value of the parameter over a 2π-period of θ as has been explained in Chapter 2.

4.6. First-Order Approximate Solution

Since (4.69)-(4.76) are the respective average values of (4.58)-(4.62) and (4.64)-(4.66) as θ ranges over 2π, the solutions thus derived give the average values of the corresponding parameters. These solutions are known as the first-order approximate solution to the problem, and the resulting values are designated by a 'bar' on the parameters.

In solving (4.69)-(4.76), the value of $\bar{X}_6$ on the right-hand side of these equations is assumed to remain constant as $\bar{\bar{X}}_{60}(\equiv r_c^2\dot{\theta}_E)$. This is permissible because of the small parameters $n_S'^2$ and $n_M'^2/\sigma$, as well as the smallness of $\bar{X}_1$ and $\bar{X}_2$ presented on the right-hand side of (4.73). Consequently, (4.74)-(4.76) yield

$$2\left[\tan^{-1}\left(\frac{(1-4e_S^2)^{1/2}}{1+2e_S}\right)\tan\tfrac{1}{2}(\bar{\alpha}-\omega_S) - \tan^{-1}\left(\frac{(1-4e_S^2)^{1/2}}{1+2e_S}\right)\times\right.$$
$$\left.\times\tan\tfrac{1}{2}(\bar{\alpha}_0-\omega_S)\right] = K_1(\theta-\theta_0) \tag{4.77a}$$

or

$$\bar{\alpha} \cong K_1(\theta-\theta_0) + \bar{\alpha}_0 , \tag{4.77b}$$

$$2\left[\tan^{-1}\left(\frac{(1-4e_M^2)^{1/2}}{1+2e_M}\right)\tan\tfrac{1}{2}\bar{\nu} - \tan^{-1}\left(\frac{(1-4e_M^2)^{1/2}}{1+2e_M}\right)\tan\tfrac{1}{2}\bar{\nu}_0\right]$$
$$= K_2(\theta-\theta_0) \tag{4.78a}$$

or

$$\bar{\nu} \cong K_2(\theta-\theta_0) + \bar{\nu}_0 , \tag{4.78b}$$

and

$$\bar{\Omega}_M = K_3(\theta-\theta_0) + \bar{\Omega}_{M0} , \tag{4.79a}$$

$$\bar{\omega}_M = K_4(\theta-\theta_0) + \bar{\omega}_{M0} , \tag{4.79b}$$

where

$$\left.\begin{aligned} K_1 &\equiv n_S\bar{\bar{X}}_{60}^3/\mu_E^2 = \text{constant},\\ K_2 &\equiv c\bar{n}_M\bar{\bar{X}}_{60}^3/\mu_E^2 = \text{constant},\\ K_3 &\equiv \dot{\Omega}_M\bar{\bar{X}}_{60}^3/\mu_E^2 = \text{constant},\\ \text{and}\quad K_4 &\equiv \dot{\omega}_M\bar{\bar{X}}_{60}^3/\mu_E^2 = \text{constant}. \end{aligned}\right\} \tag{4.80}$$

$\bar{\alpha}_0$, $\bar{\nu}_0$, $\bar{\Omega}_{M0}$, and $\bar{\omega}_{MO}$ are the integration constants, and θ_0 is an initial condition. Equations (4.77a) and (4.78a) are used to compute the positions of the Sun and Moon in their partial orbits, respectively, while (4.77b) and (4.78b) are for computing the corresponding positions at an integral orbit. However, in view of the small eccentricity of both orbits, the use of (4.77b) and (4.78b) to compute $\bar{\alpha}$ and $\bar{\nu}$ for partial orbits will not introduce serious errors in the following analyses.

With $\bar{X}_6 \cong \bar{\bar{X}}_{60}$ (= constant) and (4.77)-(4.79), Equations (4.69)-(4.72) constitute two sets of differential equations of mono-frequency oscillations in a nonlinear system with two degrees of freedom subject to external periodic forces (Bogoliubov and Mitropolsky, 1961). Each set of these equations has a dominant frequency (mono-frequency) appearing in its solution. As will be seen later, the set, (4.81) and (4.82), has a mono-frequency of ω_1 (= $(\omega_{11}\omega_{12})^{1/2}$), while another set, (4.83) and (4.84), has its corresponding value of ω_3 (= $(\omega_{31}\omega_{32})^{1/2}$). Hence, (4.69)-(4.73) can be written as

$$\frac{d\bar{X}_1}{d\theta} - \omega_{12}\bar{X}_2 = \frac{3}{4}\,\frac{K_1^2}{(1-e_S^2)^{3/2}}\left\{\frac{1}{2}\sin 2\lambda[1-\cos 2\bar{\alpha} +\right.$$
$$+ e_S\cos(-\omega_S+\bar{\alpha}) - \tfrac{1}{2}e_S\cos(-\omega_S+3\bar{\alpha}) -$$
$$- \tfrac{1}{2}e_S\cos(\omega_S+\bar{\alpha})] + \bar{X}_1\cos\lambda\sin 2\bar{\alpha} -$$

$$- \bar{X}_2 \cos 2\lambda \cos 2\bar{\alpha}\Big\} + \frac{3}{4} \frac{K_2^2}{\sigma(1 - e_M^2)^{3/2}}\Big\{\frac{1}{2} \sin 2\lambda \times$$

$$\times [1 - \cos(2\bar{\nu} + 2\bar{\omega}_M + 2\bar{\Omega}_M) + e_M \cos \nu -$$

$$- \frac{1}{2}e_M \cos(3\bar{\nu} + 2\bar{\omega}_M + 2\bar{\Omega}_M) - \frac{1}{2}e_M \cos(\bar{\nu} + 2\bar{\omega}_M + 2\bar{\Omega}_M)] -$$

$$- \cos 2\lambda \sin I[\cos \bar{\Omega}_M - \cos(2\bar{\nu} + 2\bar{\omega}_M + \bar{\Omega}_M) +$$

$$+ \frac{1}{2}e_M \sin(\bar{\nu} + \bar{\Omega}_M) - \frac{1}{2}e_M \sin(\bar{\nu} - \bar{\Omega}_M) +$$

$$+ \frac{1}{2}e_M \cos(3\bar{\nu} + 2\bar{\omega}_M + \bar{\Omega}_M) + \frac{1}{2}e_M \cos(\bar{\nu} + 2\bar{\omega}_M + \bar{\Omega}_M)] +$$

$$+ \bar{X}_1[\cos \lambda \sin(2\bar{\nu} + 2\bar{\omega}_M + 2\bar{\Omega}_M) -$$

$$- \sin \lambda \sin I\big(\sin \bar{\Omega}_M - \sin(2\bar{\nu} + 2\bar{\omega}_M + \bar{\Omega}_M)\big)] +$$

$$+ \bar{X}_2[- \cos 2\lambda \cos(2\bar{\nu} + 2\bar{\omega}_M + 2\bar{\Omega}_M) +$$

$$+ \sin 2\lambda \sin I\big(\cos \bar{\Omega}_M - \cos(2\bar{\nu} + 2\bar{\omega}_M + \bar{\Omega}_M)\big)]\Big\} , \tag{4.81}$$

$$\frac{d\bar{X}_2}{d\theta} + \omega_{11}\bar{X}_1 = - \frac{3}{4} \frac{K_1^2}{(1 - e_S^2)^{3/2}}\Big\{\sin \lambda[\sin 2\bar{\alpha} + \frac{1}{2}e_S \sin(- \omega_S + 3\bar{\alpha}) +$$

$$+ \frac{1}{2} e_S \sin(\omega_S + \bar{\alpha})] + \bar{X}_1(1 + \sin^2 \lambda)\cos 2\bar{\alpha} +$$

$$+ \bar{X}_2 \cos \lambda \sin 2\bar{\alpha}\Big\} - \frac{3}{4} \frac{K_2^2}{\sigma(1 - e_M^2)^{3/2}} \times$$

$$\times \Big\{\sin \lambda \sin(2\bar{\nu} + 2\bar{\omega}_M + 2\bar{\Omega}_M) + \cos \lambda \sin I\big(\sin \bar{\Omega}_M -$$

$$- \sin(2\bar{\nu} + 2\bar{\omega}_M + \bar{\Omega}_M)\big) + \frac{1}{2}e_M \sin \lambda \sin(3\bar{\nu} + 2\bar{\omega}_M +$$

$$+ 2\bar{\Omega}_M) + \frac{1}{2}e_M \sin \lambda \sin(\bar{\nu} + 2\bar{\omega}_M + 2\bar{\Omega}_M) +$$

$$+ \bar{X}_1[(1 + \sin^2 \lambda)\cos(2\bar{\nu} + 2\bar{\omega}_M + 2\bar{\Omega}_M) +$$

$$+ \sin 2\lambda \sin I\big(\cos \bar{\Omega}_M - \cos(2\bar{\nu} + 2\bar{\omega}_M + \bar{\Omega}_M)\big)] +$$

$$+ \bar{X}_2[\cos \lambda \sin(2\bar{\nu} + 2\bar{\omega}_M + 2\bar{\Omega}_M) +$$

$$+ \sin \lambda \sin I\big(- \sin \bar{\Omega}_M + \sin(2\bar{\nu} + 2\bar{\omega}_M + \bar{\Omega}_M)\big)]\Big\} , \tag{4.82}$$

$$\frac{d\bar{X}_3}{d\theta} + \omega_{31}\bar{X}_4 = - \frac{3}{4} \frac{K_1^2}{(1 - e_S^2)^{3/2}}[5\bar{X}_3 \cos \lambda \sin 2\bar{\alpha} - \bar{X}_4(1 + 4 \cos^2 \lambda) \times$$

$$\times \cos 2\bar{\alpha}] - \frac{3}{4} \frac{K_2^2}{\sigma(1 - e_M^2)^{3/2}}\Big\{5\bar{X}_3 \times$$

$$\times [\cos \lambda \sin 2(\bar{\nu} + \bar{\omega}_M + \bar{\Omega}_M) - \sin \lambda \sin I\big(\sin \bar{\Omega}_M -$$

$$- \sin(2\bar{\nu} + 2\bar{\omega}_M + \bar{\Omega}_M)\big)] - \bar{X}_4 [(1 + 4 \cos^2 \lambda) \times$$

$$\times \cos 2(\bar{\nu} + \bar{\omega}_M + \bar{\Omega}_M) - 4 \sin 2\lambda \sin I\big(\cos \bar{\Omega}_M -$$

$$- \cos(2\bar{\nu} + 2\bar{\omega}_M + \bar{\Omega}_M)\big)] + \frac{15}{16} \frac{K_2^2}{\sigma(1 - e_M^2)^3} \frac{1}{a_M} \times$$

$$\times \Big\{\cos \lambda [(1 - \frac{15}{4} \sin^2 \lambda)\sin(\bar{\omega}_M + \bar{\nu} + \bar{\Omega}_M) +$$

$$+ \frac{5}{4} \sin^2 \lambda \sin 3(\bar{\omega}_M + \bar{\nu} + \bar{\Omega}_M)] +$$

$$+ \sin\lambda \sin I[(8 - \frac{15}{2}\sin^2\lambda)\sin(\bar\omega_M + \bar\nu) +$$
$$+ \frac{3}{4}(-2 + 5\sin^2\lambda)\left(\sin(3\bar\omega_M + 3\bar\nu + 2\bar\Omega_M) -\right.$$
$$\left.- \sin(\bar\omega_M + \bar\nu + 2\bar\Omega_M)\right)] + e_M \cos\lambda[(1 - \frac{15}{4}\sin^2\lambda) \times$$
$$\times \left(\sin(\bar\omega_M + 2\bar\nu + \bar\Omega_M) + \sin(\bar\omega_M + \bar\Omega_M)\right) + \frac{5}{4}\sin^2\lambda \times$$
$$\times \left(\sin(3\bar\omega_M + 4\bar\nu + 3\bar\Omega_M) + \sin(3\bar\omega_M + 2\bar\nu + 3\bar\Omega_M)\right)] +$$
$$+ e_M \sin\lambda \sin I[(8 - \frac{15}{2}\sin^2\lambda)\left(\sin(\bar\omega_M + 2\bar\nu) +\right.$$
$$\left.+ \sin\bar\omega_M\right) - \frac{3}{4}(2 - 5\sin^2\lambda)\left(\sin(3\bar\omega_M + 4\bar\nu + 2\bar\Omega_M) +\right.$$
$$+ \sin(3\bar\omega_M + 2\bar\nu + 2\bar\Omega_M) - \sin(\bar\omega_M + 2\bar\nu + 2\bar\Omega_M) -$$
$$\left.\left. - \sin(\bar\omega_M + 2\bar\Omega_M)\right)]\right\}, \qquad (4.83)$$

$$\frac{d\bar X_4}{d\theta} - \omega_{32}\bar X_3 = \frac{3}{4}\frac{K_1^2}{(1 - e_S^2)^{3/2}}[\bar X_3(4 + \cos^2\lambda)\cos 2\bar\alpha +$$
$$+ 5\bar X_4 \cos\lambda \sin 2\bar\alpha] + \frac{3}{4}\frac{K_2^2}{\sigma(1 - e_M^2)^{3/2}} \times$$
$$\times\left\{\bar X_3[(4 + \cos^2\lambda)\cos(2\bar\nu + 2\bar\omega_M + 2\bar\Omega_M) -\right.$$
$$- \sin 2\lambda \sin I\left(\cos\bar\Omega_M - \cos(2\bar\nu + 2\bar\omega_M - \bar\Omega_M)\right)] +$$
$$+ 5\bar X_4[\cos\lambda \sin(2\bar\nu + 2\bar\omega_M + 2\bar\Omega_M) +$$
$$+ \sin\lambda \sin I\left(- \sin\bar\Omega_M + \sin(2\bar\nu + 2\bar\omega_M + \Omega_M)\right)] +$$
$$+ \frac{15}{16}\frac{K_2^2}{\sigma(1 - e_M^2)^3}\frac{1}{a_M}\left\{(\frac{5}{4}\sin^2\lambda - 1)\cos(\bar\omega_M + \bar\nu + \bar\Omega_M) -\right.$$
$$- \frac{3}{4}\sin^2\lambda \cos 3(\bar\omega_M + \bar\nu + \bar\Omega_M) + \frac{5}{4}\sin 2\lambda \sin I \times$$
$$\times [\cos(3\bar\omega_M + 3\bar\nu + 2\bar\Omega_M) - \cos(\bar\omega_M + \bar\nu + 2\bar\Omega_M)] +$$
$$+ e_M(\frac{5}{4}\sin^2\lambda - 1)[\cos(\bar\omega_M + 2\bar\nu + \bar\Omega_M) +$$
$$+ \cos(\bar\omega_M + \bar\Omega_M)] - \frac{3}{4}e_M \sin^2\lambda[\cos(3\bar\omega_M + 4\bar\nu + 3\bar\Omega_M) +$$
$$+ \cos(3\bar\omega_M + 2\bar\nu + 3\bar\Omega_M)] + \frac{5}{4}e_M \sin 2\lambda \sin I \times$$
$$\times [\cos(3\bar\omega_M + 4\bar\nu + 2\bar\Omega_M) + \cos(3\bar\omega_M + 2\bar\nu + 2\bar\Omega_M) -$$
$$\left. - \cos(\bar\omega_M + 2\bar\nu + 2\bar\Omega_M) - \cos(\bar\omega_M + 2\bar\Omega_M)]\right\}, \qquad (4.84)$$

and

$$\frac{\bar{\bar X}{}_{60}^6}{\bar X_6^7}\frac{d\bar X_6}{d\theta} = \frac{3}{4}\frac{K_1^2}{(1 - e_S^2)^{3/2}}[\bar X_2 \sin\lambda \sin 2\bar\alpha -$$
$$- \bar X_1 \sin\lambda \cos\lambda(1 - \cos 2\bar\alpha)] + \frac{3}{4}\frac{K_2^2}{\sigma(1 - e_M^2)^{3/2}} \times$$
$$\times \left\{\bar X_1\left(\frac{-\sin 2\lambda}{2}[1 - \cos 2(\bar\nu + \bar\omega_M + \bar\Omega_M)] -\right.\right.$$
$$\left. - \cos 2\lambda \sin I\left(\cos\bar\Omega_M - \cos(2\bar\nu + 2\bar\omega_M + \bar\Omega_M)\right)\right) +$$
$$+ \bar X_2[\sin\lambda \sin 2(\bar\nu + \bar\omega_M + \bar\Omega_M) -$$
$$\left. - \cos\lambda \sin I\left(- \sin\bar\Omega_M + \sin(2\bar\nu + 2\bar\omega_M + \bar\Omega_M)\right)]\right\}, \qquad (4.85)$$

where

$$\left.\begin{aligned}
\omega_{11} &\equiv \frac{3}{2}J_2\mu_E^2R_E^2\frac{1}{\bar{\bar{X}}_{60}^4} + \frac{3}{4}\cos^2\lambda\left(\frac{K_1^2}{(1-e_S^2)^{3/2}} + \frac{K_2^2}{\sigma(1-e_M^2)^{3/2}}\right)\\
&= \text{constant},\\
\omega_{12} &\equiv \frac{3}{2}J_2\mu_E^2R_E^2\frac{1}{\bar{\bar{X}}_{60}^4} + \frac{3}{4}\cos 2\lambda\left(\frac{K_1^2}{(1-e_S^2)^{3/2}} + \frac{K_2^2}{\sigma(1-e_M^2)^{3/2}}\right)\\
&= \text{constant},\\
\omega_{31} &\equiv \frac{3}{2}J_2\mu_E^2R_E^2\frac{1}{\bar{\bar{X}}_{60}^4} + \frac{3}{4}(1-4\sin^2\lambda)\left(\frac{K_1^2}{(1-e_S^2)^{3/2}} + \frac{K_2^2}{\sigma(1-e_M^2)^{3/2}}\right)\\
&= \text{constant},\\
\text{and}\quad \omega_{32} &\equiv \frac{3}{2}J_2\mu_E^2R_E^2\frac{1}{\bar{\bar{X}}_{60}^4} + \frac{3}{4}(1+\sin^2\lambda)\left(\frac{K_1^2}{(1-e_S^2)^{3/2}} + \frac{K_2^2}{\sigma(1-e_M^2)^{3/2}}\right)\\
&= \text{constant}.
\end{aligned}\right\} \tag{4.86}$$

In obtaining (4.81)-(4.85), terms containing the products of e_S and e_M with X's have been omitted as they are of the higher order of magnitude. Attention should be paid that $K_1^2 << 1$ and $K_2^2/\sigma << 1$; these two constants will be considered associating with a small parameter, ε.

We shall proceed to solve the first set of differential equations, (4.81) and (4.82), in our mono-frequency nonlinear oscillation system. According to the method of solving a system of differential equations of mono-frequency oscillations in nonlinear mechanics, the first-order approximate solutions to (4.81) and (4.82) are

$$\begin{aligned}
\bar{X}_k &= u_k^{(0)}(a, \Psi) + \varepsilon u_k^{(1)}(a, \Psi, \bar{\alpha}, \bar{\nu}, \bar{\Omega}_M, \bar{\omega}_M)\\
&= u_k^{(0)}(a, \Psi) + \varepsilon u_k^{(1)}(a, \Psi, K_1\theta, K_2\theta, K_3\theta, K_4\theta) \quad (k = 1, 2),
\end{aligned} \tag{4.87}$$

where a and Ψ are defined by the differential equations

$$\frac{da}{d\theta} = \varepsilon C_1(a) \tag{4.88}$$

and

$$\frac{d\Psi}{d\theta} = \omega_1 + \varepsilon C_2(a) . \tag{4.89}$$

In these equations, ω_1 is the mono-frequency of the nonlinear system, and ε is a small parameter which equals $\frac{3}{4}K_1^2/(1 - e_S^2)^{3/2}$ in the present case. As is seen from (4.87), $u_k^{(0)}$ contributes to the zeroth-order solution to $\bar{X}_k$, while $u_k^{(1)}$ constitutes the first-order magnitude part of the solution. The functions $u_k^{(0)}$, $u_k^{(1)}$, $C_1(a)$, and $C_2(a)$ are determined as follows.

The solution to the set of homogeneous differential equations of (4.81) and (4.82) is $u_k^{(0)}$:

$$u_1^{(0)} = \omega_1 a \sin\Psi \tag{4.90}$$

and

$$u_2^{(0)} = \omega_{11} a \cos\Psi , \tag{4.91}$$

where $\omega_1 \equiv (\omega_{11}\omega_{12})^{1/2}$, $\Psi = \omega_1\theta + \Phi$, and a and Φ are two arbitrary constants that are later considered as functions of θ. This can be seen from (4.88) and (4.89).

With (4.88) and (4.89), the derivative of $\bar{X}_k$ with respect to θ in (4.87) is

$$\frac{d\bar{x}_k}{d\theta} = \omega_1 \frac{\partial u_k^{(0)}}{\partial \Psi} + \varepsilon \left(\frac{\partial u_k^{(0)}}{\partial a} C_1(a) + \frac{\partial u_k^{(0)}}{\partial \Psi} C_2(a) + \frac{\partial u_k^{(1)}}{\partial \theta} \right) . \qquad (4.92)$$

Substituting (4.87) and (4.90)-(4.92) into (4.81) and (4.82) results in

$$\begin{aligned}\frac{\partial u_1^{(1)}}{\partial \theta} - \omega_{12} u_2^{(1)} = {} & \tfrac{1}{2} \sin 2\lambda [1 - \cos 2\bar{\alpha} + \tfrac{1}{2} e_S \cos(- \omega_S + \bar{\alpha}) - \\ & - \tfrac{1}{2} e_S \cos(- \omega_S + 3\bar{\alpha})] - \tfrac{1}{2} a (\omega_1 \cos \lambda + \\ & + \omega_{11} \cos 2\lambda) \cos(2\bar{\alpha} + \Psi) + \tfrac{1}{2} a(\omega_1 \cos \lambda - \\ & - \omega_{11} \cos 2\lambda) \cos(2\bar{\alpha} - \Psi) + \kappa \Big\{ \tfrac{1}{2} \sin 2\lambda \times \\ & \times [1 - \cos(2\bar{\nu} + 2\bar{\omega}_M + 2\bar{\Omega}_M) + e_M \cos \bar{\nu} - \\ & - \tfrac{1}{2} e_M \cos(3\bar{\nu} + 2\bar{\omega}_M + 2\bar{\Omega}_M) - \tfrac{1}{2} e_M \cos(\bar{\nu} + 2\bar{\omega}_M + \\ & + 2\bar{\Omega}_M)] - \cos 2\lambda \sin I [\sin \bar{\Omega}_M - \cos(2\bar{\nu} + 2\bar{\omega}_M + \\ & + \bar{\Omega}_M) - \tfrac{1}{2} e_M \cos(\bar{\nu} + \bar{\Omega}_M) - \tfrac{1}{2} e_M \cos(\bar{\nu} - \bar{\Omega}_M) + \\ & + \tfrac{1}{2} e_M \cos(3\bar{\nu} + 2\bar{\omega}_M + \bar{\Omega}_M) + \tfrac{1}{2} e_M \cos(\bar{\nu} + 2\bar{\omega}_M + \\ & + \bar{\Omega}_M)] + \tfrac{1}{2} \omega_1 a [\cos \lambda \cos(2\bar{\nu} + 2\bar{\omega}_M + 2\bar{\Omega}_M - \Psi) - \\ & - \cos \lambda \cos(2\bar{\nu} + 2\bar{\omega}_M + 2\bar{\Omega}_M + \Psi) - \sin \lambda \sin I \times \\ & \times \big(\cos(\bar{\Omega}_M - \Psi) - \cos(\bar{\Omega}_M + \Psi) - \cos(2\bar{\nu} + 2\bar{\omega}_M + \\ & + \bar{\Omega}_M - \Psi) + \cos(2\bar{\nu} + 2\bar{\omega}_M + \bar{\Omega}_M + \Psi)\big)] + \\ & + \tfrac{1}{2} \omega_{11} a [- \cos 2\lambda \cos(2\bar{\nu} + 2\bar{\omega}_M + 2\bar{\Omega}_M - \Psi) - \\ & - \cos 2\lambda \cos(2\bar{\nu} + 2\bar{\omega}_M + 2\bar{\Omega}_M + \Psi)] + \\ & + 2 \sin 2\lambda \sin I \big(\cos(\bar{\Omega}_M + \Psi) + \cos(\bar{\Omega}_M - \Psi) - \\ & - \cos(2\bar{\nu} + 2\bar{\omega}_M + \bar{\Omega}_M + \Psi) - \cos(2\bar{\nu} + 2\bar{\omega}_M + \bar{\Omega}_M - \\ & - \Psi)\big)]\Big\} - \omega_1 C_1(a) \sin \Psi - \omega_1 C_2(a) \cos \Psi \end{aligned} \qquad (4.93)$$

and

$$\begin{aligned}\frac{\partial u_2^{(1)}}{\partial \theta} + \omega_{11} u_1^{(1)} = {} & - \sin \lambda [\sin 2\bar{\alpha} + \tfrac{1}{2} e_S \sin(- \omega_S + 3\bar{\alpha}) + \\ & + \tfrac{1}{2} e_S \sin(\omega_S + \bar{\alpha})] - \tfrac{1}{2} a [\omega_1 (1 + \sin^2 \lambda) + \\ & + \omega_{11} \cos \lambda] \sin(2\bar{\alpha} + \Psi) + \tfrac{1}{2} a [\omega_1 (1 + \sin^2 \lambda) - \\ & - \omega_{11} \cos \lambda] \sin(2\bar{\alpha} - \Psi) - \kappa \Big\{ \sin \lambda [\sin(2\bar{\nu} + 2\bar{\omega}_M + \\ & + 2\bar{\Omega}_M) + \tfrac{1}{2} e_M \sin(3\bar{\nu} + 2\bar{\omega}_M + 2\bar{\Omega}_M) + \tfrac{1}{2} e_M \sin(\bar{\nu} + \\ & 2\bar{\omega}_M + 2\bar{\Omega}_M)] + \cos \lambda \sin I [\sin \bar{\Omega}_M - \sin(2\bar{\nu} + 2\bar{\omega}_M + \\ & + \bar{\Omega}_M) + \tfrac{1}{2} e_M \sin(\bar{\nu} + \bar{\Omega}_M) - \tfrac{1}{2} e_M \sin(\bar{\nu} - \bar{\Omega}_M) - \\ & - \tfrac{1}{2} e_M \sin(3\bar{\nu} + 2\bar{\omega}_M + \bar{\Omega}_M) + \tfrac{1}{2} e_M \sin(\bar{\nu} + 2\bar{\omega}_M + \\ & + \bar{\Omega}_M)] + \tfrac{1}{2} \omega_1 a [(1 + \sin^2 \lambda) \big(\sin(2\bar{\nu} + 2\bar{\omega}_M + 2\bar{\Omega}_M + \\ & + \Psi) - \sin(2\bar{\nu} + 2\bar{\omega}_M + 2\bar{\Omega}_M - \Psi)\big) + \sin 2\lambda \sin I \times \\ & \times \big(\sin(\bar{\Omega}_M + \Psi) - \sin(\bar{\Omega}_M - \Psi) - \sin(2\bar{\nu} + 2\bar{\omega}_M + \\ & + \bar{\Omega}_M + \Psi) + \sin(2\bar{\nu} + 2\bar{\omega}_M + \bar{\Omega}_M - \Psi)\big)] + \end{aligned}$$

$$+ \tfrac{1}{2}\omega_{11} a[\cos\lambda\big(\sin(2\bar\nu + 2\bar\omega_M + 2\bar\Omega_M + \Psi) + \sin(2\bar\nu + 2\bar\omega_M + 2\bar\Omega_M - \Psi)\big) + \sin\lambda \sin I \times \big(-\sin(\bar\Omega_M + \Psi) - \sin(\bar\Omega_M - \Psi) + \sin(2\bar\nu + 2\bar\omega_M + \bar\Omega_M + \Psi) + \sin(2\bar\nu + 2\bar\omega_M + \bar\Omega_M - \Psi)\big)]\Big\} - \omega_{11} C_1(a)\cos\Psi + \omega_{11} C_2(a)\sin\Psi \;, \quad (4.94)$$

where the ε^2-terms have been omitted, and $\kappa \equiv (K_2^2/\sigma K_1^2)\big((1 - e_S^2)/(1 - e_M^2)\big)^{3/2}$. Since the terms containing $C_1(a)$ and $C_2(a)$ are associated with $\sin\Psi$ and $\cos\Psi$, the requirement that $u_1^{(1)}$ and $u_2^{(1)}$ have periodic solutions generally necessitates the disappearance of these terms in (4.93) and (4.94). In other words,

$$C_1(a) = 0 \quad \text{and} \quad C_2(a) = 0 \;. \quad (4.95)$$

Consequently, (4.88) and (4.89) yield

$$a = a_0, \quad \Phi = \Psi_0 \;,$$

and

$$\Psi = \omega_1(\theta - \theta_0) + \Psi_0 \;, \quad (4.96)$$

where a_0 and Ψ_0 are constants to be determined from the initial conditions.

As a result of (4.95) and (4.96), $u_1^{(1)}$ and $u_2^{(1)}$ have as their solutions (see Appendix B for constants $A_1, \ldots, A_{44}$):

$$\begin{aligned} u_1^{(1)} = {} & \omega_1 a_0 \sin\Psi + {}^{A_1}_{A_2} \sin(2\bar\alpha \pm \Psi) + A_3 \sin 2\bar\alpha + A_4 \sin\bar\Omega_M + \\ & + A_5 \sin(2\bar\nu + 2\bar\omega_M + \bar\Omega_M) + A_6 \sin 2(\bar\nu + \bar\omega_M + \bar\Omega_M) + \\ & + {}^{A_7}_{A_8} \sin(\bar\Omega_M \pm \Psi) + {}^{A_9}_{A_{10}} \sin(2\bar\nu + 2\bar\omega_M + \bar\Omega_M \pm \Psi) + \\ & + {}^{A_{11}}_{A_{12}} \sin(2\bar\nu + 2\bar\omega_M + 2\bar\Omega_M \pm \Psi) + {}^{A_{13}}_{A_{14}} \sin(\omega_S \pm \bar\alpha) + \\ & + A_{15} \sin(\omega_S + 3\bar\alpha) + A_{16} \sin\bar\nu + {}^{A_{17}}_{A_{18}} \sin(\bar\nu \pm \bar\Omega_M) + \\ & + A_{19} \sin(\bar\nu + 2\bar\omega_M + \bar\Omega_M) + A_{20} \sin(\bar\nu + 2\bar\omega_M + 2\bar\Omega_M) + \\ & + A_{21} \sin(3\bar\nu + 2\bar\omega_M + \bar\Omega_M) + A_{22} \sin(3\bar\nu + 2\bar\omega_M + 2\bar\Omega_M) \end{aligned} \quad (4.97)$$

and

$$\begin{aligned} u_2^{(1)} = {} & \omega_{11} a_0 \cos\Psi + {}^{A_{23}}_{A_{24}} \cos(2\bar\alpha \pm \Psi) + A_{25} \cos 2\bar\alpha + A_{26} \cos\bar\Omega_M + \\ & + A_{27} \cos(2\bar\nu + 2\bar\omega_M + \bar\Omega_M) + A_{28} \cos 2(\bar\nu + \bar\omega_M + \bar\Omega_M) + \\ & + {}^{A_{29}}_{A_{30}} \cos(\bar\Omega_M \pm \Psi) + {}^{A_{31}}_{A_{32}} \cos(2\bar\nu + 2\bar\omega_M + \bar\Omega_M \pm \Psi) + \\ & + {}^{A_{33}}_{A_{34}} \cos(2\bar\nu + 2\bar\omega_M + 2\bar\Omega_M \pm \Psi) + {}^{A_{35}}_{A_{36}} \cos(\omega_S \pm \bar\alpha) + \\ & + A_{37} \cos(\omega_S + 3\bar\alpha) + A_{38} \cos\bar\nu + {}^{A_{39}}_{A_{40}} \cos(\bar\nu \pm \bar\Omega_M) + \\ & + A_{41} \cos(\bar\nu + 2\bar\omega_M + \bar\Omega_M) + A_{42} \cos(\bar\nu + 2\bar\omega_M + 2\bar\Omega_M) + \\ & + A_{43} \cos(3\bar\nu + 2\bar\omega_M + \bar\Omega_M) + A_{44} \cos(3\bar\nu + 2\bar\omega_M + 2\bar\Omega_M) - G \;, \end{aligned} \quad (4.98)$$

where $G \equiv [(1 + \kappa)\sin 2\lambda]/(2\omega_{12})$ = constant, and Ψ is shown in (4.96). Note that the last term on the right-hand side of (4.98) is of the order of $1/\varepsilon$ because $\omega_{12} \sim O(\varepsilon)$ as is seen from (4.86). Therefore, the constant $\varepsilon G \equiv D$ is of $O(\varepsilon^0)$ and should be considered together with $u_2^{(0)}$ as the zeroth-order solution to $\bar X_2$.

The first set of differential equations, (4.81) and (4.82), has been solved. The solution is in the form of the first-order approximations of $\bar{X}_1$ and $\bar{X}_2$. It is represented by (4.87), (4.90), (4.91), (4.97), and (4.98). The two constants, a_0 and Ψ_0, are determined from the initial conditions: $\bar{X}_1 = \bar{X}_{10}$ and $\bar{X}_2 = \bar{X}_{20}$ at $\theta = \theta_0$, $\bar{\alpha} = \bar{\alpha}_0$, $\bar{\nu} = \bar{\nu}_0$, $\bar{\Omega}_M = \bar{\Omega}_{M0}$, and $\bar{\omega}_M = \bar{\omega}_{M0}$.

Following the same procedures in solving the first set of differential equations above, we proceed to solve the second set of differential equations, (4.83) and (4.84), by assuming that the first-order approximate solutions to X_3 and X_4 are in the forms of

$$\bar{X}_3 = u_3^{(0)} + \frac{\varepsilon}{4} u_3^{(1)} \tag{4.99}$$

and

$$\bar{X}_4 = u_4^{(0)} + \frac{\varepsilon}{4} u_4^{(1)} \ . \tag{4.100}$$

The results are (see Appendix B for constants $B_1, \ldots, B_{46}$)

$$u_3^{(0)} = \omega_{31} b_0 \cos \xi \ , \tag{4.101}$$

$$u_4^{(0)} = \omega_3 b_0 \sin \xi \ , \tag{4.102}$$

$$\begin{aligned} u_3^{(1)} &= \omega_{31} b_0 \cos \xi + {}^{B_1}_{B_2} \cos(2\bar{\alpha} \pm \xi) + {}^{B_3}_{B_4} \cos(2\bar{\nu} + 2\bar{\omega}_M + 2\bar{\Omega}_M \pm \xi) + \\ &+ {}^{B_5}_{B_6} \cos(2\bar{\nu} + 2\bar{\omega}_M + \bar{\Omega}_M \pm \xi) + {}^{B_7}_{B_8} \cos(\bar{\Omega}_M \pm \xi) + \\ &+ B_9 \cos \bar{\omega}_M + B_{10} \cos(\bar{\omega}_M + \bar{\Omega}_M) + B_{11} \cos(\bar{\omega}_M + 2\bar{\Omega}_M) + \\ &+ B_{12} \cos(\bar{\nu} + \bar{\omega}_M) + B_{13} \cos(\bar{\nu} + \bar{\omega}_M + \bar{\Omega}_M) + B_{14} \cos(\bar{\nu} + \bar{\omega}_M + \\ &+ 2\bar{\Omega}_M) + B_{15} \cos(2\bar{\nu} + \bar{\Omega}_M) + B_{16} \cos(2\bar{\nu} + \bar{\omega}_M + \bar{\Omega}_M) + \\ &+ B_{17} \cos(2\bar{\nu} + \bar{\omega}_M + 2\bar{\Omega}_M) + B_{18} \cos(2\bar{\nu} + 3\bar{\omega}_M + 2\bar{\Omega}_M) + \\ &+ B_{19} \cos(2\bar{\nu} + 3\bar{\omega}_M + 3\bar{\Omega}_M) + B_{20} \cos(3\bar{\nu} + 3\bar{\omega}_M + 2\bar{\Omega}_M) + \\ &+ B_{21} \cos(3\bar{\nu} + 3\bar{\omega}_M + 3\bar{\Omega}_M) + B_{22} \cos(4\bar{\nu} + 3\bar{\omega}_M + 2\bar{\Omega}_M) + \\ &+ B_{23} \cos(4\bar{\nu} + 3\bar{\omega}_M + 3\bar{\Omega}_M) \ , \end{aligned} \tag{4.103}$$

and

$$\begin{aligned} u_4^{(1)} &= \omega_3 b_0 \sin \xi + {}^{B_{24}}_{B_{25}} \sin(2\bar{\alpha} \pm \xi) + {}^{B_{26}}_{B_{27}} \sin(2\bar{\nu} + 2\bar{\omega}_M + 2\bar{\Omega}_M \pm \xi) + \\ &{}^{B_{28}}_{B_{29}} \sin(2\bar{\nu} + 2\bar{\omega}_M + \bar{\Omega}_M \pm \xi) + {}^{B_{30}}_{B_{31}} \sin(\bar{\Omega}_M \pm \xi) + B_{32} \sin \bar{\omega}_M + \\ &+ B_{33} \sin(\bar{\omega}_M + \bar{\Omega}_M) + B_{34} \sin(\bar{\omega}_M + 2\bar{\Omega}_M) + B_{35} \sin(\bar{\nu} + \bar{\omega}_M) + \\ &+ B_{36} \sin(\bar{\nu} + \bar{\omega}_M + \bar{\Omega}_M) + B_{37} \sin(\bar{\nu} + \bar{\omega}_M + 2\bar{\Omega}_M) + \\ &+ B_{38} \sin(2\bar{\nu} + \bar{\Omega}_M) + B_{39} \sin(2\bar{\nu} + \bar{\omega}_M + \bar{\Omega}_M) + \\ &+ B_{40} \sin(2\bar{\nu} + \bar{\omega}_M + 2\bar{\Omega}_M) + B_{41} \sin(2\bar{\nu} + 3\bar{\omega}_M + 2\bar{\Omega}_M) + \\ &+ B_{42} \sin(2\bar{\nu} + 3\bar{\omega}_M + 3\bar{\Omega}_M) + B_{43} \sin(3\bar{\nu} + 3\bar{\omega}_M + 2\bar{\Omega}_M) + \\ &+ B_{44} \sin(3\bar{\nu} + 3\bar{\omega}_M + 3\bar{\Omega}_M) + B_{45} \sin(4\bar{\nu} + 3\bar{\omega}_M + 2\bar{\Omega}_M) + \\ &B_{46} \sin(4\bar{\nu} + 3\bar{\omega}_M + 3\bar{\Omega}_M) \ , \end{aligned} \tag{4.104}$$

where $\omega_3 \equiv (\omega_{31}\omega_{32})^{1/2}$, $\xi = \omega_3(\theta - \theta_0) + \xi_0$, and b_0 and ξ_0 are constants determined from the initial conditions, $\bar{X}_{30}$ and $\bar{X}_{40}$, when $\theta = \theta_0$, $\bar{\alpha} = \bar{\alpha}_0$, $\bar{\nu} = \bar{\nu}_0$, $\bar{\Omega}_M = \bar{\Omega}_{M0}$, and $\bar{\omega}_M = \bar{\omega}_{M0}$.

After we solve the two sets of differential equations of mono-frequency oscillations, the remaining differential equation, (4.85), will be solved by utilizing the known zeroth-order solutions of $\bar{X}_1$ (= $u_1^{(0)}$) and $\bar{X}_2$ (= $u_2^{(0)}-D$) to replace the corresponding values in (4.85). As a result, (4.85) can be directly integrated as the first-order approximation for $\bar{X}_6$:

$$\begin{aligned}
&\frac{1}{6}\left(\frac{1}{\bar{X}_{60}^6} - \frac{1}{\bar{X}_6^6}\right)\bar{\bar{X}}_{60}^6 \\
&= \frac{\varepsilon}{2}\Bigg\{a_0 \sin 2\lambda \cos\Psi - a_0 \sin\lambda\left[\frac{\omega_{11} + \omega_1\cos\lambda}{2K_1 + \omega_1}\cos(2\bar{\alpha} + \Psi) +\right. \\
&\quad \left. + \frac{\omega_{11} - \omega_1\cos\lambda}{2K_1 - \omega_1}\cos(2\bar{\alpha} - \Psi)\right] + \frac{1}{K_1} D\sin\lambda\cos 2\bar{\alpha} + \\
&\quad + 2\kappa D\left[\frac{\sin\lambda}{2(K_2 + K_3 + K_4)}\cos 2(\bar{\nu} + \bar{\omega}_M + \bar{\Omega}_M) - \right. \\
&\quad \left. - \frac{\sin I}{2K_2 + 2K_4 + K_3}\cos(2\bar{\nu} + 2\bar{\omega}_M + \bar{\Omega}_M) + \frac{\sin I\cos\lambda}{K_3}\cos\bar{\Omega}_M\right] + \\
&\quad + \kappa a_0\left[\sin 2\lambda\cos\Psi - \frac{\omega_{11}\sin\lambda + \frac{1}{2}\omega_1\sin 2\lambda}{2K_2 + 2K_3 + 2K_4 + \omega_1}\cos(2\bar{\nu} + 2\bar{\omega}_M + \right. \\
&\quad + 2\bar{\Omega}_M + \Psi) + \frac{-\omega_{11}\sin\lambda + \frac{1}{2}\omega_1\sin 2\lambda}{2K_2 + 2K_3 + 2K_4 - \omega_1}\cos(2\bar{\nu} + 2\bar{\omega}_M + 2\bar{\Omega}_M - \Psi) + \\
&\quad + \frac{\sin I}{2K_2 + 2K_4 + K_3 + \omega_1}\big(\omega_{11}\cos(2\bar{\nu} + 2\bar{\omega}_M + \bar{\Omega}_M + \Psi) + \\
&\quad + \omega_1\cos 2\lambda\sin(2\bar{\nu} + 2\bar{\omega}_M + \bar{\Omega}_M + \Psi)\big) + \frac{\sin I}{2K_2 + 2K_4 + K_3 - \omega_1}\times \\
&\quad \times\big(\omega_{11}\cos(2\bar{\nu} + 2\bar{\omega}_M + \bar{\Omega}_M - \Psi) - \omega_1\cos 2\lambda\sin(2\bar{\nu} + 2\bar{\omega}_M + \\
&\quad + \bar{\Omega}_M - \Psi)\big) - \sin I(\omega_{11}\cos\lambda + \omega_1\cos 2\lambda)\left(\frac{\cos(\bar{\Omega}_M + \Psi)}{K_3 + \omega_1} + \right. \\
&\quad \left.\left. + \frac{\cos(\bar{\Omega}_M - \Psi)}{K_3 - \omega_1}\right)\right]\Bigg\}\Bigg|_{\theta_0}^{\theta} \\
&\equiv \frac{1}{2}\varepsilon a_0(1 + \kappa)\sin 2\lambda(\cos\Psi - \cos\Psi_0) + F(\Psi, \bar{\alpha}, \bar{\nu}, \bar{\Omega}_M, \bar{\omega}_M) - \\
&\quad - F_0(\Psi, \bar{\alpha}, \bar{\nu}, \bar{\Omega}_M, \bar{\omega}_M)\,, \qquad (4.105)
\end{aligned}$$

where $F(\Psi, \bar{\alpha}, \bar{\nu}, \bar{\Omega}_M, \bar{\omega}_M)$ includes all terms on the right side of the identity except the term of $\cos\Psi$, and $F_0(\Psi, \bar{\alpha}, \bar{\nu}, \bar{\Omega}_M, \bar{\omega}_M)$ is the value of $F(\Psi, \bar{\alpha}, \bar{\nu}, \bar{\Omega}_M, \bar{\omega}_M)$ when $\theta = \theta_0$.

In summary, (4.77)-(4.79), (4.87), (4.90), (4.91), (4.97)-(4.100), and (4.105) constitute the solutions to the system of equations (4.69)-(4.76), where $\bar{X}_1, \ldots, \bar{X}_4, \bar{X}_6, \bar{\alpha}, \bar{\nu}, \bar{\Omega}_M$, and $\bar{\omega}_M$ are dependent variables and θ is the basic independent variable with constants $\bar{X}_{10}, \ldots, \bar{X}_{40}, \bar{X}_{60}, \bar{\alpha}_0, \bar{\nu}_0, \bar{\Omega}_{M0}$, and $\bar{\omega}_{M0}$ to be determined at $\theta = \theta_0$.

We have obtained our solutions without taking the $\delta_1(\;)$'s in (4.46) into consideration. In order to incorporate the long-period terms of the solar perturbation on the Moon's motion into our solutions, the $\delta_1(\;)$ in (4.46) should be added to the corresponding orbital elements, i.e.:

$$\bar{\bar{\psi}} = \psi + \delta_1\psi \cong \bar{\nu} + \bar{\Omega}_M + \bar{\omega}_M + \delta_1\bar{\psi} \,, \tag{4.106a}$$

$$\bar{\bar{\Omega}}_M = \Omega_M + \delta_1\Omega_M \cong \bar{\Omega}_M + \delta_1\bar{\Omega}_M \,, \tag{4.106b}$$

$$\bar{\bar{\omega}}_M = \omega_M + \delta_1\omega_M \cong \bar{\omega}_M + \delta_1\bar{\omega}_M \,, \tag{4.106c}$$

and

$$\bar{\bar{I}} = I + \delta_1\bar{I} \,. \tag{4.106d}$$

In other words, $\bar{\nu}$, $\bar{\Omega}_M$, $\bar{\omega}_M$, and I in (4.69)-(4.73) should be replaced by $\bar{\bar{\psi}}$, $\bar{\bar{\Omega}}_M$, $\bar{\bar{\omega}}_M$ and $\bar{\bar{I}}$, respectively, and the resulting set of equations are to be solved.

The complicated expressions of δ_1( )'s prohibit us to solve (4.69)-(4.73) directly with parameters of $\bar{\bar{\psi}}$, $\bar{\bar{\Omega}}_M$, $\bar{\bar{\omega}}_M$, and $\bar{\bar{I}}$. We shall incorporate the δ_1( )'s into our solution shown in (4.87), (4.90), (4.91), (4.97)-(4.100), and (4.105), which do not include the effects of the δ_1( )'s, with the following arrangements:

(a) $\delta_1\psi$ and $\delta_1 I$ are considered as higher order terms because m^2, me_M, me_S, and mI_0 are one order less than m and m^2/e_M; consequently, only $\delta_1\bar{\Omega}_M$ and $\delta_1\bar{\omega}_M$ are taken into consideration in our solution.

(b) Select the terms in $u_1^{(1)}$, $u_2^{(1)}$, $u_3^{(1)}$, $u_4^{(1)}$, and $F(\Psi, \bar{\alpha}, \bar{\nu}, \bar{\Omega}_M, \bar{\omega}_M)$ containing only low frequency arguments such as ω_1, ω_3, K_3, and K_4 because the low frequency values always appear in the denominators of their coefficients and result in large values of the coefficients. We find those terms in

$u_1^{(1)}$ are with coefficients A_4, A_7, and A_8;

$u_2^{(1)}$ are with coefficients A_{26}, A_{29}, and A_{30};

$u_3^{(1)}$ are with coefficients B_7, B_8, B_9, B_{10}, and B_{11};

$u_4^{(1)}$ are with coefficients B_{30}, B_{31}, B_{32}, B_{33}, and B_{34};

and

$F(\Psi, \bar{\alpha}, \bar{\nu}, \bar{\Omega}_M, \bar{\omega}_M)$ are $\cos\bar{\Omega}_M$, $\cos(\bar{\Omega}_M + \Psi)$, and $\cos(\bar{\Omega}_M - \Psi)$.

(c) In δ_1( )'s, the mean longitudes of the Sun and Moon can be approximated by

$$\lambda_S \cong \bar{\lambda}_S \cong \bar{\alpha}$$
$$\text{and} \qquad \lambda_M \cong \bar{\lambda}_M \cong \bar{\nu} + \bar{\Omega}_M + \bar{\omega}_M \,, \tag{4.107}$$

where $\bar{\alpha}$, $\bar{\nu}$, $\bar{\Omega}_M$, and $\bar{\omega}_M$ are computed from (4.77)-(4.79).

(d) From the selected A's and B's, we shall be able to find the coefficients of the corresponding terms in (4.69)-(4.73) and carry out solutions to these equations (see Appendix C for a detailed methodology). Consequently, we have additional terms in

$$u_1^{(1)}: \; A_{4M}\sin\bar{\Omega}_M\cos(2\bar{\lambda}_S - 2\bar{\Omega}_M)$$
$$\begin{matrix} A_{7M} \\ A_{8M} \end{matrix}\sin(\bar{\Omega}_M \pm \Psi)\cos(2\bar{\lambda}_S - 2\bar{\Omega}_M) \tag{4.108}$$

$$u_2^{(1)}: \; A_{26M}\cos\bar{\Omega}_M\cos(2\bar{\lambda}_S - 2\bar{\Omega}_M)$$
$$\begin{matrix} A_{29M} \\ A_{30M} \end{matrix}\cos(\bar{\Omega}_M \pm \Psi)\cos(2\bar{\lambda}_S - 2\bar{\Omega}_M) \tag{4.109}$$

$$
\begin{aligned}
u_3^{(1)}:\ & \begin{matrix}B_{7M}\\ B_{8M}\end{matrix}\ \cos(\bar{\Omega}_M \pm \xi)\cos(2\bar{\lambda}_S - 2\bar{\Omega}_M)\\
& B_{9M1}\ \cos\bar{\omega}_M \cos(3\bar{\lambda}_M - 2\bar{\lambda}_S - \bar{\bar{\omega}}_M)\\
& B_{9M2}\ \cos\bar{\omega}_M \cos(\bar{\lambda}_M - 2\bar{\lambda}_S + \bar{\bar{\omega}}_M)\\
& B_{9M3}\ \cos\bar{\omega}_M \cos(2\bar{\lambda}_S - 2\bar{\bar{\omega}}_M)\\
& B_{9M4}\ \cos\bar{\omega}_M \cos(2\bar{\lambda}_S - 2\Omega_M)\\
& B_{10M1}\ \cos(\bar{\omega}_M + \bar{\Omega}_M)\cos(3\bar{\lambda}_M - 2\bar{\lambda}_S - \bar{\bar{\omega}}_M)\\
& B_{10M2}\ \cos(\bar{\omega}_M + \bar{\Omega}_M)\cos(\bar{\lambda}_M - 2\bar{\lambda}_S + \bar{\bar{\omega}}_M)\\
& B_{10M3}\ \cos(\bar{\omega}_M + \bar{\Omega}_M)\cos(2\bar{\lambda}_S - 2\bar{\bar{\omega}}_M)\\
& B_{11M1}\ \cos(\bar{\omega}_M + 2\bar{\Omega}_M)\cos(2\bar{\lambda}_S - 2\bar{\Omega}_M)\\
& B_{11M2}\ \cos(\bar{\omega}_M + 2\bar{\Omega}_M)\cos(3\bar{\lambda}_M - 2\bar{\lambda}_S - \bar{\bar{\omega}}_M)\\
& B_{11M3}\ \cos(\bar{\omega}_M + 2\bar{\Omega}_M)\cos(\bar{\lambda}_M - 2\bar{\lambda}_S + \bar{\bar{\omega}}_M)\\
& B_{11M4}\ \cos(\bar{\omega}_M + 2\bar{\Omega}_M)\cos(2\bar{\lambda}_S - 2\bar{\bar{\omega}}_M)
\end{aligned} \tag{4.110}
$$

$$
\begin{aligned}
u_4^{(1)}:\ & \begin{matrix}B_{30M}\\ B_{31M}\end{matrix}\ \sin(\bar{\Omega}_M \pm \xi)\cos(2\bar{\lambda}_S - 2\bar{\Omega}_M)\\
& B_{32M}\ \text{(very small coefficient, hence omitted)}\\
& B_{33M1}\ \sin(\bar{\omega}_M + \bar{\Omega}_M)\cos(3\bar{\lambda}_M - 2\bar{\lambda}_S - \bar{\bar{\omega}}_M)\\
& B_{33M2}\ \sin(\bar{\omega}_M + \bar{\Omega}_M)\cos(\bar{\lambda}_M - 2\bar{\lambda}_S + \bar{\bar{\omega}}_M)\\
& B_{33M3}\ \sin(\bar{\omega}_M + \bar{\Omega}_M)\cos(2\bar{\lambda}_S - 2\bar{\bar{\omega}}_M)\\
& B_{34M1}\ \sin(\bar{\omega}_M + 2\bar{\Omega}_M)\cos(2\bar{\lambda}_S - 2\bar{\Omega}_M)\\
& B_{34M2}\ \sin(\bar{\omega}_M + 2\bar{\Omega}_M)\cos(3\bar{\lambda}_M - 2\bar{\lambda}_S - \bar{\bar{\omega}}_M)\\
& B_{34M3}\ \sin(\bar{\omega}_M + 2\bar{\Omega}_M)\cos(\bar{\lambda}_M - 2\bar{\lambda}_S + \bar{\bar{\omega}}_M)\\
& B_{34M4}\ \sin(\bar{\omega}_M + 2\bar{\Omega}_M)\cos(2\bar{\lambda}_S - 2\bar{\bar{\omega}}_M)
\end{aligned} \tag{4.111}
$$

and the constants, A_M's and B_M's, are shown in Appendix C.

The additional terms in $F(\Psi, \bar{\alpha}, \bar{\nu}, \bar{\Omega}_M, \bar{\omega}_M)$ resulting from $\delta_1(\ )$'s have very small coefficients; consequently, we may omit these terms and $\bar{X}_6$ is not effected.

With $\delta_1(\ )$'s, $\bar{\Omega}_M$ and $\bar{\omega}_M$ in (4.79a) and (4.79b) are modified accordingly:

$$
\begin{aligned}
\bar{\Omega}_{MS} &= \bar{\Omega}_M + \delta_1\Omega_M\\
&= \bar{\Omega}_M + \frac{3}{8}m\sin(2\bar{\lambda}_S - 2\bar{\Omega}_M) + \bar{\Omega}_{MS0}
\end{aligned} \tag{4.112a}
$$

and

$$
\begin{aligned}
\bar{\omega}_{MS} &= \bar{\omega}_M + \delta_1\omega_M\\
&= \bar{\omega}_M + \frac{1}{4}\frac{m^2}{e_M}[\sin(3\bar{\lambda}_M - 2\bar{\lambda}_S - \bar{\bar{\omega}}_M) - 9\sin(\bar{\lambda}_M - 2\bar{\lambda}_S + \bar{\bar{\omega}}_M)] +\\
&\quad + \frac{15}{8}m\sin(2\bar{\lambda}_S - 2\bar{\bar{\omega}}_M) - \frac{3}{8}m\sin(2\bar{\lambda}_S - 2\bar{\Omega}_M) + \bar{\omega}_{MS0}\ .
\end{aligned} \tag{4.112b}
$$

$\bar{\alpha}$ is not effected by $\delta_1(\ )$'s, and $\bar{\nu}$ is associated with $\delta_1\psi \cong \delta_1(\bar{\nu} + \bar{\Omega}_M + \bar{\omega}_M) \cong 0$ because $\delta_1\psi$ contains higher order terms as shown in (4.46d). As a result, $\delta_1\nu = -\delta_1\Omega_M - \delta_1\omega_M$.

4.7. Improved First-Order Approximate Solution

An improvement of the first-order approximate solution can bring out the oscillatory nature of the parameters. This is achieved by noting (in accordance with the asymptotic method shown in (2.100)-(2.103)) that the improved first-order approximate solution for X_1 can be expressed as

$$X_1 = \bar{X}_1 + \int \left(\frac{dX_1}{d\theta} - \overline{\frac{dX_1}{d\theta}} \right) d\theta , \tag{4.113}$$

where $dX_1/d\theta$ is shown in (4.58), and $\overline{dX_1/d\theta}$ in (4.69), with the integration carried out by assuming that $\bar{X}_1, \ldots, \bar{X}_4, \bar{X}_6, \bar{\alpha}, \bar{\nu}, \bar{\Omega}_M$, and $\bar{\omega}_M$ remain constant. As the term $\int (dX_1/d\theta - \overline{dX_1/d\theta})\, d\theta$ is basically oscillatory in nature, its average value over a range of 2π of θ should be zero. This provides the criterion to determine the integration constant for the integral. Hence, the improved first-order approximate solution for X_1 is

$$\begin{aligned}
X_1 &= \bar{X}_1 + \delta X_1 \\
&= \bar{X}_1 - \tfrac{3}{4} J_2 \frac{R_E^2}{r_c^2} (\bar{X}_1 \cos 2\theta + \bar{X}_2 \sin 2\theta) - \frac{3}{8} \frac{K_1^2}{(1 - e_S^2)^{3/2}} \times \\
&\times \Big\{ [\bar{X}_1 \cos \lambda \sin 2\bar{\alpha} + 2\bar{X}_2 \sin^2 \bar{\alpha} \cos 2\lambda + \\
&+ (1 + e_S \cos(-\omega_S + \bar{\alpha})) \sin 2\lambda \sin^2 \bar{\alpha}] \sin 2\theta + \\
&+ [2\bar{X}_1 (\cos^2 \bar{\alpha} - \sin^2 \lambda \sin^2 \bar{\alpha}) + \bar{X}_2 \cos \lambda \sin 2\bar{\alpha} + \\
&+ (1 + e_S \cos(-\omega_S + \bar{\alpha})) \sin \lambda \sin 2\bar{\alpha}] \cos 2\theta \Big\} - \\
&- \frac{2 r_c K_1^2}{(1 - e_S^2)^{3/2}} \Big\{ \sin \lambda \sin 2\bar{\alpha} [-\bar{X}_3 \cos^3 \theta + \bar{X}_4 \sin^3 \theta] + \\
&+ \sin 2\lambda \sin^2 \bar{\alpha} [\bar{X}_3 \sin^3 \theta - \bar{X}_4 \cos \theta (\sin^2 \theta + 2)] \Big\} - \\
&- \frac{3}{4} \frac{K_2^2}{\sigma (1 - e_M^2)^{3/2}} [\bar{X}_1 (\cos \lambda \bar{M}_x \bar{M}_y + \sin \lambda \bar{M}_x \bar{M}_z) + \\
&+ \bar{X}_2 (\cos 2\lambda \bar{M}_y^2 + 2 \sin 2\lambda \bar{M}_y \bar{M}_z) + (1 + e_M \cos \bar{\nu}) \times \\
&\times (\tfrac{1}{2} \sin 2\lambda \bar{M}_y^2 - \cos 2\lambda \bar{M}_y \bar{M}_z)] \sin 2\theta + [\bar{X}_1 (\bar{M}_x^2 - \sin^2 \lambda \bar{M}_y^2 + \\
&+ \sin 2\lambda \bar{M}_y \bar{M}_z) + \bar{X}_2 (\bar{M}_x \bar{M}_y \cos \lambda + \bar{M}_x \bar{M}_z \sin \lambda) + \\
&+ (1 + e_M \cos \bar{\nu}) (\sin \lambda \bar{M}_x \bar{M}_y - \cos \lambda \bar{M}_x \bar{M}_z)] \cos 2\theta \Big\} - \\
&- 4 r_c \frac{K_2^2}{\sigma (1 - e_M^2)^{3/2}} (\sin \lambda \bar{M}_y - \cos \lambda \bar{M}_z) \Big\{ \bar{M}_x (-\bar{X}_3 \cos^3 \theta + \\
&+ \bar{X}_4 \sin^3 \theta) + (\cos \lambda \bar{M}_y + \sin \lambda \bar{M}_z) [\bar{X}_3 \sin^3 \theta - \\
&- \bar{X}_4 \cos \theta (\sin^2 \theta + 2)] \Big\} - \frac{r_c K_2^2}{\sigma a_M (1 - e_M^2)^{5/2}} (1 + \tfrac{1}{2} e_M^2 + \\
&+ 2 e_M \cos \bar{\nu}) \Big\{ -\tfrac{3}{2} \cos \theta + \tfrac{5}{2} (\sin \lambda \bar{M}_y - \cos \lambda \bar{M}_z) \times \\
&\times [\cos^3 \theta \bar{M}_x^2 + \cos \theta (\sin^2 \theta + 2) \cos^2 \lambda \bar{M}_y^2 - \\
&- 2 \sin^3 \theta \cos \lambda \bar{M}_x \bar{M}_y] +
\end{aligned}$$

$$+ 5 \sin^2 \lambda\ \bar{M}_y\bar{M}_z[\cos\theta(\sin^2\theta + 2)\cos\lambda\ \bar{M}_y - \sin^3\theta\ \bar{M}_x]\Big\}, \tag{4.114}$$

$$\begin{aligned}
X_2 &= \bar{X}_2 + \delta X_2 \\
&= \bar{X}_2 - \tfrac{3}{4}J_2 \frac{R_E^2}{r_c^2}(\bar{X}_1 \sin 2\theta - \bar{X}_2 \cos 2\theta) - \frac{3}{8}\frac{K_1^2}{(1 - e_S^2)^{3/2}} \times \\
&\times \Big\{[2\bar{X}_1(\cos^2\bar{\alpha} - \sin^2\lambda \sin^2\bar{\alpha}) + \bar{X}_2 \cos\lambda \sin 2\bar{\alpha} + \\
&+ (1 + e_S \cos(-\omega_S + \bar{\alpha}))\sin\lambda \sin 2\alpha]\sin 2\theta - \\
&- [\bar{X}_1 \cos\lambda \sin 2\bar{\alpha} + 2\bar{X}_2 \sin^2\bar{\alpha} \cos 2\lambda + \\
&+ (1 + e_S \cos(-\omega_S + \bar{\alpha}))\sin 2\lambda \sin^2\bar{\alpha}]\cos 2\theta\Big\} + \\
&+ \frac{2r_c K_1^2}{(1 - e_S^2)^{3/2}}\Big\{\sin\lambda \sin 2\bar{\alpha}[\bar{X}_3 \sin\theta(\cos^2\theta + 2) - \bar{X}_4 \cos^3\theta] + \\
&+ \sin 2\lambda \sin^2\bar{\alpha}[-\bar{X}_3 \cos^3\theta + \bar{X}_4 \sin^3\theta]\Big\} - \\
&- \frac{3K_2^2}{4\sigma(1 - e_M^2)^{3/2}}\Big\{\bar{X}_1(\bar{M}_x^2 - \sin^2\lambda\ \bar{M}_y^2 + \sin 2\lambda\ \bar{M}_y\bar{M}_z) + \\
&+ \bar{X}_2(\cos\lambda\ \bar{M}_x\bar{M}_y + \sin\lambda\ \bar{M}_x\bar{M}_z) + (1 + e_M \cos\bar{\nu})(\sin\lambda\ \bar{M}_x\bar{M}_y - \\
&- \cos\lambda\ \bar{M}_x\bar{M}_z)]\sin 2\theta - [\bar{X}_1(\cos\lambda\ \bar{M}_x\bar{M}_y + \sin\lambda\ \bar{M}_x\bar{M}_z) + \\
&+ \bar{X}_2(\cos 2\lambda\ \bar{M}_y^2 + 2\sin 2\lambda\ \bar{M}_y\bar{M}_z) + (1 + e_M \cos\bar{\nu}) \times \\
&\times (\tfrac{1}{2}\sin 2\lambda\ \bar{M}_y^2 + \cos 2\lambda\ \bar{M}_y\bar{M}_z)]\cos 2\theta\Big\} + \\
&+ 4r_c \frac{K_2^2}{\sigma(1 - e_M^2)^{3/2}}(\sin\lambda\ \bar{M}_y - \cos\lambda\ \bar{M}_z) \times \\
&\times \Big\{\bar{M}_x[\bar{X}_3 \sin\theta(\cos^2\theta + 2) - \bar{X}_4 \cos^3\theta] + (\cos\lambda\ \bar{M}_y + \\
&+ \sin\lambda\ \bar{M}_z)[-\bar{X}_3 \cos^3\theta + \bar{X}_4 \sin^3\theta]\Big\} - \frac{r_c K_2^2}{\sigma a_M(1 - e_M^2)^{5/2}} \times \\
&\times (1 + \tfrac{1}{2}e_M^2 + 2e_M \cos\bar{\nu})\Big\{-\tfrac{3}{2}\sin\theta + \tfrac{5}{2}(\sin\lambda\ \bar{M}_y - \cos\lambda\ \bar{M}_z) \times \\
&\times [\sin\theta(\cos^2\theta + 2)\bar{M}_x^2 + \sin^3\theta \cos^2\lambda\ \bar{M}_y^2 - \\
&- 2\cos^3\theta\cos\lambda\ \bar{M}_x\bar{M}_y] + 5\sin^2\lambda\ \bar{M}_y\bar{M}_z(\sin^3\theta \cos\lambda\ \bar{M}_y - \\
&- \cos^3\theta\ \bar{M}_x)\Big\},
\end{aligned} \tag{4.115}$$

$$\begin{aligned}
X_3 &= \bar{X}_3 + \delta X_3 \\
&= \bar{X}_3 + \tfrac{3}{2}J_2 \frac{R_E^2}{r_c^3}\cos\theta + \frac{K_1^2}{r_c(1 - e_S^2)^{3/2}}\Big\{[\cos^3\theta(\cos^2\bar{\alpha} - \\
&- \cos^2\lambda \sin^2\bar{\alpha}) + \sin\theta \sin 2\bar{\alpha} \cos\lambda(3 - \sin^2\theta) - \\
&- \cos\theta(3\cos^2\lambda \sin^2\bar{\alpha} - 1)](1 + e_S \cos(-\omega_S + \bar{\alpha})) - \\
&- \tfrac{1}{2}\sin\lambda \sin 2\bar{\alpha}(\bar{X}_1 \cos^3\theta - \bar{X}_2 \sin^3\theta) + \tfrac{1}{2}\sin 2\lambda \sin^2\bar{\alpha} \times \\
&\times [\bar{X}_1 \sin\theta(\sin^2\theta - 6) + \bar{X}_2 \cos\theta(\cos^2\theta + 3)]\Big\} +
\end{aligned}$$

$$+ \frac{3}{4}[J_2 \frac{R_E^2}{r_c^2} - (1 + \kappa) \frac{1}{(1 - e_S^2)^{3/2}}][\bar{X}_3 \cos 2\theta + \bar{X}_4 \sin 2\theta] +$$
$$+ \frac{3}{8} \frac{K_1^2}{(1 - e_S^2)^{3/2}}\Big\{\cos^2 \bar{\alpha}[- \bar{X}_3(\cos 4\theta + 6 \cos 2\theta) - \bar{X}_4 \sin 4\theta] +$$
$$+ \cos \lambda \sin 2\bar{\alpha} [-\bar{X}_3(\sin 4\theta + 9 \sin 2\theta) + \bar{X}_4(\cos 4\theta +$$
$$+ 3 \cos 2\theta)] + \cos^2 \lambda \sin^2 \bar{\alpha}[\bar{X}_3(\cos 4\theta + 12 \cos 2\theta) +$$
$$+ \bar{X}_4(\sin 4\theta + 6 \sin 2\theta)]\Big\} +$$
$$+ \frac{K_2^2}{\sigma r_c(1 - e_M^2)^{3/2}}\Big\{\bar{M}_x^2 \cos^3 \theta(1 + e_M \cos \bar{\nu}) +$$
$$+ \bar{M}_y^2[- \cos^2 \lambda \cos \theta(3 + \cos^2 \theta)(1 + e_M \cos \bar{\nu}) +$$
$$+ \frac{1}{2} \sin 2\lambda\big(\bar{X}_1 \sin \theta(\sin^2 \theta - 6) + \bar{X}_2 \cos \theta(\cos^2 \theta + 3)\big)] +$$
$$+ \bar{M}_x\bar{M}_y[2 \sin \theta(3 - \sin^2 \theta)\cos \lambda(1 + e_M \cos \bar{\nu}) +$$
$$+ \sin \lambda(- \bar{X}_1 \cos^3 \theta + \bar{X}_2 \sin^3 \theta)] + \bar{M}_x\bar{M}_z[2 \sin \lambda \sin \theta \times$$
$$\times (3 - \sin^2 \theta)(1 + e_M \cos \bar{\nu}) + \cos \lambda(\bar{X}_1 \cos^3 \theta -$$
$$- \bar{X}_2 \sin^3 \theta)] - \bar{M}_y\bar{M}_z[\sin 2\lambda \cos \theta(\cos^2 \theta + 3)(1 + e_M \cos \bar{\nu}) +$$
$$+ \cos 2\lambda\big(\bar{X}_1 \sin \theta(\sin^2 \theta - 6) + \bar{X}_2 \cos \theta(\cos^2 \theta + 3)\big)] +$$
$$+ \cos \theta(1 + e_M \cos \bar{\nu})\Big\} + \frac{3}{8} \frac{K_2^2}{\sigma(1 - e_M^2)^{3/2}}\Big\{\bar{M}_x^2[- \bar{X}_3(\cos 4\theta +$$
$$+ 6 \cos 2\theta) - \bar{X}_4 \sin 4\theta] + \bar{M}_y^2 \cos^2 \lambda[\bar{X}_3(\cos 4\theta + 12 \cos 2\theta) +$$
$$+ \bar{X}_4(\sin 4\theta + 6 \sin 2\theta)] + 2(\bar{M}_x\bar{M}_y \cos \lambda + \bar{M}_x\bar{M}_z \sin \lambda) \times$$
$$\times [\bar{X}_3(- \sin 4\theta - 9 \sin 2\theta) + \bar{X}_4(\cos 4\theta + 3 \cos 2\theta)] +$$
$$+ \bar{M}_y\bar{M}_z \sin 2\lambda[\bar{X}_3(\cos 4\theta + 12 \cos 2\theta) + \bar{X}_4(\sin 4\theta +$$
$$+ 6 \sin 2\theta)]\Big\} + \frac{3K_2^2}{\sigma a_M(1 - e_M^2)^{5/2}}(1 + \frac{1}{2}e_M^2 + 2 e_M \cos \bar{\nu}) \times$$
$$\times \Big\{\frac{5}{8}[(\frac{3}{8} - \cos^4 \theta)\bar{M}_x^3 - \frac{3}{8} \sin 4\theta \bar{M}_x^2(\cos \lambda \bar{M}_y + \sin \lambda \bar{M}_z) +$$
$$+ 3(\sin^4 \theta - \frac{3}{8})\bar{M}_x\bar{M}_y(\cos^2 \lambda \bar{M}_y + \sin 2\lambda \bar{M}_z) +$$
$$+ \frac{1}{4}(\cos 2\theta - 4)\sin 2\theta(\cos \lambda \bar{M}_y + 3 \sin \lambda \bar{M}_z)\cos^2 \lambda \bar{M}_y^2] +$$
$$+ \frac{3}{8}[\cos 2\theta \bar{M}_x + \sin 2\theta(\cos \lambda \bar{M}_y + \sin \lambda \bar{M}_z)]\Big\} , \qquad (4.116)$$

$$X_4 = \bar{X}_4 + \delta X_4$$
$$= \bar{X}_4 + \frac{3}{2}J_2 \frac{R_E^2}{r_c^3} \sin \theta - \frac{K_1^2}{r_c(1 - e_S^2)^{3/2}}\Big\{[\sin^3 \theta(\cos^2 \bar{\alpha} -$$
$$- \cos^2 \lambda \sin^2 \bar{\alpha}) - \cos \theta \sin 2\bar{\alpha} \cos \lambda(3 - \cos^2 \theta) +$$
$$+ \sin \theta(3 \cos^2 \bar{\alpha} - 1)]\big(1 + e_S \cos(- \omega_S + \bar{\alpha})\big) -$$
$$- \frac{1}{2} \sin \lambda \sin 2\bar{\alpha}[\bar{X}_1 \sin \theta(\sin^2 \theta + 3) +$$

$$+ \bar{X}_2 \cos\theta(\cos^2\theta - 6)] - \frac{1}{2}\sin 2\lambda \sin^2\bar{\alpha}(\bar{X}_1 \cos^3\theta -$$

$$- \bar{X}_2 \sin^3\theta)\Big\} + \frac{3}{4}[J_2 \frac{R_E^2}{r_c^2} - (1 + \kappa)\frac{K_1^2}{(1 - e_S^2)^{3/2}}] \times$$

$$\times [\bar{X}_3 \sin 2\theta - \bar{X}_4 \cos 2\theta] + \frac{3}{8}\frac{K_1^2}{(1 - e_S^2)^{3/2}} \times$$

$$\times \Big\{\cos^2\bar{\alpha}[\bar{X}_3(-\sin 4\theta + 6\sin 2\theta) + \bar{X}_4(\cos 4\theta - 12\cos 2\theta)] +$$

$$+ \cos\lambda \sin 2\bar{\alpha}[\bar{X}_3(\cos 4\theta - 3\cos 2\theta) + \bar{X}_4(\sin 4\theta -$$

$$- 9\sin 2\theta)] + \cos^2\lambda \sin^2\bar{\alpha}[\bar{X}_3 \sin 4\theta + \bar{X}_4(-\cos 4\theta +$$

$$+ 6\cos 2\theta)]\Big\} - \frac{K_2^2}{\sigma r_c(1 - e_M^2)^{3/2}}\Big\{\bar{M}_x^2 \sin\theta(\sin^2\theta + 3) \times$$

$$\times (1 + e_M \cos\bar{\nu}) + \bar{M}_y^2[-\sin^3\theta \cos^2\lambda(1 + e_M\cos\bar{\nu}) +$$

$$+ \frac{1}{2}\sin 2\lambda(-\bar{X}_1 \cos^3\theta + \bar{X}_2 \sin^3\theta)] + \bar{M}_x\bar{M}_y[-2\cos\theta \times$$

$$\times (3 - \cos^2\theta)\cos\lambda(1 + e_M\cos\bar{\nu}) - \bar{X}_1 \sin\lambda \sin\theta \times$$

$$\times (\sin^2\theta + 3) - \bar{X}_2 \sin\lambda \cos\theta(\cos^2\theta - 6)] +$$

$$+ \bar{M}_x\bar{M}_z[-2\sin\lambda\cos\theta(3 - \cos^2\theta)(1 + e_M\cos\bar{\nu}) +$$

$$+ \bar{X}_1 \cos\lambda \sin\theta(\sin^2\theta + 3) + \bar{X}_2\cos\lambda\cos\theta(\cos^2\theta - 6)] +$$

$$+ \bar{M}_y\bar{M}_z[-\sin^3\theta \sin 2\lambda(1 + e_M\cos\bar{\nu}) + 3\cos 2\lambda(-\bar{X}_1\cos^3\theta +$$

$$+ \bar{X}_2 \sin^3\theta)] - \sin\theta(1 + e_M\cos\bar{\nu})\Big\} +$$

$$+ \frac{3}{8}\frac{K_2^2}{\sigma(1 - e_M^2)^{3/2}}\Big\{\bar{M}_x^2[\bar{X}_3(-\sin 4\theta + 6\sin 2\theta) + \bar{X}_4(\cos 4\theta -$$

$$- 12\cos 2\theta)] + M_y^2 \cos^2\lambda[\bar{X}_3 \sin 4\theta + \bar{X}_4(-\cos 4\theta +$$

$$+ 6\cos 2\theta)] + 2(\bar{M}_x\bar{M}_y \cos\lambda + \bar{M}_x\bar{M}_z \sin\lambda)[\bar{X}_3(\cos 4\theta -$$

$$- 3\cos 2\theta) + \bar{X}_4(\sin 4\theta - 9\sin 2\theta)] +$$

$$+ \bar{M}_y\bar{M}_z \sin 2\lambda[\bar{X}_3 \sin 4\theta + \bar{X}_4(-\cos 4\theta + 6\cos 2\theta)]\Big\} -$$

$$- \frac{3K_2^2}{\sigma a_M(1 - e_M^2)^{5/2}}(1 + \frac{1}{2}e_M^2 + 2e_M\cos\bar{\nu}) \times$$

$$\times \Big\{\frac{5}{8}[\frac{1}{4}\sin 2\theta(4 + \cos 2\theta)\bar{M}_x^3 - \frac{3}{8}\sin 4\theta\, \bar{M}_x\bar{M}_y(\cos^2\lambda\, \bar{M}_y +$$

$$+ \sin 2\lambda\, \bar{M}_z) - 3(\cos^4\theta - \frac{3}{8})\bar{M}_x^2(\cos\lambda\, \bar{M}_y + \sin\lambda\, \bar{M}_z) +$$

$$+ (\sin^4\theta - \frac{3}{8})\cos^2\lambda\, \bar{M}_y^2(\cos\lambda\, \bar{M}_y + 3\sin\lambda\, \bar{M}_z) -$$

$$- \frac{3}{8}[\sin 2\theta\, \bar{M}_x - \cos 2\theta(\cos\lambda\, \bar{M}_y + \sin\lambda\, \bar{M}_z)]\Big\}, \qquad (4.117)$$

$$X_5 = \bar{X}_5 + \delta X_5$$

$$= \frac{\mu_E}{\bar{X}_6^2}\frac{1}{1 + \bar{X}_1^2 + \bar{X}_2^2}[1 + (\bar{X}_1\cos\theta + \bar{X}_2\sin\theta)^2]^{1/2} +$$

$$+ \frac{\mu_E}{\bar{X}_6^2}\frac{2}{1 + \bar{X}_1^2 + \bar{X}_2^2}[1 + (\bar{X}_1\cos\theta + \bar{X}_2\sin\theta)^2]^{1/2} \times$$

$$\times \left(- \frac{1}{\bar{X}_6^3} \delta X_6 - \frac{\bar{X}_1 \delta X_1 + \bar{X}_2 \delta X_2}{1 + \bar{X}_1^2 + \bar{X}_2^2} + \right.$$
$$\left. + \frac{1}{2} \frac{(\cos\theta\ \delta X_1 + \sin\theta\ \delta X_2)(\bar{X}_1 \cos\theta + \bar{X}_2 \sin\theta)}{1 + (\bar{X}_1 \cos\theta + \bar{X}_2 \sin\theta)^2} \right) , \tag{4.118}$$

$$X_6 = \bar{X}_6 + \delta X_6$$
$$= \bar{X}_6 + \frac{3}{4} \frac{\bar{X}_6 K_1^2}{(1 - e_S^2)^{3/2}} \left\{ \left(\frac{1}{2} [\sin^2 \lambda + (1 + \cos^2 \lambda) \cos 2\bar{\alpha}] \times \right. \right.$$
$$\times \left(1 + e_S \cos(-\omega_S + \bar{\alpha})\right) - \sin\lambda \sin\bar{\alpha}(\bar{X}_1 \cos\bar{\alpha} -$$
$$\left. - \bar{X}_2 \cos\lambda \sin\bar{\alpha}) \right) \cos 2\theta + [\cos\lambda \sin 2\bar{\alpha} \times$$
$$\times \left(1 + e_S \cos(-\omega_S + \bar{\alpha})\right) - \sin\lambda \sin\bar{\alpha}(\bar{X}_1 \cos\lambda \sin\bar{\alpha} +$$
$$\left. + \bar{X}_2 \cos\bar{\alpha})] \sin 2\theta \right\} - 2r_c \frac{\bar{X}_6 K_1^2}{(1 - e_S^2)^{3/2}} \times$$
$$\times \left\{ [\sin^2 \lambda + (1 + \cos^2 \lambda) \cos 2\bar{\alpha}][\bar{X}_3 \cos^3 \theta - \bar{X}_4 \sin^3 \theta] + \right.$$
$$+ \cos\lambda \sin 2\bar{\alpha} [\bar{X}_3 \sin\theta(3 - 2\sin^2\theta) -$$
$$\left. - \bar{X}_4 \cos\theta(3 - 2\cos^2\theta)] \right\} + \frac{3}{4} \bar{X}_6 \frac{K_2^2}{\sigma(1 - e_M^2)^{3/2}} \times$$
$$\times \left\{ [\bar{M}_x^2 \cos 2\theta + \bar{M}_x(\cos\lambda\ \bar{M}_y + \sin\lambda\ \bar{M}_z) \sin 2\theta + \right.$$
$$+ \bar{M}_x(\cos\lambda\ \bar{M}_y + \sin\lambda\ \bar{M}_z) \sin 2\theta - (\cos^2\lambda\ \bar{M}_y^2 +$$
$$+ \sin 2\lambda\ \bar{M}_y \bar{M}_z) \cos 2\theta](1 + e_M \cos\bar{\nu}) +$$
$$+ \bar{M}_x(\bar{X}_1 \cos 2\theta + \bar{X}_2 \sin 2\theta)(\sin\lambda\ \bar{M}_y + \cos\lambda\ \bar{M}_z) +$$
$$\left. + (\bar{X}_1 \sin 2\theta - \bar{X}_2 \cos 2\theta)(\bar{M}_y^2 \sin\lambda \cos\lambda + \bar{M}_y \bar{M}_z) \right\} -$$
$$- 4r_c \frac{\bar{X}_6 K_2^2}{\sigma(1 - e_M^2)^{3/2}} \left\{ [\bar{X}_3 \cos^3\theta - \bar{X}_4 \sin^3\theta][\bar{M}_x^2 - \bar{M}_y \bar{M}_z \sin 2\lambda - \right.$$
$$- \bar{M}_y^2 \cos^2\lambda] + \bar{M}_x(\bar{M}_y \cos\lambda + \bar{M}_z \sin\lambda) \times$$
$$\left. \times [\bar{X}_3 \sin\theta(\cos 2\theta + 2) + \bar{X}_4 \cos\theta(-\cos 2\theta + 2)] \right\} +$$
$$+ \frac{15}{2} \frac{r_c \bar{X}_6 K_2^2}{\sigma a_M (1 - e_M^2)^{5/2}} (1 + \tfrac{1}{2} e_M^2 + 2e_M \cos\bar{\nu}) \left\{ \tfrac{1}{3} \bar{M}_x^3 \cos^3\theta + \right.$$
$$+ \bar{M}_x \bar{M}_y (\bar{M}_y \cos^2\lambda + \bar{M}_z \sin 2\lambda) \sin^2\theta \cos\theta + \bar{M}_x^2 (\bar{M}_y \cos\lambda +$$
$$+ \bar{M}_z \sin\lambda) \sin\theta \cos^2\theta + \bar{M}_y^2 \cos^2\lambda (\tfrac{1}{3} \bar{M}_y \cos\lambda +$$
$$+ \bar{M}_z \sin\lambda) \sin^3\theta - \tfrac{1}{5}(\bar{M}_x \cos\theta + \bar{M}_y \cos\lambda \sin\theta +$$
$$\left. + \bar{M}_z \sin\lambda \sin\theta) \right\} , \tag{4.119}$$

$$\alpha = \bar{\alpha} + \delta\alpha$$
$$= \bar{\alpha} - 2K_1 r_c (\bar{X}_3 \sin\theta - \bar{X}_4 \cos\theta) + \alpha_0 , \tag{4.120}$$

$$\nu = \bar{\nu} + \delta_1\nu + \delta\nu$$

$$= \bar{\nu} + \delta_1\nu - 2K_2r_c(\bar{X}_3 \sin\theta - \bar{X}_4 \cos\theta) + \nu_0 , \qquad (4.121)$$

$$\Omega_M = \bar{\Omega}_{MS} + \delta\Omega_M$$

$$= \bar{\Omega}_{MS} - 2K_3r_c(\bar{X}_3 \sin\theta - \bar{X}_4 \cos\theta) + \Omega_{M0} , \qquad (4.122a)$$

and

$$\omega_M = \bar{\omega}_{MS} + \delta\omega_M$$

$$= \bar{\omega}_{MS} - 2K_4r_c(\bar{X}_3 \sin\theta - \bar{X}_4 \cos\theta) + \omega_{M0} , \qquad (4.122b)$$

where $\bar{X}_6^2/\mu_E$ in the oscillatory terms is approximated by r_c, and the terms containing M_z^2 are omitted because $\sin^2 I \cong 0$. It should be noted that $\bar{X}$'s, $\bar{\alpha}$, $\bar{\nu}$, $\bar{\Omega}_M$, and $\bar{\omega}_M$ in (4.112) and (4.114)-(4.122) can be calculated from (4.77)-(4.79), (4.87), (4.90), (4.91), (4.97)-(4.100), and (4.105), and these values are actually the secular portion of the solution. $\bar{M}_x$, $\bar{M}_y$, and $\bar{M}_z$ in the oscillatory terms have their expressions shown in (4.32) with terms having the arguments of $\bar{\nu}$, $\bar{\omega}_M$, and $\bar{\Omega}_M$.

4.8. Introducing the Time Parameter (t)

For actual tracking of a geostationary satellite, it is desirable to have the time, t, as an additional dependent variable. In other words, a relationship between t and the independent variable, θ, is required. This relationship is established as follows. Combining (4.39a), (4.39b), (4.54), (4.55), and (4.63) results in

$$\frac{dt}{d\theta} = \frac{X_6^3}{\mu_E^2}(1 + X_1^2 + X_2^2)^2[1 + (X_1 \cos\theta + X_2 \sin\theta)^2]^{-1} \times$$
$$\times \left\{1 + \frac{X_6^2}{\mu_E} \frac{X_3 \cos\theta + X_4 \sin\theta}{[1 + (X_1 \cos\theta + X_2 \sin\theta)^2]^{1/2}}(1 + X_1^2 + X_2^2)\right\}^{-2} . \qquad (4.123)$$

Following the same reasoning in obtaining (4.68), the braces on the right-hand side of this equation can be expanded into a series. Since a geostationary satellite is supposed to be in a circular orbit, this series can be approximated by the first two terms of its expansion. Thus

$$\frac{dt}{d\theta} = \frac{X_6^3}{\mu_E^2}\left\{(1 + X_1^2 + X_2^2)^2[1 + (X_1 \cos\theta + X_2 \sin\theta)^2]^{-1} - \right.$$
$$\left. - 2\frac{X_6^2}{\mu_E} \frac{X_3 \cos\theta + X_4 \sin\theta}{[1 + (X_1 \cos\theta + X_2 \sin\theta)^2]^{3/2}}(1 + X_1^2 + X_2^2)^3\right\} ,$$

which, in turn, can be written as

$$\frac{dt}{d\theta} = \frac{X_6^3}{\mu_E^2}\left\{(1 + X_1^2 + X_2^2)^2[1 + (X_1 \cos\theta + X_2 \sin\theta)^2]^{-1} - \right.$$
$$- 2\frac{X_6^2}{\mu_E}(X_3 \cos\theta + X_4 \sin\theta)[1 - \tfrac{3}{4}(X_1^2 + X_2^2) - \tfrac{3}{2}X_1X_2 \sin 2\theta -$$

$$- \frac{3}{4}(X_1^2 - X_2^2)\cos 2\theta](1 + X_1^2 + X_2^2)^3\Big\} \qquad (4.124)$$

as $[1 + (X_1 \cos\theta + X_2 \sin\theta)^2]^{-3/2}$ is expanded into a series and only two terms are retained for the reason of X_1 and X_2 being small.

The averaged value of the above equation over a θ range of 2π is

$$\frac{\overline{dt}}{d\theta} = \frac{\bar{X}_6^3}{\mu_E^2}\Big\{(1 + \bar{X}_1^2 + \bar{X}_2^2)^{3/2} - \omega_4 + \frac{3}{2}\frac{K_1^2}{(1 - e_S^2)^{3/2}}\sin^2\lambda\cos 2\bar{\alpha} +$$

$$+ \frac{3}{2}\frac{\kappa K_1^2}{(1 - e_S^2)^{3/2}}[\sin^2\lambda\cos 2(\nu + \omega_M + \Omega_M) + \qquad (4.125)$$

$$+ \sin 2\lambda \sin I(\cos\bar{\Omega}_M - \cos(2\bar{\nu} + 2\bar{\omega}_M + \bar{\Omega}_M))](1 + \frac{9}{4}(\bar{X}_1^2 + \bar{X}_2^2))\Big\},$$

where

$$\omega_4 \equiv 3J_2\mu_E^2R_E^2\frac{1}{\bar{\bar{X}}_{60}^4} + K_1^2\frac{(1 + \kappa)}{(1 - e_S^2)^{3/2}}(\frac{3}{2}\sin^2\lambda - 1) = \text{constant}.$$

It should be noted that, in the process of averaging, X_3 and X_4 are replaced by $\bar{X}_3 + \delta X_3$ and $\bar{X}_4 + \delta X_4$, respectively. $\delta X_3 \cos\theta + \delta X_4 \sin\theta$ does bring out the combined secular effect due to δX_3 and δX_4 as was explained in Section 2.5.5., while the terms containing $\bar{X}_1$, $\bar{X}_2$, $\bar{X}_3$, and $\bar{X}_4$ in δX_3 and δX_4, and the products of both X_1 and X_2 with $\bar{X}_3$ and $\bar{X}_4$ are omitted because of their smallness. Furthermore, approximations of $(1 + \bar{X}_2^2 + \bar{X}^2)^3 \cong 1 + 3(\bar{X}_1^2 + \bar{X}_2^2)$ and $\bar{X}_6^2 \cong \bar{X}_{60}^2$ are also made in the second term in the braces of (4.124) since geostationary satellites are near-equatorial.

The integration of (4.125) requires further approximations:

(a) The X_6^3 is approximated as

$$\bar{X}_6^3 \cong \bar{X}_{60}^3\Big\{1 + 3\frac{\bar{X}_{60}^6}{\bar{\bar{X}}_{60}^6}(\tfrac{1}{2}\varepsilon a_0(1 + \kappa)\sin 2\lambda(\cos\Psi - \cos\Psi_0) +$$

$$+ \frac{9}{4}\frac{\bar{X}_{60}^6}{\bar{\bar{X}}_{60}^6}[\tfrac{1}{2} a_0(1 + \kappa)\sin 2\lambda]^2(\cos 2\Psi - \cos 2\Psi_0) + F - F_0\Big\} \qquad (4.105a)$$

by means of a binomial expansion of (4.105) when only two terms are retained since both F and $[\frac{1}{2}\varepsilon a_0(1 + \kappa)\sin 2\lambda]^2$ have their order of magnitude of $10^{-4}(\cong \varepsilon)$.

(b) The first term in the braces of (4.125) is replaced by $1 + \frac{3}{2}(\bar{X}_1^2 + \bar{X}_2^2)$ as a result of a series expansion.

(c) X_1 and X_2 are replaced by their respective first-order solutions

$$\bar{X}_1 = u_1^{(0)} + \varepsilon u_1^{(1)} \quad \text{and} \quad \bar{X}_2 = u_1^{(0)} - D + \varepsilon(u_2^{(1)} + G).$$

With the above approximations, (4.125) is then ready to be integrated to the order of ε as:

$$\bar{t} - \bar{t}_0 = \frac{\bar{X}_{60}^3}{\mu_E^2}(1 - 2\omega_4)\Big\{f_1(\theta, \Psi) + \varepsilon f_2(\theta, \Psi, \bar{\alpha}, \bar{\nu}, \bar{\Omega}_M, \bar{\omega}_M)\Big\}\Big|_{\theta_0}^{\theta}, \qquad (4.126)$$

where $\bar{t} = \bar{t}_0$ at $\theta = \theta_0$ is the given initial condition, and the terms containing e_S and e_M in $u_1^{(1)}$ and $u_2^{(1)}$ are omitted as they are of a higher order of magnitude. The $f_1(\theta, \Psi)$ and $f_2(\theta, \Psi, \bar{\alpha}, \bar{\nu}, \bar{\Omega}_M, \bar{\omega}_M)$ have their expressions as (see Appendix D for constants $E_1, \ldots, E_4$ and $D_1, \ldots, D_{39}$):

$$f_1(\theta, \Psi) = E_1\theta + E_2 \sin \Psi + E_3 \sin 2\Psi + E_4 \sin 3\Psi \ , \tag{4.127}$$

and

$$\begin{aligned} f_2(\theta, \Psi, \bar{\alpha}, \bar{\nu}, \bar{\Omega}_M, \bar{\omega}_M) = \\ & D_0\theta + D_1 \sin \Psi + D_2 \sin 2\Psi + D_3 \sin 3\Psi + D_4 \sin 4\Psi + D_5 \sin 2\bar{\alpha} + \\ & + {D_6 \atop D_7} \sin(2\bar{\alpha} \pm \Psi) + {D_8 \atop D_9} \sin(2\bar{\alpha} \pm 2\Psi) + {D_{10} \atop D_{11}} \sin(2\bar{\alpha} \pm 3\Psi) + \\ & + D_{12} \sin \bar{\Omega}_M + {D_{13} \atop D_{14}} \sin(\bar{\Omega}_M \pm \Psi) + {D_{15} \atop D_{16}} \sin(\bar{\Omega}_M \pm 2\Psi) + \\ & + {D_{17} \atop D_{18}} \sin(\bar{\Omega}_M \pm 3\Psi) + D_{19} \sin(2\bar{\nu} + 2\bar{\omega}_M + \bar{\Omega}_M) + \\ & + D_{20} \sin(2\bar{\nu} + 2\bar{\omega}_M + 2\bar{\Omega}_M) + {D_{21} \atop D_{22}} \sin(2\bar{\nu} + 2\bar{\omega}_M + \bar{\Omega}_M \pm \Psi) + \\ & + {D_{23} \atop D_{24}} \sin(2\bar{\nu} + 2\bar{\omega}_M + \bar{\Omega}_M \pm 2\Psi) + {D_{25} \atop D_{26}} \sin(2\bar{\nu} + 2\bar{\omega}_M + \bar{\Omega}_M \pm 3\Psi) + \\ & + {D_{27} \atop D_{28}} \sin(2\bar{\nu} + 2\bar{\omega}_M + 2\bar{\Omega}_M \pm \Psi) + {D_{29} \atop D_{30}} \sin(2\bar{\nu} + 2\bar{\omega}_M + 2\bar{\Omega}_M \pm 2\Psi) + \\ & + {D_{31} \atop D_{32}} \sin(2\bar{\nu} + 2\bar{\omega}_M + 2\bar{\Omega}_M \pm 3\Psi) + D_{33} \cos(2\bar{\nu} + 2\bar{\omega}_M + \bar{\Omega}_M) + \\ & + {D_{34} \atop D_{35}} \cos(2\bar{\nu} + 2\bar{\omega}_M + \bar{\Omega}_M \pm \Psi) + {D_{36} \atop D_{37}} \cos(2\bar{\nu} + 2\bar{\omega}_M + \bar{\Omega}_M \pm 2\Psi) + \\ & + {D_{38} \atop D_{39}} \cos(2\bar{\nu} + 2\bar{\omega}_M + \bar{\Omega}_M \pm 3\Psi) \ . \end{aligned} \tag{4.128}$$

It should be noted that $f_1(\theta, \Psi)$ is the zeroth-order solution of $\bar{t}$, and $f_1(\theta, \Psi) + \varepsilon f_2(\theta, \Psi, \bar{\alpha}, \bar{\nu}, \bar{\Omega}_M, \bar{\omega}_M)$ is the first-order solution. Additional terms should be added in $f_2(\theta, \Psi, \xi, \bar{\alpha}, \bar{\nu}, \bar{\Omega}_M, \bar{\omega}_M)$ as a result of the long-period terms of solar perturbation on the Moon's motion appearing in $\delta_1\bar{\Omega}_M$ and $\delta_1\bar{\omega}_M$, and these long-period terms should play a role on the terms with low-frequency arguments such as $\bar{\Omega}_M$, $(\bar{\Omega}_M \pm \Psi)$, $(\bar{\Omega}_M \pm 2\Psi)$, and $(\bar{\Omega}_M \pm 3\Psi)$. However, the resulting terms carry a coefficient of m/K_1, which is small compared with the corresponding coefficients $1/K_3$, $1/(K_3 \pm \omega_1)$, etc. in the D's. Consequently, we drop all terms related to $\delta_1\bar{\Omega}_M$ and $\delta_1\bar{\omega}_M$ in $f_2(\theta, \Psi, \xi, \bar{\alpha}, \bar{\nu}, \bar{\Omega}_M, \bar{\omega}_M)$.

Examination of $E_1\theta$ in $f_1(\theta, \Psi)$ of (4.127) shows that the secular effect of the term $(1 - \omega_4)$ can be eliminated if $\bar{X}_{60} \cong \bar{\bar{X}}_{60} \equiv r_c'\dot{\theta}_E^2$ and r_c' is so chosen that $\bar{\bar{X}}_{60}^3(1 - \omega_4)/\mu_E^2 = 1/\dot{\theta}_E$; thus,

$$\frac{\bar{\bar{X}}_{60}^3}{\mu_E^2}(1 - \omega_4) = \frac{1}{\dot{\theta}_E}\left(\frac{r_c'}{r_s}\right)^6 (1 - \omega_4) = \frac{1}{\dot{\theta}_E} \ ,$$

where the relationship $\mu_E = r_s^3\dot{\theta}_E^2$ has been employed. Consequently, the synchronous radius (r_c') can be computed from

$$r_c' = r_s(1 + \tfrac{1}{6}\omega_4) \ . \tag{4.129}$$

Equation (4.129) represents the radius of an equatorial-circular geostationary satellite orbit, which has taken into account the effects of the oblateness of the Earth (J_2), and the Sun and Moon attractions.

When only the oblateness effect (J_2) of the Earth is considered, (4.129) becomes

$$r_c = r_s\left(1 + \frac{1}{2}J_2\frac{R_E^2}{r_s^2}\right) .$$

This expression is in agreement with the derivation shown in Section 3.3.

Equation (4.126) gives the first-order approximate relation between t and θ. The improved first-order approximate solution of t can be obtained by employing the corresponding relationship of (4.113) for t with (4.124) and its averaged value (4.125) as the integrand. Thus

$$\begin{aligned} t &= \bar{t} + \delta t \\ &= \bar{t} + \frac{\bar{X}_6^3}{\mu_E^2}(1 + \bar{X}_1^2 + \bar{X}_2^2)^2\left[\frac{1}{(1 + \bar{X}_1^2 + \bar{X}_2^2)^{1/2}} \times\right. \\ &\quad \times \left\{\tan^{-1}\left[(1 + \bar{X}_1^2 + \bar{X}_2^2)^{1/2}\frac{\bar{X}_2 \sin\theta + \bar{X}_1 \cos\theta}{\bar{X}_2 \cos\theta - \bar{X}_1 \sin\theta}\right] -\right. \\ &\quad \left.- \left[\theta + \tan^{-1}\left(\frac{\bar{X}_1}{\bar{X}_2}\right)\right]\right\} + \frac{2\bar{X}_6^2}{\mu_E} \times \\ &\quad \left.\times \frac{\bar{X}_4 \cos\theta - \bar{X}_3 \sin\theta + (\bar{X}_1\bar{X}_4 - \bar{X}_2\bar{X}_3)(\bar{X}_2 \sin\theta + \bar{X}_1 \cos\theta)}{[1 + (\bar{X}_1 \cos\theta + \bar{X}_2 \sin\theta)^2]^{1/2}}\right], \end{aligned} \tag{4.130}$$

where $\bar{t}$ is the first-order approximation as shown in (4.126), and the second term on the right-hand side depicts the oscillatory nature of t.

4.9. Recapitulation

The derivation of the necessary equations for the problem is now completed, and a recapitulation of the work is in order. The following chart shows what steps have been taken:

Basic Equations of Motion: Equations (4.36), (4.37), (4.38), (4.43), (4.44), (4.49)	→	Equations in H, Q, and S System: Equations (4.40), (4.41), (4.42), (4.50), (4.51), (4.52)	→
First-Order Differential Equations (4.58)-(4.62) and (4.64)-(4.66)	→	Averaged First-Order Differential Equations (4.69)-(4.76)	→
Solutions to Equations (4.69)-(4.76): Equations (4.77), (4.78), (4.79), (4.87), (4.90), (4.91), (4.97)-(4.105)	→	Improved Solutions to Equations (4.69)-(4.76): Equations (4.114)-(4.122)	→

→ Introducing Time Parameter: (4.126) and (4.130) .

Our basic equations relating the X-parameters and the H, Q and S may be written by combining (4.39) and (4.53)-(4.55) as

$$S = \tan\phi = X_1 \cos\theta + X_2 \sin\theta , \tag{4.131}$$

$$Q = \frac{1}{r\cos\phi} = X_5 + X_3 \cos\theta + X_4 \sin\theta , \tag{4.132}$$

and $$H = X_6 = r^2\dot{\theta}\cos^2 , \tag{4.133}$$

where θ and X_1, ..., X_4, X_6 are related by (4.114)-(4.119), in which the values of $\bar{X}_1$, ..., $\bar{X}_4$, $\bar{X}_6$, $\bar{\alpha}$, $\bar{\nu}$, $\bar{\Omega}_M$, and $\bar{\omega}_M$ are related by (4.77)-(4.79), (4.87), (4.90), (4.91), and (4.97)-(4.105). Equations (4.131)-(4.133) are the solutions of (4.36)-(4.38), which show the relations among the three basic geometric parameters, r, θ, and ϕ, and the angular momentum 'H' about the polar axis; the time 't' is tied to the solutions by (4.130).

The integration constants, a_0, Ψ_0, b_0, ξ_0, $\bar{\alpha}_0$, $\bar{\nu}_0$, $\bar{\Omega}_{M0}$, $\bar{\omega}_{M0}$, and $\bar{t}_0$ used in obtaining (4.77)-(4.79), (4.90), (4.91), (4.97)-(4.100), (4.105), and (4.126) are related to the quantities of $\bar{X}_{10}$, ..., $\bar{X}_{40}$, $\bar{X}_{60}$, $\bar{\alpha}_0$, $\bar{\nu}_0$, $\bar{\Omega}_{M0}$, and $\bar{\omega}_{M0}$. These quantities are derived from the given initial conditions r_0, θ_0, ϕ_0, $\dot{r}_0$, $\dot{\theta}_0$, $\dot{\phi}_0$, α_0, ν_0, Ω_{M0}, and ω_{M0}. With these ten given values, the values of X_{10}, ..., X_{40}, X_{60} can be determined from (4.131)-(4.133) and their first derivatives with respect to θ.

As a remark on the practical computation, a substitution of the expressions

$$\left.\frac{dX_1}{d\theta}\right|_0 , \ldots , \left.\frac{dX_4}{d\theta}\right|_0 , \left.\frac{dX_6}{d\theta}\right|_0 , \left.\frac{d\alpha}{d\theta}\right|_0 , \left.\frac{d\nu}{d\theta}\right|_0 , \left.\frac{d\Omega_M}{d\theta}\right|_0 \text{ and } \left.\frac{d\omega_M}{d\theta}\right|_0$$

into the first derivatives of (4.131)-(4.133) results in the same expressions as if X_1, ..., X_4, X_6, α, ν, Ω_M, and ω_M were treated as constants. With known values of X_{10}, ..., X_{40}, X_{60}, α_0^*, ν_0^*, Ω_{M0}^*, and ω_{M0}^*, the constants, $\bar{X}_{10}$, ..., $\bar{X}_{40}$, $\bar{X}_{60}$, $\bar{\alpha}_0$, $\bar{\nu}_0$, $\bar{\Omega}_{M0}$, and $\bar{\omega}_{M0}$ can be calculated from (4.114)-(4.122). Owing to the implicit nature of these equations, approximating $\bar{X}_{10}$, ..., $\bar{X}_{40}$, $\bar{X}_{60}$, $\bar{\alpha}_0$, $\bar{\nu}_0$, $\bar{\Omega}_{M0}$, and $\bar{\omega}_{M0}$ in the oscillatory terms by their respective X_{10}, ..., X_{40}, X_{60}, α_0, ν_0, Ω_{M0}, and ω_{M0} is suggested for the first round of computation , and an iteration may be made if any further refinement is necessary.

The constant $\bar{t}_0$ in (4.126) is calculated from (4.126) and (4.130) as all parameters in these equations are already determined at $\theta = \theta_0$ and $t = 0$.

For actual trajectory calculations, the relationship $r = \bar{r} + \delta r$ and $\phi = \bar{\phi} + \delta\phi$ should be employed, where $\bar{r}$ and $\bar{\phi}$ are obtained by inserting $\bar{X}_1$, $\bar{X}_4$, ..., $\bar{X}_6$, $\bar{\alpha}$, $\bar{\nu}$, $\bar{\Omega}_M$, and $\bar{\omega}_M$ into (4.131) and (4.133), and δr and $\delta\phi$ are derived from the variational quantities of these two equations along $\bar{r}$ and $\bar{\phi}$. The variational values of X_1, ..., X_4, X_6, α, ν, Ω_M, and ω_M are shown by the oscillatory terms in (4.114)-(4.122).

---

*If a date is known, the position of the Sun (α_0), the Moon (ν_0), the line of ascending node (Ω_{M0}), and the argument of perigee (ω_{M0}) of the Moon's orbit can be found in the American Ephemeris and Nautical Almanac published yearly by the U.S. Government Printing Office, Washington, D.C.

4.10. Osculating Orbital Elements

Geometrically, the trajectory of a near-equatorial geostationary satellite under the influences of the Earth oblateness, the Sun, and the Moon is a space curve, which is described by (4.131)-(4.133). However, if the concept of an osculating orbit is introduced (as it was in Section 2.5.5.), the orbital elements can be expressed in terms of the X's (i.e. $\bar{X} + \delta X$) in (4.114)-(4.119) of the previous analysis. The orbit is defined by (see Figure 2.1.)

$$r = \frac{\ell}{1 + e \cos(\eta - \omega)} ,$$

where e is the eccentricity, ℓ is the semilatus rectum, η is the argument of latitude, and ω is the argument of perigee.

After combining (4.39b) and (4.54) and comparing the resulting expression with the above equation, we obtain

$$\begin{aligned}\frac{1}{r} &= \frac{\mu_E}{X_6^2} \frac{1}{1 + X_1^2 + X_2^2}\left\{1 + \frac{X_6^2}{\mu_E}(1 + X_1^2 + X_2^2) \times \right. \\ &\quad \left. \times \frac{X_3 \cos\theta + X_4 \sin\theta}{[1 + (X_1 \cos\theta + X_2 \sin\theta)^2]^{1/2}}\right\} \\ &\cong \frac{\mu_E}{X_6^2} \frac{1}{1 + X_1^2 + X_2^2}\left[1 + \frac{X_6^2}{\mu_E}(X_3 \cos\theta + X_4 \sin\theta)(1 + X_1^2 + X_2^2)\right] \\ &= \frac{1}{\ell}[1 + e \cos(\eta - \omega)] \end{aligned} \tag{4.134}$$

as $[1 + (X_1 \cos\theta + X_2 \sin\theta)^2]^{1/2}$ is approximated by 1.

Since δX_3 and δX_4 depict the respective oscillatory nature of X_3 and X_4 within an orbit, the resulting expression of $\delta X_3 \cos\theta + \delta X_4 \sin\theta$ in (4.134) will give terms of constant, secular (containing only X's), long-period (containing combinations of X's and trigonometric terms with $\bar{\alpha}$, $\bar{\nu}$, $\bar{\Omega}_M$, and $\bar{\omega}_M$ as their arguments), and short-period (containing combinations of X's and trigonometric terms with $\bar{\alpha}$, $\bar{\nu}$, $\bar{\Omega}_M$, $\bar{\omega}_M$, θ, 2θ, 3θ, and 4θ as their arguments) in nature. In the short-period terms, the ones containing $\sin\theta$ and $\cos\theta$ are particularly sorted out for defining $\delta\bar{X}_3'$ and $\delta\bar{X}_4'$ later in (4.139) relating to $e \cos(\eta - \omega)$, since both θ and η have a single orbital period of 2π. The remaining terms of $\delta X_3 \cos\theta + \delta X_4 \sin\theta$ are then defined as a part of the semilatus rectum. As a result of various terms described above, the semilatus rectum (ℓ) will have the corresponding secular (ℓ_S), long-period (ℓ_L), and short-period (ℓ_P) components. Therefore,

$$\ell \equiv \bar{\ell} + \ell_P \equiv \ell_s + \ell_L + \ell_P , \tag{4.135}$$

$$\begin{aligned}\ell_s &\equiv \frac{\bar{X}_6^2}{\mu_E}(1 + \bar{X}_1^2 + \bar{X}_2^2)\left[1 - \frac{3}{2}J_2 \frac{R_E^2}{r_c^2} - \frac{1}{4} \frac{1 + \kappa}{(1 - e_S^2)^{3/2}} \times \right. \\ &\quad \left. \times K_1^2(1 - 3\cos^2\lambda + \tfrac{3}{2}\bar{X}_2 \sin 2\lambda)\right] ,\end{aligned} \tag{4.136}$$

$$\begin{aligned}\ell_L &\equiv \frac{\bar{X}_6^2}{\mu_E}(1 + \bar{X}_1^2 + \bar{X}_2^2)\left\{- \frac{3}{4} \frac{1}{(1 - e_S^2)^{3/2}} K_1^2 \sin^2\lambda \sin 2\bar{\alpha} \times \right. \\ &\quad \times \left(1 + e_S \cos(- \omega_S + \bar{\alpha})\right) -\end{aligned}$$

$$
\begin{aligned}
& - \frac{1}{4}\frac{1+\kappa}{(1-e_S^2)^{3/2}} K_1^2 (1 - 3\cos^2\lambda) e_S \cos(-\omega_S + \bar\alpha) - \\
& - \frac{3}{4}\frac{K_1^2}{(1-e_S^2)^{3/2}}[(\bar X_1 \sin\lambda + 2\bar X_4 r_c \cos\lambda)\sin 2\bar\alpha - \\
& - \frac{1}{2}\bar X_2 \sin 2\lambda \cos 2\bar\alpha] - \frac{3}{4}\frac{K_2^2}{\sigma(1-e_M^2)^{3/2}}[(1 + e_M\cos\bar\nu) \times \\
& \times \big(-\sin^2\lambda\cos 2(\bar\nu + \bar\omega_M + \bar\Omega_M) - 2\sin 2\lambda \sin I \sin(\bar\nu + \bar\omega_M)\times \\
& \times \sin((\bar\nu + \bar\omega_M + \bar\Omega_M)\big) + \bar X_1 \sin\lambda \sin 2(\bar\nu + \bar\omega_M + \bar\Omega_M) - \\
& - \frac{1}{2}\bar X_2 \sin 2\lambda \cos 2(\bar\nu + \bar\omega_M + \bar\Omega_M) - 2\sin I \sin(\bar\nu + \bar\omega_M) \times \\
& \times \big(X_1 \cos\lambda \cos(\bar\nu + \bar\omega_M + \bar\Omega_M) + \frac{5}{3}\bar X_2 \cos 2\lambda \sin(\bar\nu + \bar\omega_M + \bar\Omega_M)\big)]\Big\}
\end{aligned}
\tag{4.137}
$$

$$
\begin{aligned}
\ell_P \equiv\ & \frac{\bar X_6^2}{\mu_E}(1 + \bar X_1^2 + \bar X_2^2)\Big\{\frac{-K_1^2}{(1-e_S^2)^{3/2}}\Big[(1 + e_S\cos(-\omega_S + \bar\alpha) \times \\
& \times [\big(\frac{5}{4}\sin^2\lambda + \frac{5}{4}(1 + \cos^2\lambda)\cos 2\bar\alpha\big)\cos 2\theta + \\
& + \cos\lambda \sin 2\bar\alpha \sin 2\theta] + \\
& + \frac{5}{8}[-2\bar X_1 \sin\lambda \sin 2\bar\alpha + \bar X_2 \sin 2\lambda(1 - \cos 2\bar\alpha)]\cos 2\theta - \\
& - \frac{5}{8}\bar X_1 \sin 2\lambda(1 - \cos 2\bar\alpha)\sin 2\theta\Big] - \frac{K_2^2}{\sigma(1-e_M^2)^{3/2}} \times \\
& \times \Big[(1 + e_M\cos\bar\nu)\Big([\frac{5}{4}\sin^2\lambda + \frac{5}{4}(1 + \cos^2\lambda)\cos 2(\bar\nu + \bar\omega_M + \bar\Omega_M) - \\
& - \frac{5}{2}\sin 2\lambda \sin I \sin(\bar\nu + \bar\omega_M)\sin(\bar\nu + \bar\omega_M + \bar\Omega_M)]\cos 2\theta + \\
& + [\frac{5}{2}\cos\lambda \sin 2(\bar\nu + \bar\omega_M + \bar\Omega_M) + \frac{5}{2}\bar X_2 \cos\lambda \sin I \sin(\bar\nu + \bar\omega_M)\times \\
& \times \cos(\bar\nu + \bar\omega_M + \bar\Omega_M)]\sin 2\theta\Big) + \cos 2\theta\Big(\frac{5}{8}\bar X_2 \sin 2\lambda \times \\
& \times [1 - \cos 2(\bar\nu + \bar\omega_M + \bar\Omega_M)] - \frac{5}{4}\bar X_1 \sin\lambda \sin 2(\bar\nu + \bar\omega_M + \bar\Omega_M)\Big) + \\
& + \sin 2\theta\Big(\frac{5}{8}\bar X_1 \sin 2\lambda[-1 + \cos 2(\bar\nu + \bar\omega_M + \bar\Omega_M)] - \\
& - \frac{5}{4}\bar X_2 \sin\lambda \sin 2(\bar\nu + \bar\omega_M + \bar\Omega_M)\Big) + \frac{1}{2}\sin I \sin(\bar\nu + \bar\omega_M) \times \\
& \times \sin(\bar\nu + \bar\omega_M + \bar\Omega_M)\cos 2\lambda\Big(\bar X_1(\sin 4\theta + 7\sin 2\theta) - \\
& - \bar X_2(\cos 4\theta + \cos 2\theta)\Big)\Big]\Big\} + \\
& + \frac{21}{8}\frac{K_1^2}{(1-e_S^2)^{3/2}}[(-\cos^2\bar\alpha + \cos^2\lambda \sin^2\bar\alpha)(\bar X_3 \cos 3\theta + \\
& + \bar X_4 \sin 3\theta) - \cos\lambda \sin 2\bar\alpha(\bar X_3 \cos 3\theta - \bar X_4 \sin 3\theta)] + \\
& + \frac{21}{8}\frac{K_2^2}{\sigma(1-e_M^2)^{3/2}}\Big\{[-\cos^2(\bar\nu + \bar\omega_M + \bar\Omega_M) + \\
& + \sin^2(\bar\nu + \bar\omega_M + \bar\Omega_M)\cos^2\lambda + \sin(\bar\nu + \bar\omega_M + \bar\Omega_M)\sin(\bar\nu + \bar\omega_M) \times \\
& \times \sin I \sin 2\lambda](\bar X_3 \cos 3\theta + \bar X_4 \sin 3\theta) + 2\cos(\bar\nu + \bar\omega_M + \bar\Omega_M)\times \\
& \times [\sin(\bar\nu + \bar\omega_M + \bar\Omega_M)\cos\lambda + \sin(\bar\nu + \bar\omega_M)\sin I \sin\lambda] \times \\
& \times (\bar X_4 \cos 3\theta - \bar X_3 \sin 3\theta)\Big\} +
\end{aligned}
$$

$$+ \frac{15}{64} \frac{K_2^2}{\sigma a_M (1 - e_M^2)^{5/2}} (1 + \tfrac{1}{2} e_M^2 + 2e_M \cos \bar{\nu}) \times$$
$$\times \Big\{ [+ \cos^3 (\bar{\nu} + \bar{\omega}_M + \bar{\Omega}_M) - \tfrac{3}{2} \cos^2 \lambda \sin 2(\bar{\nu} + \bar{\omega}_M + \bar{\Omega}_M) \times$$
$$\times \sin(\bar{\nu} + \bar{\omega}_M + \bar{\Omega}_M) - \tfrac{3}{2} \sin 2\lambda \sin 2(\bar{\nu} + \bar{\omega}_M + \bar{\Omega}_M) \times$$
$$\times \sin(\bar{\nu} + \bar{\omega}_M) \sin I] \cos 3\theta + [\tfrac{3}{2} \cos \lambda \sin 2(\bar{\nu} + \bar{\omega}_M + \bar{\Omega}_M) \times$$
$$\times \cos(\bar{\nu} + \bar{\omega}_M + \bar{\Omega}_M) - \tfrac{3}{2} \sin \lambda \sin 2(\bar{\nu} + \bar{\omega}_M + \bar{\Omega}_M) \times$$
$$\times \cos(\bar{\nu} + \bar{\omega}_M + \bar{\Omega}_M) \sin I - \cos^3 \lambda \sin^3 (\bar{\nu} + \bar{\omega}_M + \bar{\Omega}_M) -$$
$$- \tfrac{1}{2} \cos \lambda \sin 2\lambda \sin^2 (\bar{\nu} + \bar{\omega}_M + \bar{\Omega}_M) \sin(\bar{\nu} + \bar{\omega}_M) \sin I] \sin 3\theta \Big\} +$$
$$+ \frac{2\bar{X}_6}{\mu_E} (1 + \bar{X}_1^2 + \bar{X}_2^2) \delta X_6 + \frac{2\bar{X}_6^2}{\mu_E} (\bar{X}_1 \delta X_1 + \bar{X}_2 \delta X_2) , \tag{4.138}$$

and

$$e \cos(\eta - \omega) = \ell (\bar{X}_3' \cos \theta + \bar{X}_4' \sin \theta) , \tag{4.139}$$

where

$$\bar{X}_3' \equiv \bar{X}_3 + \delta \bar{X}_3'$$
$$= \bar{X}_3 + \frac{3}{4} \left[J_2 \frac{R_E^2}{r_c^2} - \frac{1}{(1 - e_S^2)^{3/2}} K_1^2 (1 + \kappa - 3 \cos^2 \lambda \sin^2 \bar{\alpha}) \right] \bar{X}_3 -$$
$$- \frac{9}{8} \frac{1}{(1 - e_S^2)^{3/2}} K_1^2 \cos \lambda \sin 2\bar{\alpha}\, \bar{X}_4 + \frac{9}{4} \frac{1}{\sigma (1 - e_M^2)^{3/2}} K_2^2 \times$$
$$\times \Big\{ [\sin^2 (\bar{\nu} + \bar{\omega}_M + \bar{\Omega}_M) \cos^2 \lambda + \sin I \sin 2\lambda \sin(\bar{\nu} + \bar{\omega}_M + \bar{\Omega}_M) \times$$
$$\times \sin(\bar{\nu} + \bar{\omega}_M)] \bar{X}_3 - [\tfrac{1}{2} \sin 2(\bar{\nu} + \bar{\omega}_M + \bar{\Omega}_M) \cos \lambda +$$
$$+ \sin I \sin \lambda \cos(\bar{\nu} + \bar{\omega}_M + \bar{\Omega}_M) \sin(\bar{\nu} + \bar{\omega}_M)] \bar{X}_4 \Big\} +$$
$$+ \frac{9}{8} \frac{K_2^2}{\sigma a_M (1 - e_M^2)^{5/2}} (1 + \tfrac{1}{2} e_M^2 + 2e_M \cos \bar{\nu}) \times$$
$$\times [- \tfrac{15}{8} \cos^3 (\bar{\nu} + \bar{\omega}_M + \bar{\Omega}_M) - \tfrac{5}{8} \sin(\bar{\nu} + \bar{\omega}_M + \bar{\Omega}_M) \times$$
$$\times \sin 2(\bar{\nu} + \bar{\omega}_M + \bar{\Omega}_M) \cos^2 \lambda - \tfrac{5}{8} \sin 2(\bar{\nu} + \bar{\omega}_M + \bar{\Omega}_M) \times$$
$$\times \sin(\bar{\nu} + \bar{\omega}_M) \sin 2\lambda \sin I + \cos(\bar{\nu} + \bar{\omega}_M + \bar{\Omega}_M)] \tag{4.140}$$

and

$$\bar{X}_4' \equiv \bar{X}_4 + \delta \bar{X}_4'$$
$$= \bar{X}_4 + \frac{3}{4} \left[J_2 \frac{R_E^2}{r_c^2} - \frac{1}{(1 - e_S^2)^{3/2}} K_1^2 (1 + \kappa - 3 \cos^2 \bar{\alpha}) \right] \bar{X}_4 -$$
$$- \frac{9}{8} \frac{1}{(1 - e_S^2)^{3/2}} K_1^2 \cos \lambda \sin 2\bar{\alpha}\, \bar{X}_3 + \frac{9}{4} \frac{1}{\sigma (1 - e_M^2)^{3/2}} K_2^2 \times$$
$$\times \Big\{ [- \tfrac{1}{2} \cos \lambda \sin 2(\bar{\nu} + \bar{\omega}_M + \bar{\Omega}_M) -$$
$$- \sin I \sin \lambda \sin(\bar{\nu} + \bar{\omega}_M + \bar{\Omega}_M) \sin(\bar{\nu} + \bar{\omega}_M)] \bar{X}_3 +$$
$$+ \cos^2 (\bar{\nu} + \bar{\omega}_M + \bar{\Omega}_M) \bar{X}_4 \Big\} + \frac{9}{8} \frac{K_2^2}{\sigma a_M (1 - e_M^2)^{5/2}} (1 + \tfrac{1}{2} e_M^2 + 2e_M \cos \bar{\nu}) \times$$
$$\times [- \tfrac{15}{8} \cos \lambda \sin 2(\bar{\nu} + \bar{\omega}_M + \bar{\Omega}_M) \cos(\bar{\nu} + \bar{\omega}_M + \bar{\Omega}_M) -$$

$$- \frac{5}{4} \sin \lambda \sin I \cos^2(\bar{\nu} + \bar{\omega}_M + \bar{\Omega}_M)\sin(\bar{\nu} + \bar{\omega}_M) -$$
$$- \frac{5}{4} \cos^3 \lambda \sin^3(\bar{\nu} + \bar{\omega}_M + \bar{\Omega}_M) - \frac{15}{8} \cos \lambda \sin 2\lambda \sin^2(\bar{\nu} + \bar{\omega}_M +$$
$$+ \bar{\Omega}_M)\sin(\bar{\nu} + \bar{\omega}_M)\sin I + \cos \lambda \sin(\bar{\nu} + \bar{\omega}_M + \bar{\Omega}_M) +$$
$$+ \sin \lambda \sin I \sin(\bar{\nu} + \bar{\omega}_M)] \quad . \tag{4.141}$$

In order to observe some physical aspects of the problem from our previous analytical results, we only consider the zeroth-order solutions of X's to examine the characteristics of various orbital elements. However, in the computation of the first-order approximation of the orbital elements, the full expression of X's in (4.87), (4.99), (4.100), and (4.105) should be used. We shall first deal with the simple case that the Moon lies in the ecliptic plane. After exploring the simple case, the case involving the movement of the Moon's orbit will be examined in detail in Section 4.10.6.

The zeroth-order expressions for the X's are

$$\left.\begin{aligned}
\bar{X}_1 &\cong \omega_1 a_0 \sin \Psi , \\
\bar{X}_2 &= \omega_{11} a_0 \cos \Psi - D , \\
\bar{X}_3 &\cong \omega_{31} b_0 \cos \xi , \\
\bar{X}_4 &\cong \omega_3 b_0 \sin \xi , \\
\text{and}\quad & \\
\bar{X}_6 &\cong \bar{X}_{60}\left[1 + \frac{\bar{X}_{60}^6}{\bar{\bar{X}}_{60}^6} \tfrac{1}{2}\varepsilon a_0 (1 + \kappa)\sin 2\lambda(\cos \Psi - \cos \Psi_0)\right] \\
&\cong \bar{\bar{X}}_{60}[1 + \tfrac{1}{2}\varepsilon a_0 (1 + \kappa)\sin 2\lambda(\cos \Psi - \cos \Psi_0)]
\end{aligned}\right\} \tag{4.142}$$

as $\bar{X}_{60} \cong \bar{\bar{X}}_{60}$.

Consequently, the zeroth-order approximation for $\bar{t}$ is

$$\bar{t} - \bar{t}_0 = \frac{\bar{X}_{60}^3}{\mu_E^2}(1 - 2\omega_4)(E_1\theta + E_2 \sin \Psi + E_3 \sin 2\Psi + E_4 \sin 3\Psi)\Big|_{\theta_0}^{\theta} \tag{4.143}$$

as is obtained from (4.126).

For a satellite initially in the equatorial plane of the Earth, with $\theta_0 = 0$, it is found that

$$a_0 = D/\omega_{11} \quad \text{and} \quad \Psi_0 = 0 \tag{4.144}$$

to the zeroth-order approximation, and furthermore, if the satellite is at the synchronous altitude (r_c), then

$$\left.\begin{aligned}
\omega_1 &= 0.5064 \times 10^{-4}, & \omega_3 &= 0.5023 \times 10^{-4}, \\
\omega_{11} &= 0.5206 \times 10^{-4}, & \omega_{31} &= 0.4366 \times 10^{-4}, \\
\omega_{12} &= 0.4925 \times 10^{-4}, & \omega_{32} &= 0.5779 \times 10^{-4}, \\
\varepsilon &= 0.0559 \times 10^{-4}, & \omega_4 &= 0.5603 \times 10^{-4}, \\
D &= -0.1313, & \text{and}\quad \kappa &= 2.1678.
\end{aligned}\right\} \tag{4.145}$$

These constants are computed from their definitions described in the previous sections.

As is seen from above,

$$\omega_1 \cong \omega_{11} \cong \omega_{12} \; . \tag{4.146}$$

Equations (4.142)-(4.146) will be used in the following derivations of various orbital elements and their physical interpretations.

4.10.1. Time (t)

As a result of (4.144) and (4.146), (4.143) may be simplified as

$$\bar{t} = \frac{1}{\dot{\theta}_E}\left\{\theta + \frac{3D^2}{\omega_1(1-\omega_4)}[\omega_4 + 6D^2(1 - 2\omega_4)](\sin\Psi - \omega_1\theta) + \right.$$
$$\left. + \frac{9}{4}\frac{D^4}{\omega_1(1-\omega_4)}(1 - 2\omega_4)(2\omega_1\theta - \sin 2\Psi)\right\} , \tag{4.147}$$

where the initial condition $\bar{t}_0 = 0$, $\theta_0 = 0$ has been employed. When $\omega_1\theta$ is small, (4.147) gives

$$\bar{t} = \theta/\dot{\theta}_E \quad (\dot{\theta}_E = 2\pi/\text{sidereal day}) \tag{4.148}$$

as is expected. For long time durations (i.e., $\sin\Psi \neq \omega_1\theta$ and $\sin 2\Psi \neq 2\omega_1\theta$), the synchronism cannot be maintained.

4.10.2. Longitude of Ascending Node (Ω)

The line of nodes of the osculating orbit can be determined by setting $\phi = 0$ in (4.131). This results in

$$\theta = \Omega = \tan^{-1}\left(-\frac{X_1}{X_2}\right) , \tag{4.149}$$

which, in turn, gives

$$\Omega = \bar{\Omega} + \delta\Omega$$
$$= \tan^{-1}\left(-\frac{\bar{X}_1}{\bar{X}_2}\right) + \frac{\bar{X}_1\delta X_1 - \bar{X}_2\delta X_2}{\bar{X}_1^2 + \bar{X}_2^2} . \tag{4.150}$$

With (4.142), (4.144), and (4.146), the zeroth-order approximation for Ω is

$$\bar{\Omega} = \tan^{-1}\left(\frac{-\sin\Psi}{\cos\Psi - 1}\right) = \tan^{-1}(\cot\frac{\Psi}{2})$$
$$= \frac{\pi}{2} - \frac{\Psi}{2} . \tag{4.151}$$

As $\Psi = \omega_1\theta + \Psi_0$, which shows that the average rate of the regression of the line of nodes over a 2π-period of θ is $\frac{1}{2}\omega_1$ ($= 1.595 \times 10^{-4}$ radian/sidereal day), which is less than that due to the Earth oblateness alone (2.393×10^{-4} radian/sidereal day). It is seen then that the attractions of the Sun and Moon reduce the rate of nodal regression (due to Earth oblateness alone) by approximately one third.

4.10.3. Inclination of Orbital Plane (i)

The inclination of the osculating orbit has been derived in (2.117) and (2.118) as

$$\tan i = (X_1^2 + X_2^2)^{1/2} \tag{4.152}$$

and

$$i = \bar{i} + \delta i$$
$$= \tan^{-1}[(\bar{X}_1^2 + \bar{X}_2^2)^{1/2}] + \frac{\bar{X}_1\delta X_1 + \bar{X}_2\delta X_2}{(\bar{X}_1^2 + \bar{X}_2^2)^{1/2}} \ . \tag{4.153}$$

Substituting (4.142), (4.144), and (4.146) into (4.152) yields*

$$\tan \bar{i} = (\bar{X}_1^2 + \bar{X}_2^2)^{1/2}$$
$$= 2D \sin \frac{\Psi}{2} \tag{4.154}$$

as the zeroth-order approximation of the orbital inclination. Consequently, the average rate of change of the inclination is

$$\frac{d\bar{i}}{d\theta} = D\omega_1 \cos \frac{\Psi}{2}(1 - 2D^2 \sin^2 \frac{\Psi}{2}) \tag{4.155}$$

when (4.154) is differentiated with respect to θ. From (4.145), (4.147), and (4.155), it can be seen that when $\Psi = \pi$, then $\bar{t}$ = 26.85 years, and $\bar{i}$ has its maximum of 14 degrees 41.2 minutes. When $\bar{i} = 0$ and $\Psi = 0$, $d\bar{i}/d\theta = D\omega_1$, which gives the initial rate of change of orbital inclination as 0.853 degrees per year. It should be noted that the maximum value of $\bar{i}$ and the initial rate of change of $\bar{i}$ described above are for an orbit lying initially in the Earth equatorial plane. The ambiguity of $\bar{\Omega}_0$ for an initial zero-inclination orbit needs some explanation. $\bar{\Omega}_0$ is indeterminate when (4.149) is employed for $\bar{X}_1 = 0$ and $\bar{X}_2 = 0$. An application of the L'Hospital rule in the limiting process to (4.149) results in $\bar{\Omega}_0 = 270°$. In other words, the zero-inclination orbit plane begins to rotate at a line located at $\bar{\Omega}_0 = 270°$, while the ascending node regresses from the same line.

If the orbit has its initial orientation ($\bar{i}_0$ and $\bar{\Omega}_0$) other than $\bar{i}_0 = 0$, the maximum inclination and the initial rate of change of inclination will be different. In the following, we shall study the rotation of the orbital plane as affected by its initial orientation.

Since the principal effect of the Sun and the Moon is to produce a rotation and regression of the orbital plane, a generalized study of this rotation and regression can be achieved by employing (4.131), (4.144), (4.149), and the first two equations of (4.142). If the initial orientation of the orbital plane has an inclination, $\bar{i}_0$, and a longitude of ascending node, $\bar{\Omega}_0$, then

$$\tan \bar{\Omega}_0 = - \frac{\bar{X}_{10}}{\bar{X}_{20}} \ , \tag{4.156}$$

$$\tan \bar{i}_0 = \tan \phi_0 = \bar{X}_{10} \cos(\bar{\Omega}_0 + 90°) + \bar{X}_{20} \sin(\bar{\Omega}_0 + 90°) \ , \tag{4.157}$$

$$\bar{X}_{10} = a_0\omega_1 \sin \Psi_0 \ , \tag{4.158}$$

and
$$\bar{X}_{20} = a_0\omega_{11} \cos \Psi_0 - D \ . \tag{4.159}$$

After obtaining a_0 and Ψ_0 from the above equations, we have

$$\bar{X}_1 = a_0\omega_1 \sin[\omega_1(\theta - \bar{\Omega}_0 - 90°) + \Psi_0] \ , \tag{4.160}$$

*It is understood that the resulting orbital plane lies between the ecliptic and the equatorial plane of the Earth.

$$\bar{X}_2 = a_0\omega_{11}\cos[\omega_1(\theta - \bar{\Omega}_0 - 90°) + \Psi_0] - D\ , \tag{4.161}$$

and

$$\tan^2 \bar{i} = \bar{X}_1^2 + \bar{X}_2^2$$
$$= \left\{a_0\ \omega_1 \sin[\omega_1(\theta - \bar{\Omega}_0 - 90°) + \Psi_0]\right\}^2 +$$
$$+ \left\{a_0\omega_{11}\cos[\omega_1(\theta - \bar{\Omega}_0 - 90°) + \Psi_0] - D\right\}^2 . \tag{4.162}$$

The above equations together with

$$\bar{t} - \bar{t}_0 = \frac{\bar{X}_{60}^3}{\mu^2}(1 - 2\omega_4)(E_1\theta + E_2 \sin\Psi + E_3 \sin 2\Psi + + E_4 \sin 3\Psi)\Big|_{\bar{\Omega}_0 + 90°}^{\theta} \tag{4.163}$$

and

$$\Psi = \omega_1(\theta - \bar{\Omega}_0 - 90°) + \Psi_0 \tag{4.164}$$

yield the rotation of the orbital plane for an arbitrary initial orientation (both inclination, $\bar{i}_0$, and the location of the line of ascending node, $\bar{\Omega}_0$) as shown in Figures 4.2 and 4.3.

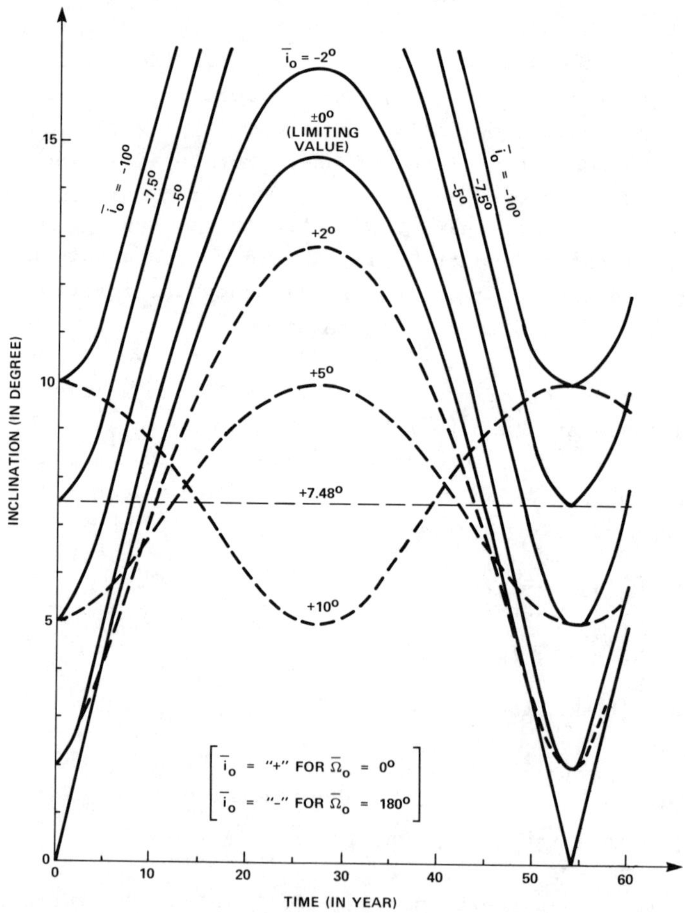

Fig. 4.2. Drift of the geostationary orbital plane for $\bar{\Omega}_0$ = 0° and $\bar{\Omega}_0$ = 180°.

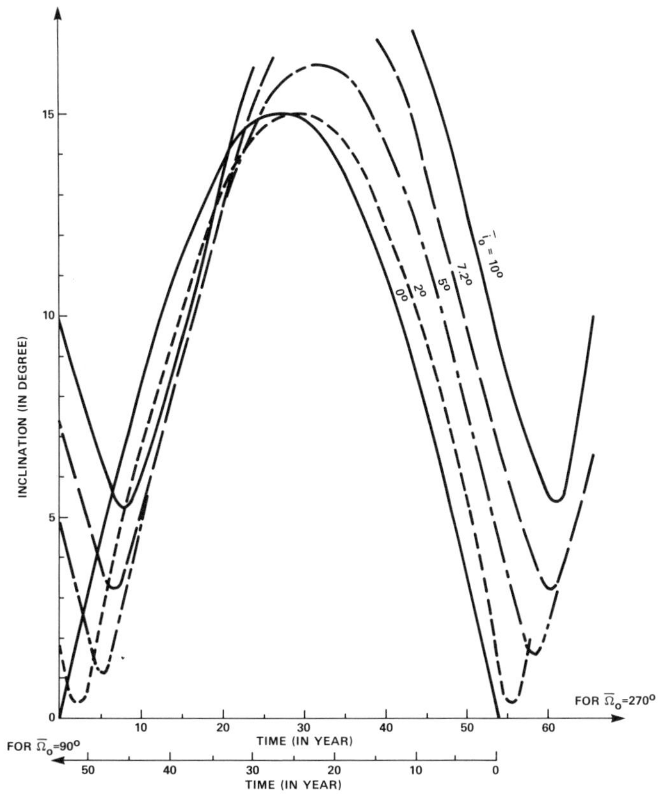

Fig. 4.3. Drift of the geostationary orbital plane for $\bar{\Omega}_0 = 90°$ and $\bar{\Omega}_0 = 270°$.

It can be seen that for a passive satellite, the drift of the orbital plane can be minimized by placing the satellite in a near-equatorial orbit with $\bar{\Omega}_0$ in the neighborhood of 270°. The orbital plane initially drifts toward and then moves away from the equatorial plane of the Earth; consequently, the satellite can remain in the neighborhood of the equatorial plane for a long period of time. This characteristic has its practical importance in the station-keeping of a geostationary satellite. We shall further explore this characteristic.

The time required for the drift of the orbital plane from an initial inclination ($\bar{i}_0$) (with $\bar{\Omega}_0$ being in the neighborhood of 270°) toward its minimum value ($\bar{i}_m = 0$) can be obtained by differentiating (4.162) with respect to θ and setting $d\bar{i}/d\theta = 0$. Consequently,

$$\Psi = \omega_1(\theta - \bar{\Omega}_0 - 90°) + \Psi_0 = 0 \quad \text{or}$$

$$\theta = \bar{\Omega}_0 + 90° - \frac{1}{\omega_1}\Psi_0 \equiv \theta_m ,$$

and

$$\tan i_m = a_0\omega_{11} - D = 0,$$

where θ_m is the value of θ (hence, t) when $\bar{i}_0$ reaches $\bar{i}_m$ (= 0). From the above two equations, it can be seen that θ_m is a function of $\bar{\Omega}_0$ and Ψ_0 (or $\bar{i}_0$) when $\bar{i}_m$ reaches 0. In order to maintain the orbital plane within the bound of $\bar{i}_0$ for a long duration of time, we shall try to find an $\bar{\Omega}_0$ to maximize θ_m or to minimize Ψ_0. The expression of Ψ_0 in terms of $\bar{i}_0$ and $\bar{\Omega}_0$ can be obtained by substituting both (4.158) and (4.159) into (4.157) as

$$\tan \Psi_0 = - \frac{\omega_1}{\omega_{11}} \frac{\sin \bar{\Omega}_0 \tan \bar{i}_0}{D + \cos \bar{\Omega}_0 \tan \bar{i}_0} . \tag{4.165}$$

By differentiating (4.165) with respect to $\bar{\Omega}_0$ and setting $\partial\Psi_0/\partial\bar{\Omega}_0 = 0$, we obtain

$$D \cos \bar{\Omega}_0 + \tan \bar{i}_0 = 0 . \tag{4.166}$$

This equation gives the $\bar{\Omega}_0$ which minimizes Ψ_0 in (4.165). In turn, the maximized θ_m and the corresponding maximum duration of time (from 4.163) can be found. Figure 4.4. gives the maximum duration of time for orbits drifting toward the equatorial plane with initial inclinations of $\frac{1}{2}°$, 1°, $1\frac{1}{2}°$, 2°, and $2\frac{1}{2}°$ together with their corresponding $\bar{\Omega}_0$'s.

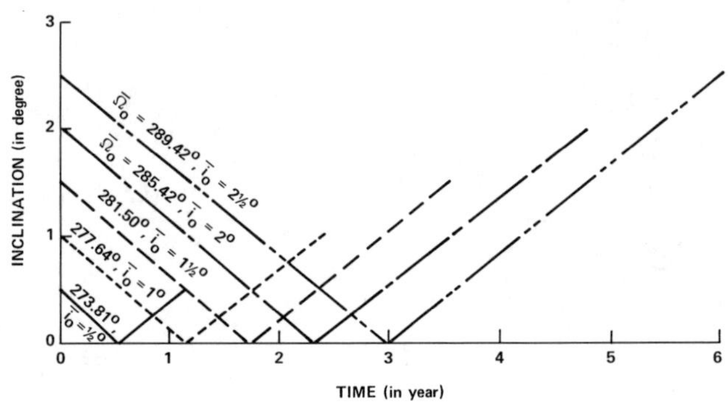

Fig. 4.4. Maximum drifting duration of the geostationary orbital plane toward the equatorial plane of the Earth.

4.10.4. Semilatus Rectum (ℓ) and Semimajor Axis (a)

The semilatus rectum has been derived in (4.135)-(4.138), and the secular part of ℓ can be considered as the zeroth-order approximation. With the $\bar{X}$'s shown in (4.142), we have

$$\begin{aligned} \ell_s &= \frac{\bar{\bar{X}}_{60}^2}{\mu_E}[1 + \tfrac{1}{2}\varepsilon a_0(1 + \kappa)\sin 2\lambda(\cos \Psi - 1)][1 + D^2 \sin^2 \Psi + \\ &\quad + D^2(\cos \Psi - 1)^2][1 - \tfrac{3}{2}J_2 \frac{R_E^2}{r_c^2} - \frac{1}{4} \frac{K_1^2(1 + \kappa)}{(1 - e_S^2)^{3/2}}\big(1 - 3 \cos^2 \lambda + \\ &\quad + \tfrac{3}{2} D \sin 2\lambda(\cos \Psi - 1)\big)] \\ &= \frac{\bar{\bar{X}}_{60}^2}{\mu_E}\big(1 - \tfrac{3}{2}J_2 \frac{R_E^2}{r_c^2} - \frac{1}{4} \frac{K_1^2(1 + \kappa)}{(1 - e_S^2)^{3/2}}[1 - 3 \cos^2 \lambda + \\ &\quad + \tfrac{3}{2} D \sin 2\lambda(\cos \Psi - 1)]\big) \times \end{aligned}$$

$$\times [1 - \varepsilon a_0(1 + \kappa)\sin 2\lambda \ \sin^2 \frac{\Psi}{2}][1 + 4D_2 \ \sin^2 \frac{\Psi}{2}]$$

$$\cong \frac{\bar{\bar{X}}_{60}^2}{\mu_E}\left(1 - \frac{3}{2}J_2 \frac{R_E^2}{r_c^2} - \frac{1}{4} \frac{K_1^2(1 + \kappa)}{(1 - e_S^2)^{3/2}} [1 - 3 \cos^2 \lambda + \right.$$
$$\left. + \frac{3}{2} D \sin 2\lambda(\cos \Psi - 1)]\right)\left(1 + [- \varepsilon a_0(1 + \kappa) \times \right.$$
$$\left. \times \sin 2\lambda + 4D^2]\sin^2 \frac{\Psi}{2}\right) . \tag{4.167}$$

As the term $\frac{3}{8}\left((1 + \kappa)/(1 - e_S^2)^{3/2}\right) K_1^2 D \sin 2\lambda(\cos \Psi - 1)$ in the first brackets is of the order of magnitude of 10^{-4}, and the term $4D^2 - \varepsilon a_0(1 + \kappa)\sin 2\lambda$ in the second brackets is greater than zero and has the order of magnitude of 10^{-2}, (4.167) shows that the semilatus rectum reaches its maximum value when $\Psi = \pi$. As is seen from the last section, when $i = 0$ and $\Psi = \pi$, $\bar{t} =$ 26.84 years and $\bar{i}$ has its maximum value; this fact gives a general notion that the semilatus rectum increases as the inclination of the orbital plane increases.

From the general definition $a \equiv \ell/(1 - e^2)$, since the eccentricity of a geostationary satellite orbit is very small, 'a' can be approximated by ℓ, and the behavior of the semimajor axis, a, can be inferred from (4.167).

4.10.5. Eccentricity (e) and Argument of Perigee (ω)

The eccentricity and the argument of perigee of an osculating orbit in terms of X's have been derived as shown in (2.121) and (2.122). They are

$$\tan \omega = \frac{X_1\bar{X}_3' + X_2\bar{X}_4'}{X_2\bar{X}_3' - X_1\bar{X}_4'} \frac{1}{(1 + X_1^2 + X_2^2)^{1/2}} \tag{4.168}$$

and

$$e = [\bar{X}_3'^2 + \bar{X}_4'^2 + (X_2\bar{X}_3' - X_1\bar{X}_4')^2]^{1/2} \ \ell . \tag{4.169}$$

The average values of e and ω (i.e. $\bar{e}$ and $\bar{\omega}$) are obtained by inserting $\bar{X}$'s into (4.168) and (4.169) with ℓ replaced by $\bar{\ell}$, and the oscillatory natures of e and ω over a θ-cycle of 2π can be brought out by applying the variational technique to (4.168) and (4.169) in terms of $\bar{X}$'s, ℓ, δX's, and $\delta\ell$. The complete variational equations can be easily written out when computations need to be performed.

Equation (4.168) gives the zeroth-order approximation for ω as

$$\tan \bar{\omega} = \frac{\omega_1\omega_{31}a_0 \ \sin \Psi \cos \xi + \omega_3(\omega_{11}a_0 \ \cos \Psi - D)\sin \xi}{\omega_{31}(\omega_{11}a_0 \ \cos \Psi - D)\cos \xi - \omega_1\omega_3a_0 \ \sin \Psi \sin \xi} \times$$
$$\times \frac{1}{(1 + 4D^2 \ \sin^2 \frac{\Psi}{2})^{1/2}} , \tag{4.170}$$

where (4.142) and (4.146) have been used. Equation (4.170) may be reduced to

$$\tan \bar{\omega} = \frac{\omega_{31} \cos \frac{\Psi}{2} \cos \xi - \omega_3 \sin \frac{\Psi}{2} \sin \xi}{- \omega_{31} \sin \frac{\Psi}{2} \cos \xi - \omega_3 \cos \frac{\Psi}{2} \sin \xi} \tag{4.171}$$

when the approximations, $1/[1 + 4D^2 \sin^2 \frac{\Psi}{2}]^{1/2} \cong 1$ and $\omega_1 \cong \omega_{11}$, are made. If the further approximation, $\omega_{31} \cong \omega_3$, is made, then (4.170) becomes

$$\bar{\omega} = -\frac{\pi}{2} + \xi + \frac{1}{2}\Psi = -\frac{\pi}{2} + (\omega_3 + \frac{1}{2}\omega_1)\theta + \xi_0 . \tag{4.172}$$

However, this is a rather coarse approximation because actually $\omega_3/\omega_{31} = 1.15$. Since $\omega_3 \cong \omega_1$, a comparison of (4.151) and (4.172) shows that the line of apsides rotates three times faster than does the regression of the line of nodes. When only Earth oblateness is taken into account, the former rotates two times faster than the latter.

For a circular orbit, the average value of e is zero because both $\bar{X}_3'$ and $\bar{X}_4'$ are zero. If the initial orbit has a small eccentricity, the zeroth-order approximation for e is

$$\begin{aligned}\bar{e} &= \omega_{31} b_0 \frac{\bar{\bar{X}}_{60}^2}{\mu_E}\left\{1 - \frac{3}{2}J_2 \frac{R_E^2}{r_c^2} - \frac{1}{4}\frac{K_1^2(1+\kappa)}{(1-e_S^2)^{3/2}}[1 - 3\cos^2\lambda + \right.\\ &\left. + \frac{3}{2}D\sin 2\lambda(\cos\Psi - 1)]\right\}\left(1 - [\varepsilon a_0(1+\kappa)\sin 2\lambda - 4D^2]\times\right.\\ &\left.\times \sin^2\frac{\Psi}{2}\right)\left(1 + \left(\frac{\omega_3 - \omega_{31}}{\omega_{31}}\right)^2 \sin^2\xi + D^2\cos^2(\Psi + \xi)\right)^{1/2}\\ &\cong \omega_{31} b_0 \frac{\bar{\bar{X}}_{60}^2}{\mu_E}\left\{1 - \frac{3}{2}J_2 \frac{R_E^2}{r_c^2} - \frac{1}{4}\frac{K_1^2(1+\kappa)}{(1-e_S^2)^{3/2}}[1 - 3\cos^2\lambda + \right.\\ &\left. + \frac{3}{2}D\sin 2\lambda(\cos\Psi - 1)]\right\}\left(1 + [-\varepsilon a_0(1+\kappa)\sin 2\lambda + 4D^2]\times\right.\\ &\left.\times \sin^2\frac{\Psi}{2} + \frac{1}{2}\left(\frac{\omega_3 - \omega_{31}}{\omega_{31}}\right)^2 \sin^2\xi + \frac{1}{2}D^2\cos^2(\Psi + \xi)\right) ,\end{aligned} \tag{4.173}$$

which is a result of inserting (4.142) and (4.167) into (4.169). The approximation, $\omega_3 \cong \omega_{31}$, has also been made in the term $(\bar{X}_2\bar{X}_3' - \bar{X}_1\bar{X}_4')^2$ in order to simplify the resulting expression. Using the constants in (4.145), the expression

$$-\varepsilon a_0(1+\kappa)\sin 2\lambda + 4D^2$$

in (4.173) may be computed as 0.0364, a positive value. Consequently, (4.173) indicates that all terms in the second brackets are positive. As with the semilatus rectum, the eccentricity increases with increasing inclination.

4.10.6. Effect of Movement of the Moon's Orbit

The effect of the movement of the Moon's orbit on a geostationary satellite orbit is exhibited by terms containing Ω_M and ω_M in the X's. Since the zeroth-order solutions of the X's do not have those terms, the first-order solutions should be examined, and the dominant terms in the solutions are selected to approximate the X's. As is seen in Sections 4.10.2., 4.10.3., and 4.10.5., X_1 and X_2 are mainly related to the orientation of the orbital plane, while X_3 and X_4 are basically associated with the eccentricity and the argument of perigee of the orbit. We shall be interested in the orientation of the orbital plane as affected by the movement of the Moon's orbit because it has a practical application in the station-keeping of a geostationary satellite. It will be seen in Chapter 5 that the solar radiation pressure plays a dominant role on the changes of eccentricity and argument of perigee. Furthermore, the complicated expressions of ω and e in terms of X's by including the dominant terms containing Ω_M and ω_M prohibit us from extracting some meaningful information.

Examination of the coefficients of the first-order terms of X_1 and X_2 shows that A_4 and A_{26} are of the order of 10^3, while the rest are of the order of 10^1. We shall include these two dominant terms in our study of the effect of movement of the Moon's orbit in the zeroth-order solution. Therefore, we have

$$\bar{X}_1 \cong \omega_1 a_0 \sin \Psi + \frac{\kappa\varepsilon}{\omega_1^2 - K_3^2} \sin I(-\omega_{12} \cos \lambda + K_3 \cos 2\lambda) \sin \bar{\Omega}_M \tag{4.174}$$

and

$$\bar{X}_2 \cong \omega_{11} a_0 \cos \Psi - D + \frac{\kappa\varepsilon}{\omega_1^2 - K_3^2} \times$$

$$\times \sin I(-K_3 \cos \lambda + \omega_{11} \cos 2\lambda) \cos \bar{\Omega}_M . \tag{4.175}$$

With the known constants in (4.145) and the constants related to the Moon's orbit and the equatorial plane of the Earth,

$$I = 5°09', \quad \lambda = -23°27', \quad K_3 = -1.4709 \times 10^{-4} ,$$

(4.174) and (4.175) become

$$\bar{X}_1 = \omega_1 a_0 \sin \Psi + 0.0083 \sin \bar{\Omega}_M \tag{4.176}$$

and

$$\bar{X}_2 = \omega_{11} a_0 \cos \Psi - D - 0.0097 \cos \bar{\Omega}_M . \tag{4.177}$$

For a near-equatorial geostationary satellite orbit having an initial inclination of zero (with $\theta_0 = 0$), i.e., $\bar{X}_1 = \bar{X}_2 = 0$, (4.176) and (4.177) yield*

$$a_0 = -\left[\left(\frac{-0.0083 \sin \bar{\Omega}_{M0}}{\omega_1}\right)^2 + \left(\frac{D + 0.0097 \cos \bar{\Omega}_{M0}}{\omega_{11}}\right)^2\right]^{1/2} \tag{4.178}$$

and

$$\sin \Psi_0 = \frac{-0.0083 \sin \bar{\Omega}_{M0}}{\omega_1 a_0} \tag{4.179}$$

for an initial longitude of lunar ascending node $\bar{\Omega}_{M0}$. The initial ascending node for the zero inclination orbit is obtained from (4.156) by an application of the L'Hospital rule in the limiting process for the case of $\bar{X}_{10} = \bar{X}_{20} = 0$ as was mentioned previously.

With known values of a_0 and Ψ_0, the inclination ($\bar{i}$) and the longitude of ascending node ($\bar{\Omega}$) of a geostationary satellite orbit can be computed from

$$\tan \bar{i} = [(\omega_1 a_0 \sin \Psi + 0.0083 \sin \bar{\Omega}_M)^2 +$$

$$+ (\omega_{11} a_0 \cos \Psi - D - 0.0097 \cos \bar{\Omega}_M)^2]^{1/2} \tag{4.180}$$

and

$$\tan \bar{\Omega} = -\frac{\omega_1 a_0 \sin \Psi + 0.0083 \sin \bar{\Omega}_M}{\omega_{11} a_0 \cos \Psi - D - 0.0097 \cos \bar{\Omega}_M} . \tag{4.181}$$

The rotational rate of the orbital plane is

---

*The ' - ' sign is chosen for a_0 because $D = a_0 \omega_1 = -0.1313$ and $D < 0.0097 \cos \bar{\Omega}_{M0}$.

$$\frac{d\bar{i}}{d\theta} = \frac{1}{[1+\bar{x}_1^2+\bar{x}_2^2][\bar{x}_1^2+\bar{x}_2^2]^{1/2}}\left(x_1\frac{d\bar{x}_1}{d\theta}+x_2\frac{d\bar{x}_2}{d\theta}\right), \tag{4.182}$$

which is obtained by differentiating (4.152) with respect to θ. The rate of regression of the line of nodes is obtained from (4.149) as

$$\frac{d\bar{\Omega}}{d\theta} = \frac{1}{\bar{x}_1^2+\bar{x}_2^2}\left(x_1\frac{d\bar{x}_2}{d\theta}-\bar{x}_2\frac{d\bar{x}_1}{d\theta}\right). \tag{4.183}$$

With (4.176)-(4.179), (4.182), and a known $\bar{\Omega}_{M0}$, the maximum inclination ($\bar{i}_M$) is calculated by finding θ_M when $d\bar{i}/d\theta$ vanishes. As is seen from (4.182), $d\bar{i}/d\theta = 0$ requires that

$$\bar{x}_1\frac{d\bar{x}_1}{d\theta}+\bar{x}_2\frac{d\bar{x}_2}{d\theta} = 0 \quad \text{or}$$

$$\begin{aligned}&(\omega_1 a_0 \sin\Psi + 0.0083\sin\bar{\Omega}_M)(\omega_1^2 a_0 \cos\Psi + 0.0083K_3\cos\bar{\Omega}_M) + \\ &+(\omega_{11}a_0\cos\Psi - D - 0.0093\cos\bar{\Omega}_M)\times \\ &\times(-\omega_1\omega_{11}a_0\sin\Psi + 0.0097K_3\sin\bar{\Omega}_M) = 0.\end{aligned} \tag{4.184}$$

The value of Ψ_M (hence, θ_M) should satisfy the above equation.

The time, $\bar{t}_M$, required to attain to $\bar{i}_M$ and the corresponding $\bar{\Omega}_M$ can then be determined from (4.147) and (4.181), respectively. The initial rate of rotation of the orbital plane, $(d\bar{i}/d\theta)_0$, and that of the regression of the line of nodes, $(d\bar{\Omega}/d\theta)_0$, and the initial location of the line of nodes ($\bar{\Omega}_0$) for any known $\bar{\Omega}_{M0}$ are evaluated from (4.182), (4.183), and (4.181), respectively, when $\theta = 0$ and (4.147) is used. It should be noted that in obtaining $(d\bar{i}/dt)_0$, $(d\bar{\Omega}/dt)_0$, and $\bar{\Omega}_0$, the L'Hospital rule in the limiting process has been employed.

Table III. shows the values of $\bar{\Omega}_0$, $\bar{i}_M$, $\bar{\Omega}$ at $\bar{t}_M$, $(d\bar{i}/dt)_0$, and $(d\bar{\Omega}/dt)_0$ for different $\bar{\Omega}_{M0}$ for an initial zero inclination geostationary satellite orbit. Readers will be able to study the behavior of a geostationary satellite for different $\bar{\Omega}_{M0}$ with the initial condition other than $\bar{i}_0 = 0$ by following the same procedures described above in studying the case of $\bar{i}_0 = 0$.

Table III. <u>Characteristics of the geostationary satellite orbital plane as affected by the movement of the Moon's orbit for initially zero-inclination orbit</u>.

$\bar{\Omega}_{M0}$	$\bar{\Omega}_0$	$\bar{i}_M$	$\bar{\Omega}$ at $\bar{t}_M$	$\bar{t}_M$(yrs)	$(d\bar{i}/dt)_0$ (deg/yr)	$(d\bar{\Omega}/dt)_0$ (deg/yr)
0°	270°	14.69°	180.49°	26.97	- 0.951	- 4.770
90°	278.74°	14.59°	181.55°	26.47	- 0.863	- 3.595
180°	270°	14.74°	179.62°	26.85	- 0.755	- 1.748
270°	268.74°	14.89°	178.42°	27.48	- 0.863	- 3.595
none	270°	14.71°	180°	26.85	- 0.853	- 3.432

The bottom line labelled 'none', referred to as the zeroth-order solution in Section 4.10.3., is that where the motion of the Moon is assumed to be constrained to the ecliptic plane (see also Zee, 1972). The numerical values presented in Table III. are checked reasonably well with those obtained by means of a direct numerical integration of the equations of motion of the problem (Allan, 1964). However, it should be mentioned that in Allan's integration, his initial ascending node (Ω_0) of the orbit is located along the line from the Earth's center to the first point of Aries (i.e., $\Omega_0 = 0$) for all cases ($\Omega_{M0} = 0°$, $90°$, $180°$, and $270°$). In fact, for an initially equatorial geostationary orbit, Ω_0 should be a function of Ω_{M0} as shown in Table III., and their values are in the neighborhood of 270° and are not equal to 0°.

A special case of interest occurs when the constant 'a_0' in (4.176) and (4.177) is equal to zero. The resulting inclination equation, then, may be approximated by

$$\tan \bar{i} = 0.1316 - 0.0097 \cos \bar{\Omega}_M . \tag{4.185}$$

It can be seen that $\bar{i}$ will vary from 6.950°, when $\bar{\Omega}_M = 0°$, to 8.043°, when $\bar{\Omega}_M = 180°$, while the case of the Moon's orbit remaining in the ecliptic plane gives a constant value of 7.480°.

In order to study the drift of the orbital plane toward the equatorial plane of the Earth from an initial inclination, $\bar{i}_0$, we shall follow the same procedures described in Section 4.10.3. by utilizing (4.156), (4.157), and the initial values, $\bar{X}_{10}$ and $\bar{X}_{20}$, of (4.176) and (4.177) to obtain two equations similar to (4.165) and (4.166).

From (4.156) and (4.157), we find

$$\bar{X}_{10} = -\sin \bar{\Omega}_0 \tan \bar{i}_0 = -\tan \phi_0 \sin \bar{\Omega}_0$$

and

$$\bar{X}_{20} = -\tan \bar{i}_0 \cos \bar{\Omega}_0 = -\tan \phi_0 \cos \bar{\Omega}_0 .$$

Therefore,

$$\bar{X}_{10} = -\sin \bar{\Omega}_0 \tan \bar{i}_0$$

$$= \omega_1 a_0 \sin \Psi_0 + 0.0083 \sin \bar{\Omega}_{M0} \tag{4.186}$$

and

$$\bar{X}_{20} = \cos \bar{\Omega}_0 \tan \bar{i}_0$$

$$= \omega_{11} a_0 \cos \Psi_0 - D - 0.0097 \cos \bar{\Omega}_{M0} . \tag{4.187}$$

From the above two equations, we have

$$\tan \Psi_0 = \frac{-\omega_{11}}{\omega_1} \; \frac{\sin \bar{\Omega}_0 \tan \bar{i}_0 + 0.0083 \sin \bar{\Omega}_{M0}}{\cos \bar{\Omega}_0 \tan \bar{i}_0 + D + 0.0097 \cos \bar{\Omega}_{M0}} . \tag{4.188}$$

In the above equation, it is required to find an $\bar{\Omega}_0$ to minimize Ψ_0. Therefore, we differentiate (4.188) with respect to $\bar{\Omega}_0$ and set $d\Psi_0/d\bar{\Omega}_0 = 0$. The resulting expression is

$$\tan \bar{i}_0 + (D + 0.0097 \cos \bar{\Omega}_{M0}) \cos \bar{\Omega}_0 + 0.0083 \sin \bar{\Omega}_{M0} \sin \bar{\Omega}_0 = 0 , \tag{4.189}$$

which gives the $\bar{\Omega}_0$ required to minimize Ψ_0.

The minimum $\bar{i}$ can be obtained from $d\bar{i}/d\theta = 0$ when (4.180) is differentiated with respect to θ; thus,

$$(\omega_1 a_0 \sin \Psi + 0.0083 \sin \bar{\Omega}_M)(\omega_1^2 a_0 \cos \Psi + 0.0083 K_3 \cos \bar{\Omega}_M) +$$
$$+ (\omega_{11} a_0 \cos \Psi - D - 0.0097 \cos \bar{\Omega}_M) \times$$
$$\times (-\omega_1 \omega_{11} a_0 \sin \Psi + 0.0097 K_3 \sin \bar{\Omega}_M) = 0 . \tag{4.190}$$

With known $\bar{\Omega}_0$, Ψ_0, and $\bar{\Omega}_{M0}$, hence a_0, the solution to the above equation depicts the θ_M (included in Ψ) in (4.180) for the minimum $\bar{i}$. In view of the smallness of the numerical coefficients of the terms containing $\sin \bar{\Omega}_M$ and $\cos \bar{\Omega}_M$ in (4.190), θ_M is approximately equal to $\bar{\Omega}_0 + 90° - \frac{1}{\omega_1} \Psi_0$ as is shown in Section 4.10.3. Hence, t_M may be computed from (4.147).

Table IV. shows the $\bar{\Omega}_0$'s, Ψ_0's, and t_M's for various $\bar{i}_0$ at different $\bar{\Omega}_{M0}$'s. The column marked 'none' is the case when the Moon's orbit coincides with the ecliptic plane as shown in Figure 4.4.

From the tabulated values in Table IV., it can be seen that when $\bar{\Omega}_{M0}$ is in the neighborhood of 270°, the synchronous orbit will remain close to the equatorial plane for a long time.

4.11. Canonical Methods of Solution

We shall present canonical methods of solution to the problem by referring to our descriptions on the subject in Chapter 2, notably Kozai's solution from solving the Lagrange's equations and the solution of Voshkov'yak and Lidov based on the Delaunay equations.

4.11.1. Direct Solution from Lagrange's Equations (Kozai, 1973)

Based on the disturbing function, R_d, in (2.22) and the constants described in (2.22a) for the case of geostationary satellites with the approximations of $e = e_S = e_M = 0$, R_d is reduced to

$$\begin{aligned}R_d = {} & 2.46 \times 10^{-5}\, n^2a^2(\tfrac{1}{2} - \tfrac{3}{4}\sin^2 i) + 1.63 \times 10^{-5}\, n^2a^2 \times \\ & \times [\tfrac{1}{4}(1 - \tfrac{3}{2}\sin^2 i)(1 - \tfrac{3}{2}\sin^2 i_M) + \\ & + \tfrac{3}{16}\sin 2i \sin 2i_M \cos(\Omega - \Omega_M) + \\ & + \tfrac{3}{16}\sin^2 i \sin^2 i_M \cos 2(\Omega - \Omega_M)] + 0.75 \times 10^{-5}\, n^2a^2 \times \\ & \times [\tfrac{1}{4}(1 - \tfrac{3}{2}\sin^2 i)(1 - \tfrac{3}{2}\sin^2 \lambda) - \tfrac{3}{16}\sin 2i \sin 2\lambda \cos\Omega + \\ & + \tfrac{3}{16}\sin^2 i \sin^2 \lambda \cos 2\Omega]\,. \end{aligned} \tag{4.191}$$

If the variables

$$p = \sin i \cos \Omega \quad \text{and} \quad q = \sin i \sin \Omega \tag{4.192}$$

are introduced instead of i and Ω, then

$$\left.\begin{aligned}\frac{dp}{dt} &= \cos i \cos \Omega \frac{di}{dt} - \sin i \sin \Omega \frac{d\Omega}{dt}, \\ \frac{dq}{dt} &= \cos i \sin \Omega \frac{di}{dt} + \sin i \cos \Omega \frac{d\Omega}{dt}, \\ \frac{\partial R_d}{\partial p} &= \frac{\cos \Omega}{\cos i}\frac{\partial R_d}{\partial i} - \frac{\sin \Omega}{\sin i}\frac{\partial R_d}{\partial \Omega}, \\ \text{and} \quad \frac{\partial R_d}{\partial q} &= \frac{\sin \Omega}{\cos i}\frac{\partial R_d}{\partial i} + \frac{\cos \Omega}{\sin i}\frac{\partial R_d}{\partial \Omega}.\end{aligned}\right\} \tag{4.193}$$

A substitution of (2.21) into (4.193) results in

$$\frac{dp}{dt} = -\frac{\cos i}{na^2\sqrt{1 - e^2}}\frac{\partial R_d}{\partial q} \cong -\frac{1}{na^2}\frac{\partial R_d}{\partial q}$$

and

Table IV. Maximum drifting duration of the synchronous orbital plane toward the equatorial plane of the Earth as affected by the movement of the Moon's orbit.

$\bar{i}_0$	$\bar{\Omega}_{M0} = 0°$			$\bar{\Omega}_{M0} = 90°$			$\bar{\Omega}_{M0} = 180°$			$\bar{\Omega}_{M0} = 270°$			None		
	$\bar{\Omega}_0$ deg.	Ψ_0 deg.	t_M yr.	$\bar{\Omega}_0$ deg.	Ψ_0 deg.	t_M yr.	$\bar{\Omega}_0$ deg.	Ψ_0 deg.	t_M yr.	$\bar{\Omega}_0$ deg.	Ψ_0 deg.	t_M yr.	$\bar{\Omega}_0$ deg.	Ψ_0 deg.	t_M yr.
$\frac{1}{2}°$	274.12	- 4.23	0.63	270.19	- 0.19	0.03	273.55	- 3.65	0.55	277.42	- 7.63	1.14	273.81	- 3.92	0.59
1°	278.25	- 8.48	1.27	274.01	- 4.12	0.62	277.11	- 7.31	1.09	281.24	-11.55	1.73	277.64	- 7.85	1.18
$1\frac{1}{2}°$	282.44	-12.13	1.91	277.86	- 8.08	1.21	280.70	-11.00	1.65	285.10	-15.50	2.32	281.50	-11.82	1.77
2°	286.69	-17.13	2.56	281.78	-12.10	1.81	284.34	-14.73	2.21	289.01	-19.50	2.92	285.42	-15.84	2.37
$2\frac{1}{2}°$	291.04	-21.58	3.23	285.77	-16.18	2.42	288.04	-18.51	2.77	293.00	-23.57	3.53	289.42	-19.93	2.98

$$\frac{dq}{dt} = \frac{\cos i}{na^2\sqrt{1-e^2}} \frac{\partial R_d}{\partial p} \cong \frac{1}{na^2} \frac{\partial R_d}{\partial p} , \tag{4.194}$$

where $\cos i \cong 1$ and $e \cong 0$ are used because of the geostationary satellite orbit.

With (4.191), (4.192), (4.193), (2.31), and $\cos i \cong 1$, (4.194) becomes

$$\frac{dp}{dt} = n \times 10^{-5}(5.20q - 0.096 \sin \Omega_M)$$

and (4.195)

$$\frac{dq}{dt} = n \times 10^{-5}(- 4.92p + 0.646 + 0.078 \cos \Omega_M) .$$

As $\dot{\Omega}_M = - 1.47 \times 10^{-4}$ n, the solutions to (4.195) are obtained by introducing two integration constants, γ_1 and t_0, as

$$p = - 1.03\gamma_1 \cos[5.06 \times 10^{-5} n(t - t_0)] + 0.1313 - 0.0095 \cos \Omega_M$$

and (4.196)

$$q = \gamma_1 \sin[5.06 \times 10^{-5} n(t - t_0)] - 0.0085 \sin \Omega_M .$$

These two equations are equivalent to our findings in (4.176) and (4.177) because (2.113), (2.117), and (4.192) show that

$$\tan \bar{\Omega} = - \frac{\bar{X}_1}{\bar{X}_2} = \frac{p}{q}$$

and (4.197)

$$\tan \bar{i} = (\bar{X}_1^2 + \bar{X}_2^2)^{1/2} \cong \sin \bar{i} = (p^2 + q^2)^{1/2}$$

as geostationary satellites are in near-equatorial orbits. Consequently,

$$q = - \bar{X}_1 \quad \text{and} \quad p = \bar{X}_2 . \tag{4.198}$$

Kozai (1973) did not further explore the geostationary satellites problem because his interest was in the computational method of the luni-solar perturbations on satellite motions in general.

4.11.2. Solution from Delaunay Equations (Vashkov'yak and Lidov, 1973)

The analytical solution of Vashkov'yak and Lidov for geostationary satellites was based on a simplified Hamiltonian function, which consists of three separate parts related to:

(1) the eccentricity (e) and longitude of perigee ($\tilde{\omega}$),

(2) the inclination (i) and longitude of ascending node (Ω), and

(3) the geodetic longitude (λ_A) and the radial deviation (Δr_s) from the reference geostationary orbit (r_s) resulting from a spherical Earth.

These three 'sub-Hamiltonians' then form three sets of Delaunay equations and the three sets equations are not cross-coupled. Consequently, each set of Delaunay equations independently gives the solutions of its related orbital parameters. Since the simplified Hamiltonian does include the solar radiation pressure, we shall give the presentation of their solutions in Chapter 5.

4.12. Summary

The dominant effect of the Sun and Moon and an oblate Earth on a geostationary satellite is exhibited by the rotation and regression of the orbital plane. The rotation can be computed by the relationships of

$$\bar{X}_1 = \omega_1 a_0 \sin\Psi + \varepsilon[\omega_1 a_0 \sin\Psi + A_3 \sin 2\bar{\alpha} + A_4 \sin\bar{\Omega}_M + A_7 \sin(\bar{\Omega}_M + \Psi) + A_8 \sin(\bar{\Omega}_M - \Psi)] , \tag{4.199}$$

$$\bar{X}_2 = \omega_{11} a_0 \cos\Psi + \varepsilon[\omega_{11} a_0 \cos\Psi + A_{25} \cos 2\bar{\alpha} + A_{26} \cos\bar{\Omega}_M + A_{29} \cos(\bar{\Omega}_M + \Psi) + A_{30} \cos(\bar{\Omega}_M - \Psi) - G] \tag{4.200}$$

and

$$\tan\bar{i} = (\bar{X}_1^2 + \bar{X}_2^2)^{1/2} ; \tag{4.201}$$

while the regression of the line of node is calculated from

$$\bar{\Omega} = \tan^{-1}\left(-\frac{\bar{X}_1}{\bar{X}_2}\right) . \tag{4.202}$$

The changes of eccentricity and argument of perigee are less prominent. However, they may be shown by

$$\begin{aligned}\bar{X}_3' &\cong \omega_{31} b_0 \cos\xi \\ &+ \frac{\varepsilon}{4}\Big\{B_7 \cos(\bar{\Omega}_M + \xi) + B_8 \cos(\bar{\Omega}_M - \xi) + B_9 \cos\bar{\omega}_M + \\ &+ \cos(\bar{\omega}_M + \bar{\Omega}_M)[B_{10} + B_{10M2} \cos(\bar{\lambda}_M - 2\bar{\lambda}_S + \bar{\omega}_M + \bar{\Omega}_M) + \\ &+ B_{10M3} \cos 2(\bar{\lambda}_S - \bar{\omega}_M - \bar{\Omega}_M)] + B_{11} \cos(\bar{\omega}_M + 2\bar{\Omega}_M)\Big\} ,\end{aligned} \tag{4.203}$$

$$\begin{aligned}\bar{X}_4' &\cong \omega_3 b_0 \sin\xi \\ &+ \frac{\varepsilon}{4}\Big\{B_{30} \sin(\bar{\Omega}_M + \xi) + B_{31} \sin(\bar{\Omega}_M - \xi) + B_{32} \sin\bar{\omega}_M + \\ &+ \sin(\bar{\omega}_M + \bar{\Omega}_M)[B_{33} + B_{33M2} \cos(\bar{\lambda}_M - 2\bar{\lambda}_S + \bar{\omega}_M + \bar{\Omega}_M) + \\ &+ B_{33M3} \cos 2(\bar{\lambda}_S - \bar{\omega}_M - \bar{\Omega}_M)] + \sin(\bar{\omega}_M + 2\bar{\Omega}_M) \times \\ &\times [B_{34} + B_{34M4} \cos 2(\bar{\lambda}_S - \bar{\omega}_M - \bar{\Omega}_M)]\Big\}\end{aligned} \tag{4.204}$$

$$\tan\bar{\omega} = \frac{\bar{X}_1\bar{X}_3' + \bar{X}_2\bar{X}_4'}{\bar{X}_2\bar{X}_3' - \bar{X}_1\bar{X}_4'} \frac{1}{(1 + \bar{X}_1^2 + \bar{X}_2^2)^{1/2}} , \tag{4.205}$$

$$\bar{e} = \bar{\ell}[\bar{X}_3'^2 + \bar{X}_4'^2 + (\bar{X}_2\bar{X}_3' - \bar{X}_1\bar{X}_4')^2]^{1/2} , \tag{4.206}$$

and

$$\bar{\ell} \cong \frac{\bar{X}_6^2}{\mu_E}(1 + \bar{X}_1^2 + \bar{X}_2^2)\left[1 - \frac{3}{2}\frac{R_E^2}{r_c^2} - \frac{1}{4}\frac{1 + \kappa}{(1 - e_S^2)^{3/2}} K_1^2 \times (1 - 3\cos^2\lambda + \frac{3}{2}\bar{X}_2 \sin 2\lambda)\right] , \tag{4.207}$$

where

$$\bar{X}_6 \cong \bar{X}_{60}\left\{1 + \frac{\bar{X}_{60}^6}{\bar{\bar{X}}_{60}^6}[\tfrac{1}{2}\varepsilon a_0(1 + \kappa)\sin 2\lambda](\cos\Psi - \cos\Psi_0) + \frac{9}{4}\left(\frac{\bar{X}_{60}^6}{\bar{\bar{X}}_{60}^6}\right)^2 [\tfrac{1}{2}\varepsilon a_0(1 + \kappa)\sin 2\lambda]^2(\cos 2\Psi - \cos 2\Psi_0)\right\} , \tag{4.208}$$

and

$$\bar{X}_{60} = r_c^2\dot{\theta}_E/(1 + \bar{X}_{10}^2 + \bar{X}_{20}^2)^{1/2} . \tag{4.209}$$

All parameters in the above equations are related to the time by

$$\bar{t} - \bar{t}_0 \cong \frac{\bar{X}_{60}^3}{\mu_E^2}(1 - 2\omega_4)[E_1(\theta - \theta_0) + E_2(\sin\Psi - \sin\Psi_0) + E_3(\sin 2\Psi - \sin 2\Psi_0) + E_4(\sin 3\Psi - \sin 3\Psi_0)] . \tag{4.210}$$

As the mean values of osculating orbital elements over a 2π-period of θ are known from (4.199)-(4.210), the 'true value' of orbital elements in-between the 2π-period are then obtained by replacing $X_1 = \bar{X}_1 + \delta X_1$, $X_2 = \bar{X}_2 + \delta X_2$, and $X_6 = \bar{X}_6 + \delta X_6$ for $\bar{X}_1$, $\bar{X}_2$, and $\bar{X}_6$, respectively, wherein the δX's are developed in the previous sections.

It should be noted that the terms in $\bar{X}_1$, $\bar{X}_2$, $\bar{X}_3'$, $\bar{X}_4'$, $\bar{X}_6$ and $\bar{t}$ in the above equations are the dominant ones. A complete expression of these parameters is referred to the sections wherein these parameters are derived from their method of solutions.

Attention should be paid that a refinement of all $\bar{X}$'s can be made if all terms shown in (4.108)-(4.111) are included in the $\bar{X}$'s as a result of the Sun's perturbation on the Moon's motion.

Bibliography

The perturbations of the Sun and Moon on satellites have been receiving attention as a result of Kozai's study (Kozai, 1959), which shows their effects of a possible reduction in lifetime for satellites in highly eccentric orbits. He presented the time rate equations for the secular terms of the ascending node and argument of perigee. This was followed by a large number of investigations, notably those by Moe, Sehnal, Musen, and Lidov. Moe (1960) first estimated their effects on satellites in a single orbit and then deduced the long-term effects by incorporating the movements of the Sun and Moon into the estimate; this was accomplished by means of a Taylor series expansion, and the case of highly eccentric orbits was studied. Sehnal (1961) numerically integrated the equations of orbital elements for a duration of one month with perturbing forces such as the Sun, Moon, and the Earth's potential (J_2, J_{22}, and J_{30}). Musen (1961a, b, c) developed equations for the problem, and they are readily numerically integrated. Lidov (1963a) recognized that the third harmonic (parallactic term) of the lunar attraction will change the eccentricity of a circular orbit if the orbit is sufficiently large. He (Lidov, 1963b) also derived expressions of the rate change of the doubly-averaged values of orbital elements over a period of the satellite as well as that of the Sun and Moon. Since he was dealing with a general class of orbits, no particular result was expected to be deduced; however, his approach of doubly-averaging technique is later adopted for all subsequent investigations by approximate analytical methods.

As geostationary satellites are in a special class of orbit with both nearly-zero eccentricity and inclination, the treatment of this class of orbit, when the perturbation of the Sun and Moon and Earth oblateness are taken into consideration, falls into two categories. The first category gives a linearized version of the problem (Frick and Garber, 1962; Cook, 1963), and the second category considers the problem by determining the time rate of the osculating orbital elements (Kozai, 1959; Musen, 1961a, b, c; Allan, 1962, 1963). The former approach loses its predictive accuracy with increasing time, and the latter requires numerical integration to obtain useful solutions. In a later study, Allan (1964) obtained the analytical expressions of the long-period motion of the orbital plane for circular orbits having a radius of 3-10 Earth radii by assuming that the lunar orbit lies in the ecliptic plane. While the geostationary orbit is within the range of his study, the pole of an initially equatorial geostationary orbit performs a precessional motion with a period of about 52 years, and the inclination reaches a maximum of about 15° at the middle of the period. When the inclination of the lunar orbit is taken into consideration, his study is limited to the equatorial synchronous satellites with an assumed initial ascending node line pointing to the vernal equinox (see comments in Section 4.10.6.), and results were achieved by numerical integration for different initial locations of the lunar ascending node. Frick (1967) also further generalized the problem by studying the motion of satellites in near-circular orbits, at any inclination, when the orbital radius is less than 10 Earth radii. However, both Allan's and Frick's studies are limited to the motion of the orbital plane, and the satellite's in-plane motion was not investigated.

In 1969, Voshkov'yak proposed a numerical scheme to compute the geostationary orbit for a time interval of a few days based on an extrapolation from an averaged orbit with the lunisolar perturbations in terms of time dependent polynomials (Voshkov'yak, 1969). His basic approach is similar to that Moe suggested in 1960. As an extension of his work, he further proposed a method to compute lunisolar perturbations in geostationary satellite motions by a semi-analytical and semi-numerical procedure; the secular and long-period effects are computed by means of numerical integration with large integration steps, and short-period oscillatory effects are obtained analytically (Voshkov'yak, 1972). A similar proposal was suggested by Kozai in 1973 for a general class of orbital computation, while the motion for the orbital plane of a geostationary satellite was briefly discussed (Kozai, 1973).

In 1968, Zee attempted to solve the problem by the asymptotic method in nonlinear mechanics (Zee, 1968), and later he obtained an integrated solution (Zee, 1972) to the problem of a near-equatorial synchronous satellite. The resulting zeroth-order solution brought out all the main features; however, it should be noted that the motion of the Moon's orbit was not included in his study. Similar results comparable to Zee's first-order approximate solution to the problem (Zee, 1971, 1972) were obtained by Voshkov'yak with the inclusion of the movement of the Moon's orbital plane (Voshkov'yak, 1973)

Along the line of further analytical developments of the problem, Flury used a set of nonsingular Poincare variables to develop a long-period Hamiltonian for the motion of a geostationary satellite (Flury, 1973a, b); Giacaglia (1974) used the orbital elements of the Moon in relation to the ecliptic plane and that of a satellite with respect to the equatorial plane of the Earth to present an algorithm for computing the lunar perturbation on a satellite; and Graf presented an extended phase-space canonical treatment to determine the lunisolar effects on the long-period behavior of the orbital inclination and ascending node (Graf, 1975).

With minimizing station-keeping efforts for a variety of geostationary satellite missions as their objectives, Kamel and Tibbits obtained a linearized solution for the longitude and semimajor axis deviations in the neighborhood of a prespecified station as a result of the perturbations of the Sun, Moon, and the Earth's potential (J_2, J_{22}, J_{31} and J_{33}) (Kamel, 1973a). In order to keep the inclination of the geostationary orbit within a certain bound, a second-order long-period solution for the inclination and longitude of ascending node was also derived (Kamel, 1973b).

As developments of computer programs progress, the application of computers not only achieves the numerical results of the problems in celestial mechanics, but also aids in obtaining the analytical approximate solutions by means of computerized symbolic algebraic manipulations. Richardson (1976) utilized computer programs in developing a theory for the long-period motion of a geosynchronous satellite using extended phase-space canonical formalism, while Kamel (1978) used computer algebraic manipulation to the fullest extent to develop algebraic expressions for the calculation of synchronous satellite ephemeris due to the Earth's triaxiality and lunisolar effects.

References

Allan, R.R.; 'Satellite Orbit Perturbations Due to Radiation Pressure and Luni-Solar Forces', Technical Note No. Space 7, Royal Aircraft Establishment, Farnborough, England (1962).

Allan, R.R.; 'Perturbation of a Geostationary Satellite - 2. Lunisolar Effects', Technical Note No. Space 47, Royal Aircraft Establishment, Farnborough, England (1963).

Allan, R.R. and Cook, G.E.; 'The Long-Period Motion of the Plane of a Distant Circular Orbit', Proceedings of Royal Society (London) A280, 97-109 (1964).

Bogoliubov, H. and Mitropolsky, J.; Asymptotic Method in the Theory of Nonlinear Oscillations, Hindustan Publishing Corporation, Delhi-6, India, 1961.

Brouwer, D. and Clemence, G.M.; Methods of Celestial Mechanics, Academic Press, New York and London, 1961.

Cook, G.E.; 'The Long-Term Evolution of Circular Near-Equatorial Orbits at Medium Distances from the Earth', Technical Note No. Space 46, Royal Aircraft Establishment, Farnborough, England (1963).

Escobal, P.R.; Method of Orbit Determination, p. 366, John Wiley & Sons, Inc., New York, 1965.

Flury, W.; 'Zur Bahnberechnung Von Geostationarer Satelliten', Celestial Mechanics 7, 376-383 (1973a).

Flury, W.; 'Perturbation Theory for Geostationary Satellites - Application of Bohlin von Zeipel Method', 10th International Symposium on Space

Science and Technology, Tokyo, Japan (1973b).

Flury, W.; 'Station-Keeping of a Geostationary Satellite', Review of Scientific Technique CECLES/CERS 5, 131-155 (1973c).

Frick, R.H. and Garber, T.B.; 'Perturbations of a Synchronous Satellite', R-399-NASA, RAND Corporation, Santa Monica, California (1962).

Frick, R.H.; 'Orbital Regression of Synchronous Satellites Due to the Combined Gravitational Effects of the Sun, the Moon and the Oblate Earth', R-454-NASA, RAND Corporation, Santa Monica, California (1967).

Giacaglia, G.E.; 'Lunar Perturbations of Artificial Satellite of the Earth', Celestial Mechanics 9, 239-267 (1975).

Graf, A; 'Lunar and Solar Perturbations on the Orbit of a Geostationary Satellite', AAS paper No. 75-023 (1975).

Kamel, A. and Tibbits, R.; 'Some Useful Results on Initial Node Locations for Near-Equatorial Circular Satellite Orbits', Celestial Mechanics 8, 45-73 (1973a).

Kamel, A., Ekman, D., and Tibbits, R.; 'East-west Stationkeeping Requirements of Nearly Synchronous Satellites Due to Earth's Triaxiality and Luni-Solar Effects', Celestial Mechanics 8, 129-148 (1973b).

Kamel, A.; 'Synchronous Satellite Ephemeris Due to Earth's Triaxiality and Luni-Solar Effects', AIAA/AAS Astrodynamics Conference, Palo Alto, California (1978).

Kozai, Y.; 'On the Effects of the Sun and Moon upon the Motion of a Close Earth Satellite', Special Report 22, Smithsonian Astrophysical Observatory, Cambridge, Mass. (1959).

Kozai, Y.; 'A New Method to Compute Lunisolar Perturbations in Satellite Motions', Special Report 349, Smithsonian Astrophysical Observatory, Cambridge, Mass. (1973).

Lidov, M.L.; 'On the Approximate Analysis of the Orbit of Evolution of Artificial Satellites', Dynamics of Satellites, M. Roy (ed.), Springer-Verlag, Berlin, 1963a.

Lidov, M.L.; 'Evolution of the Orbits of Artificial Satellites of Planets as Affected by Gravitational Perturbations from External Bodies', AIAA J. 1, 1985-2002 (1963b).

Moe, M.M.; 'Solar-Lunar Perturbation of the Orbit of an Earth Satellite', ARS J. 30, 485-487 (1960).

Musen, P., Bailie, A.E., and Upton, E.; 'Development of the Lunar and Solar Perturbations in the Motion of an Artificial Satellite', NASA TND-494 (1961a).

Musen, P.; 'On the Long-Period Luni-Solar Effect in the Motion of the Artificial Satellite', J. Geophys. Research 66, 1659-1666 (1961b).

Musen, P.; 'On the Long-Term Luni-Solar Effect in the Motion of the Artificial Satellite', J. Geophys. Research 66, 2797-2805 (1961c).

Richardson, D.L.; 'The Long-Period Motion of 24-hr Satellites', AIAA Paper No. 76-828 (1976).

Sehnal, L.; 'The Stability of the 24-hr Satellite', Astronautica Acta 7, 445-454 (1961).

Voshkov'yak, M.A.; 'Perturbed Motion of a Stationary Earth Satellite During a Short Time Interval', Cosmic Research 7, 756-765 (1969).

Voshkov'yak, M.A.; 'On the Method of Approximate Computation of the Motion of a Synchronous Artificial Satellite', Cosmic Research 10, 131-140 (1972).

Voshkov'yak, M.A. and Lidov, M.L.; 'An Approximate Description of the Evolution of the Orbit of a Stationary Artificial Earth Satellite', Cosmic Research 11, 305-316 (1973).

Zee, C.H.; 'Effects of the Sun and the Moon on a Synchronous Satellite', Report No. ADR 06-06-68.2, Grumman Aircraft Engineering Corporation, Bethpage, New York (1968).

Zee, C.H.; 'Effect of Earth Oblateness and Equator Ellipticity on a Synchronous Satellite', Astronautica Acta 16, 143-153 (1971).

Zee, C.H.; 'Effects of the Sun and the Moon on a Near-Equatorial Synchronous Satellite', Astronautica Acta 17, 891-906 (1972).

Chapter 5

COMBINED EFFECTS OF THE SUN, MOON, AND SOLAR RADIATION PRESSURE

5.1. General

Before studying the effect of solar radiation pressure on a geostationary satellite, we shall examine the Sun-Earth eclipse relationships. The Sun-Earth radius vector, lying in the ecliptic plane, rotates once a year. The Earth continually casts a circular shadow of its own diameter (neglecting the slightly larger penumbra annulus and smaller umbra circle) and creates an eclipse cylinder on the side opposite the Sun (as shown in Figure 5.1.). This eclipse cylinder has a diameter of 6,880 nm (12,742 km).

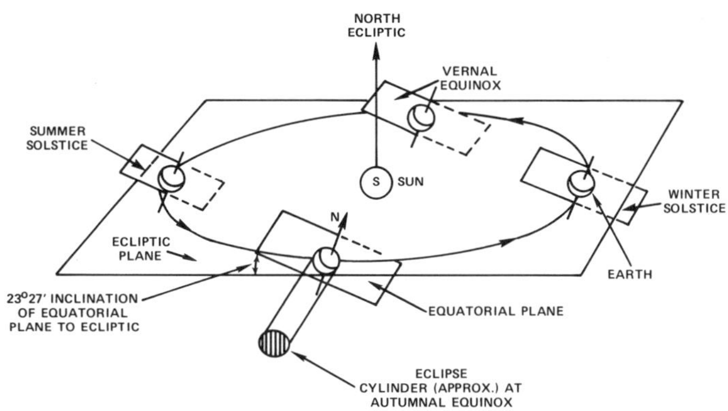

Fig. 5.1. Sun, Earth and eclipse relations.

In a year's time, every satellite in a synchronous radius (r_c) orbit crosses the eclipse cylinder with different durations. The maximum duration of eclipsing is 70 minutes in a single day, while the total number of eclipsed days in which a satellite in a synchronous radis (r_c) orbit experiences eclipse from the Sun, depends on orbit inclination and the right ascension (relative to the orbit ascending node). For an orbit having its ascending node coinciding with the equinoxes, the total number of eclipsed days varies from a maximum of 90 in equatorial to 36 in polar orbits. The relationships of orbit inclination, ascending node, and the days of eclipsing per year are shown in Figure 5.2. The maximum eclipse occurs on March 21 and September 21 for orbits lying in the equatorial plane of the Earth; inclining the orbit from the equatorial plane does shift away the maximum

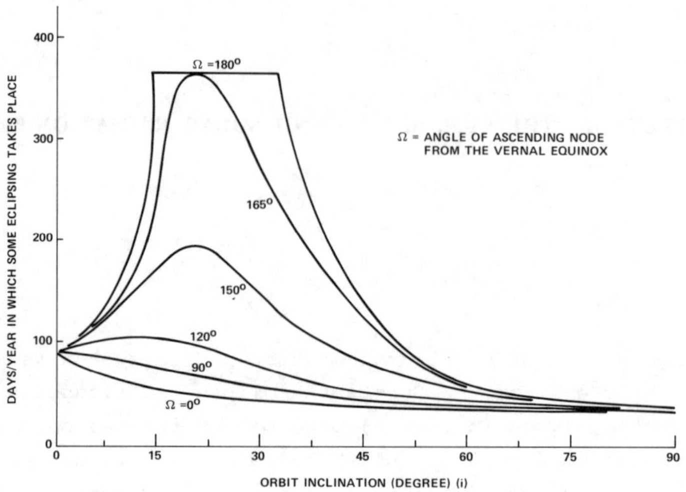

Fig. 5.2. Eclipse frequencies of satellite in synchronous-radius orbit.

eclipse date from these two dates as shown in Figure 5.3. It can be seen that a 10-degree inclination would reduce 26 days per year in eclipse but would shift the 70-min. maximum dark time from occurring at the equinoxes to a maximum of 25 days later depending on the position of ascending node. The duration of darkness during the individual eclipsed day decreases from

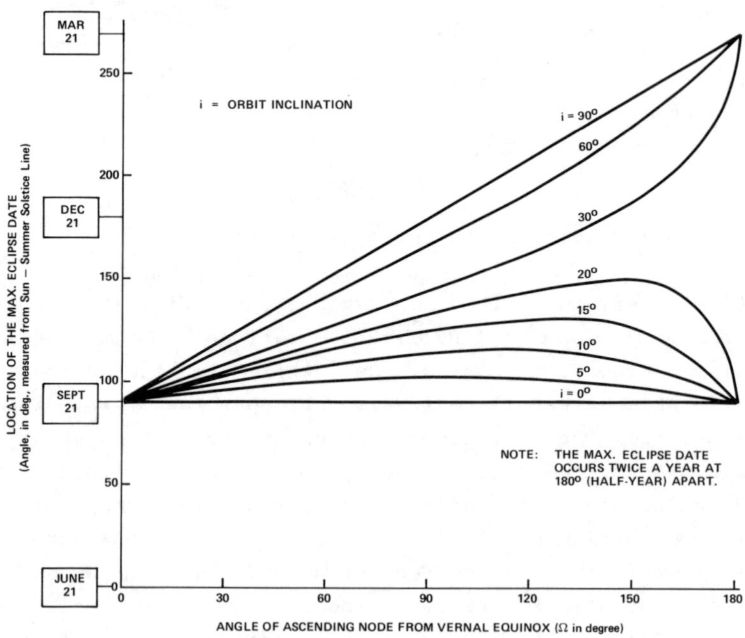

Fig. 5.3. Maximum eclipse date of a satellite in synchronous-radius orbit.

the date of maximum eclipse towards both ends of the eclipsed period. Presented in Figure 5.4. is a plot showing the darkness duration in relation to the orbit inclination and the position of ascending node. It should be noted that Figures 5.2.-5.4. are obtained from the geometric relationships between the eclipse cylinder, synchronous orbits, and the motion of the Earth.

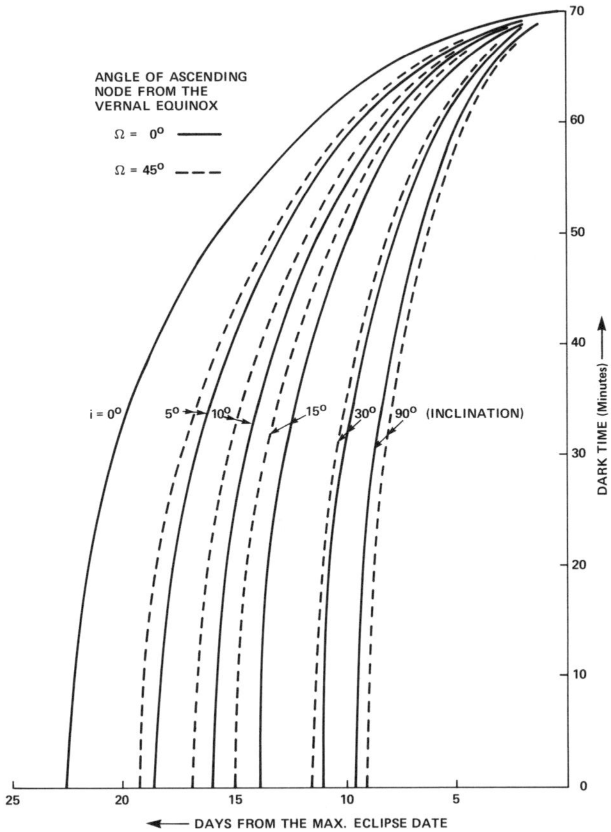

Fig. 5.4. Eclipse duration related to the orientation of the synchronous-radius orbit.

Based on the above observations of the characteristics of the eclipse, the Earth shadow effect on a spherical-geostationary satellite over a long period is minimal, and it will be neglected in the following study. Furthermore, a spherical shape satellite will be assumed to eliminate the satellite orientation problem. However, for a satellite equipped with solar panels, it can be assumed that the solar panels continually lie normal to the Sun-satellite radius vector because of the operational requirement to receive a maximum amount of solar power.

The addition of solar radiation pressure to the problem of the effects of the Sun and Moon on the near-equatorial, synchronous satellite treated in the last chapter creates an additional analytic complication. However, in dealing with this new problem, we shall follow the same format used in the earlier chapters, and a major portion of the previous derivations will be utilized. Consequently, the amount of work requird to solve this new problem will be considerably reduced.

5.2. Basic Equations of Motion

We shall proceed with our approach to the problem with the basic equations of motion. Having assumed both a spherical satellite and the absence of an eclipse cylinder, the equations of motion for a satellite subject to an oblate Earth, perturbations from solar radiation pressure, and gravitational perturbations due to the Sun and Moon, are (see (4.36)-(4.38) and Figure 4.1.)

$$\ddot{r} - r\dot{\theta}^2 \cos^2 \phi - r\dot{\phi}^2 = \text{Terms in (4.36)} - - P(a_x \cos \alpha + a_y \sin \alpha) , \tag{5.1}$$

$$\frac{1}{r \cos \phi} \frac{d}{dt}(r^2\dot{\theta} \cos^2 \phi) = \text{Terms in (4.37)} - - P(b_x \cos \alpha + b_y \sin \alpha) , \tag{5.2}$$

and

$$\frac{1}{r} \frac{d}{dt}(r^2\dot{\theta}) + r\dot{\theta}^2 \cos \phi \sin \phi = \text{Terms in (4.38)} - - P(c_x \cos \alpha + c_y \sin \alpha) , \tag{5.3}$$

where the solar radiation force (P) is previously defined in (2.26).

After carrying out the transformations (4.39a)-(4.39c):

$$S = \tan \phi = X_1 \cos \theta + X_2 \sin \theta , \tag{5.4}$$

$$\begin{aligned} Q &= \frac{1}{r \cos \phi} = X_5 + X_3 \cos \theta + X_4 \sin \theta \\ &= \frac{\mu_E}{X_6^2} \frac{1}{1 + X_1^2 + X_2^2} [1 + (X_1 \cos \theta + X_2 \sin \theta)^2]^{1/2} + \\ &\quad + X_3 \cos \theta + X_4 \sin \theta , \end{aligned} \tag{5.5}$$

$$X_6 = r^2\dot{\theta} \cos^2 \phi , \tag{5.6}$$

and the approximations used to obtain (4.58)-(4.63), Equations (5.1)-(5.3) can be resolved into the following system of first-order differential equations:

$$\begin{aligned} \frac{dX_1}{d\theta} &= \text{Terms in (4.58)} - \frac{P}{X_6^2 Q^3} \sin \theta [S \cos \theta \cos \alpha + \\ &\quad + (S \sin \theta \cos \lambda + \sin \lambda) \sin \alpha] , \end{aligned} \tag{5.7}$$

$$\begin{aligned} \frac{dX_2}{d\theta} &= \text{Terms in (4.59)} + \frac{P}{X_6^2 Q^3} \cos \theta [S \cos \theta \cos \alpha + \\ &\quad + (S \sin \theta \cos \lambda + \sin \lambda) \sin \alpha] , \end{aligned} \tag{5.8}$$

$$\frac{dX_3}{d\theta} = \text{Terms in (4.60)} - \frac{P}{X_6^2 Q^2} \sin\theta[\cos\theta\cos\alpha + $$
$$+ (\sin\theta\cos\lambda - S\sin\lambda)\sin\alpha] , \tag{5.9}$$

$$\frac{dX_4}{d\theta} = \text{Terms in (4.61)} + \frac{P}{X_6^2 Q^2} \cos\theta[\cos\theta\cos\alpha + $$
$$+ (\sin\theta\cos\lambda - S\sin\lambda)\sin\alpha] , \tag{5.10}$$

and

$$\frac{dX_6}{d\theta} = \text{Terms in (4.62)} - \frac{P}{X_6 Q^3}(-\sin\theta\cos\alpha + \cos\theta\cos\lambda\sin\alpha) . \tag{5.11}$$

This system of equations together with (4.64)-(4.66) constitute the basic differential equations to be solved for the present problem.

Following the same procedures used in obtaining (4.69)-(4.76), the application of the asymptotic method (Bogoliubov and Mitropolsky, 1961) to (5.7)-(5.11) and (4.64)-(4.66) results in

$$\frac{d\bar{X}_1}{d\theta} = \text{Terms in (4.69)} + \frac{3P}{2\mu_E^4} \bar{X}_6^6 \bar{X}_4 \sin\lambda\sin\bar{\alpha} , \tag{5.12}$$

$$\frac{d\bar{X}_2}{d\theta} = \text{Terms in (4.70)} - \frac{3P}{2\mu_E^4} \bar{X}_6^6 \bar{X}_3 \sin\lambda\sin\bar{\alpha} , \tag{5.13}$$

$$\frac{d\bar{X}_3}{d\theta} = \text{Terms in (4.71)} - \frac{\bar{X}_6^2}{2\mu_E^2} P\sin\bar{\alpha}(3\cos\lambda - \bar{X}_2\sin\lambda) , \tag{5.14}$$

$$\frac{d\bar{X}_4}{d\theta} = \text{Terms in (4.72)} + \frac{\bar{X}_6^2}{2\mu_E^2} P(3\cos\bar{\alpha} - \bar{X}_1\sin\lambda\sin\bar{\alpha}) , \tag{5.15}$$

$$\frac{d\bar{X}_6}{d\theta} = \text{Terms in (4.73)} + \frac{3P}{2\mu_E^4} \bar{X}_6^7(\bar{X}_3\cos\lambda\sin\bar{\alpha} - \bar{X}_4\cos\bar{\alpha}) , \tag{5.16}$$

$$\frac{d\bar{\alpha}}{d\theta} = \text{Same as (4.74)} , \tag{5.17}$$

$$\frac{d\nu}{d\theta} = \text{Same as (4.75)} , \tag{5.18}$$

$$\frac{d\bar{\Omega}_M}{d\theta} = \text{Same as (4.76a)} , \tag{5.19a}$$

and

$$\frac{d\bar{\omega}_M}{d\theta} = \text{Same as (4.76b)} . \tag{5.19b}$$

This set of first-order differential equations will be solved by the same methodology adopted in Chapter 4.

5.3. First-Order Approximate Solution

For completeness of our presentation, we shall write out the full set of equations which make up the solution, although some of them are identical to those shown in Chapter 4.

Following the same method of solution used in Chapter 4, the solutions to (5.12)-(5.19) are:

$$\bar{X}_1 = u_1^{(0)} \text{ (see (4.90))} + \varepsilon u_1^{(1)} \text{ (see (4.97))} + \\ + \varepsilon\left\{\frac{3}{8}\frac{1}{\mu_E^2}\bar{X}_6^4 \chi \sin\lambda\, b_0\left[\frac{\omega_{12}\omega_{31} - \omega_3(K_1+\omega_3)}{\omega_1^2 - (K_1+\omega_3)^2}\sin(\bar{\alpha}+\xi) + \right.\right. \\ \left.\left. + \frac{\omega_{12}\omega_{31} + \omega_3(K_1-\omega_3)}{\omega_1^2 - (K_1-\omega_3)^2}\sin(\bar{\alpha}-\xi)\right]\right\}, \tag{5.20}$$

$$\bar{X}_2 = u_2^{(0)} \text{ (see (4.91))} + \varepsilon u_2^{(1)} \text{ (see (4.98)} + \\ + \varepsilon\left\{\frac{3}{8}\frac{1}{\mu_E^2}\bar{X}_6^4 \chi \sin\lambda\, b_0\left[\frac{-\omega_{11}\omega_3 + \omega_{31}(K_1+\omega_3)}{\omega_1^2 - (K_1+\omega_3)^2}\cos(\bar{\alpha}+\xi) + \right.\right. \\ \left.\left. + \frac{\omega_{11}\omega_3 + \omega_{31}(K_1-\omega_3)}{\omega_1^2 - (K_1-\omega_3)^2}\cos(\bar{\alpha}-\xi)\right]\right\}, \tag{5.21}$$

$$\bar{X}_3 = u_3^{(0)} \text{ (see (4.101))} + \frac{\varepsilon}{4} u_3^{(1)} \text{ (see (4.103))} + \\ + \frac{\varepsilon}{4}\chi\left\{\frac{1}{\omega_3^2 - K_1^2}[3\omega_{31} + K_1(3\cos\lambda + D\sin\lambda)]\cos\bar{\alpha} - \right. \\ - \frac{1}{2}\frac{a_0\sin\lambda}{\omega_3^2 - (K_1+\omega_1)^2}[\omega_{11}(K_1+\omega_1) - \omega_1\omega_{31}]\cos(\bar{\alpha}+\Psi) - \\ \left. - \frac{1}{2}\frac{a_0\sin\lambda}{\omega_3^2 - (K_1-\omega_1)^2}[\omega_{11}(K_1-\omega_1) + \omega_1\omega_{31}]\cos(\bar{\alpha}-\Psi)\right\}, \tag{5.22}$$

$$\bar{X}_4 = u_4^{(0)} \text{ (see (4.102))} + \frac{\varepsilon}{4} u_4^{(1)} \text{ (see (4.104))} + \\ + \frac{\varepsilon}{4}\chi\left\{\frac{1}{\omega_3^2 - K_1^2}[3K_1 + \omega_{32}(3\cos\lambda + D\sin\lambda)]\sin\bar{\alpha} + \right. \\ + \frac{1}{2}\frac{a_0\sin\lambda}{\omega_3^2 - (K_1+\omega_1)^2}[\omega_1(K_1+\omega_1) - \omega_1\omega_{32}]\sin(\bar{\alpha}+\Psi) + \\ \left. + \frac{1}{2}\frac{a_0\sin\lambda}{\omega_3^2 - (K_1-\omega_1)^2}[\omega_1(K_1-\omega_1) + \omega_1\omega_{32}]\sin(\bar{\alpha}-\Psi)\right\}, \tag{5.23}$$

$$\frac{1}{6}\bar{\bar{X}}_{60}^6\left(\frac{1}{\bar{X}_{60}^6} - \frac{1}{\bar{X}_6^6}\right) = \frac{1}{2}\varepsilon a_0(1+\kappa)\sin 2\lambda(\cos\Psi - \cos\Psi_0) + \\ + [F(\Psi, \bar{\alpha}, \bar{\nu}, \bar{\Omega}_M, \bar{\omega}_M) - F_0(\Psi, \bar{\alpha}, \bar{\nu}, \bar{\Omega}_M, \bar{\omega}_M)] \\ \text{(see (4.105))} + \frac{3}{8}\varepsilon r_c^2\chi b_0\left\{\frac{\omega_{31}\cos\lambda + \omega_3}{K_1 - \omega_3}\times\right. \\ \times[\cos(\bar{\alpha}-\xi) - \cos(\bar{\alpha}_0 - \xi_0)] + \frac{\omega_{31}\cos\lambda - \omega_3}{K_1+\omega_3}\times \\ \left.\times[\cos(\bar{\alpha}+\xi) - \cos(\bar{\alpha}_0+\xi_0)]\right\}, \tag{5.24}$$

$$\bar{\alpha} = K_1(\theta - \theta_0) + \bar{\alpha}_0, \tag{5.25}$$

$$\bar{\nu} = K_2(\theta - \theta_0) + \bar{\nu}_0, \tag{5.26}$$

$$\bar{\Omega}_M = K_3(\theta - \theta_0) + \bar{\Omega}_{M0}, \tag{5.27a}$$

and

$$\bar{\omega}_M = K_4(\theta - \theta_0) + \bar{\omega}_{M0}, \tag{5.27b}$$

where

$$\chi \equiv -P\bar{\bar{X}}_{60}^{2}/(\tfrac{1}{2}\varepsilon\mu_E^2) \ . \tag{5.28}$$

The above presentation of the first-order solutions to $\bar{X}_1$, ..., $\bar{X}_4$, and $\bar{X}_6$ in (5.20)-(5.24) has been shortened because the derivation of the corresponding expressions was presented in detail in Chapter 4. However, the reader is reminded that these expressions are obtained by following the same procedures used in Chapter 4, and they constitute an integrated first-order approximate solution of the combined effects of the solar radiation pressure and the attractions of the oblate Earth, the Sun, and the Moon on a geostationary satellite.

Readers are reminded that the above first-order solution does not include the long-period terms of the Sun's perturbation on the Moon's motion, i.e., the $\delta_1(\ )$'s in (4.46). Since the effects of the $\delta_1(\ )$'s and the solar radiation pressure on the motions of geostationary satellites are not related, we can, therefore, combine these two effects by superimposition. Consequently, with our findings in Section 4.6., the first-order solutions shown in (5.20)-(5.23) are modified accordingly:

$u_1^{(1)}$: Terms in (4.97) and (4.108) ,

$u_2^{(1)}$: Term in (4.98) and (4.109) ,

$u_3^{(1)}$: Terms in (4.103) and (4.110) ,

and

$u_4^{(1)}$: Terms in (4.104) and (4.111) .

In addition, (5.27a) and (5.27b) are replaced by (4.112a) and (4.112b), respectively, and (5.26) is modified by adding $\delta_1\nu$ to $\bar{\nu}$.

5.4. Improved First-Order Approximate Solution

The improved first-order approximate solutions of various parameters are:

$$\begin{aligned} X_1 &= \bar{X}_1 + \delta X_1 \\ &= \bar{X}_1 \text{ [in (5.20) with additional terms of (4.108) in } u_1^{(1)}] + \\ &\quad + \text{terms of } \delta X_1 \text{ in (4.114)} - \\ &\quad - P\frac{1}{\mu_E^3}\bar{X}_6^4\Big\{\tfrac{1}{3}(\bar{X}_2 \sin\bar{\alpha}\cos\lambda - \bar{X}_1\cos\bar{\alpha})\cos^3\theta + \\ &\quad + \tfrac{1}{3}(\bar{X}_2\cos\bar{\alpha} + \bar{X}_1\sin\bar{\alpha}\cos\lambda)\sin^3\theta - \\ &\quad - (\bar{X}_2\cos\lambda + \sin\lambda)\sin\bar{\alpha}\cos\theta\Big\} - \tfrac{3}{4}P\frac{1}{\mu_E^4}\bar{X}_6^6 \times \\ &\quad \times \sin\lambda\sin\bar{\alpha}(\bar{X}_3\cos 2\theta + \bar{X}_4\sin 2\theta) \ , \end{aligned} \tag{5.29}$$

$$\begin{aligned} X_2 &= \bar{X}_2 + \delta X_2 \\ &= \bar{X}_2 \text{ [in (5.21) with additional terms of (4.109) in } u_2^{(1)}] + \\ &\quad + \text{terms of } \delta X_2 \text{ in (4.115)} - \\ &\quad - P\frac{1}{\mu_E^3}\bar{X}_6^4\Big\{\tfrac{1}{3}(\bar{X}_2\sin\bar{\alpha}\sin\lambda - \bar{X}_1\cos\bar{\alpha})\sin^3\theta - \\ &\quad - \tfrac{1}{3}(\bar{X}_2\cos\bar{\alpha} + \bar{X}_1\sin\bar{\alpha}\cos\lambda)\cos^3\theta + \end{aligned}$$

$$+ (\bar{X}_1 \cos \bar{\alpha} + \sin \lambda \sin \bar{\alpha}) \sin \theta\Big\} -$$

$$- \frac{3}{4} P \frac{1}{\mu_E^4} \bar{X}_6^6 \sin \lambda \sin \bar{\alpha} (\bar{X}_3 \sin 2\theta - \bar{X}_4 \cos 2\theta) , \tag{5.30}$$

$$\begin{aligned} X_3 &= \bar{X}_3 + \delta X_3 \\ &= \bar{X}_3 \text{ (in (5.22) with additional terms of (4.110) in } u_3^{(1)}) + \\ &\quad + \text{terms of } \delta X_3 \text{ in (4.116)} - \\ &\quad - \frac{1}{4\mu_E^2} P\bar{X}_6^2 [(\bar{X}_1 \sin \lambda \sin \bar{\alpha} + \cos \bar{\alpha}) \cos 2\theta + \\ &\quad + (\cos \lambda + \bar{X}_2 \sin \lambda) \sin \bar{\alpha} \sin 2\theta] - \\ &\quad - \frac{2}{3} \frac{1}{\mu_E^3} P\bar{X}_6^4 [\bar{X}_3 (\cos \bar{\alpha} \cos^3 \theta - \cos \lambda \sin \bar{\alpha} \sin^3 \theta) + \\ &\quad + \bar{X}_4 (- \cos \bar{\alpha} \sin^3 \theta + \cos \lambda \sin \bar{\alpha} (\sin^2 \theta + 2) \cos \theta)] , \end{aligned} \tag{5.31}$$

$$\begin{aligned} X_4 &= \bar{X}_4 + \delta X_4 \\ &= \bar{X}_4 \text{ (in (5.23) with additional terms of (4.111) in } u_4^{(1)}) + \\ &\quad + \text{terms of } \delta X_4 \text{ in (4.117)} + \\ &\quad + \frac{1}{4\mu_E^2} P\bar{X}_6^2 [(\bar{X}_2 \sin \lambda + \cos \lambda) \sin \bar{\alpha} \cos 2\theta - \\ &\quad - (\cos \bar{\alpha} + \bar{X}_1 \sin \lambda \sin \bar{\alpha}) \sin 2\theta] - \\ &\quad - \frac{2}{3} \frac{1}{\mu_E^3} P\bar{X}_6^4 [\bar{X}_3 (\cos \bar{\alpha} (\cos^2 \theta + 2) \sin \theta - \cos \lambda \sin \bar{\alpha} \cos^3 \theta) + \\ &\quad + \bar{X}_4 (- \cos \bar{\alpha} \cos^3 \theta + \cos \lambda \sin \bar{\alpha} \sin^3 \theta)] , \end{aligned} \tag{5.32}$$

$$\begin{aligned} X_5 &= \bar{X}_5 + \delta X_5 \\ &= \text{same as (4.118)} , \end{aligned} \tag{5.33}$$

$$\begin{aligned} X_6 &= \bar{X}_6 + \delta X_6 \\ &= \bar{X}_6 \text{ (in (5.24)) + terms of } \delta X_6 \text{ in (4.119)} - \\ &\quad - P \frac{1}{\mu_E^3} \bar{X}_6^5 (\cos \bar{\alpha} \cos \theta + \cos \lambda \sin \bar{\alpha} \sin \theta) + \\ &\quad + \frac{3}{4} P \frac{1}{\mu_E^4} \bar{X}_6^7 [\bar{X}_3 (\cos \bar{\alpha} \cos 2\theta + \cos \lambda \sin \bar{\alpha} \sin 2\theta) + \\ &\quad + \bar{X}_4 (\cos \bar{\alpha} \sin 2\theta - \cos \lambda \sin \bar{\alpha} \cos 2\theta)] , \end{aligned} \tag{5.34}$$

$$\begin{aligned} \alpha &= \bar{\alpha} + \delta\alpha \\ &= \bar{\alpha} - 2K_1 r_c (\bar{X}_3 \sin \theta - \bar{X}_4 \cos \theta) + \alpha_0 , \end{aligned} \tag{5.35}$$

$$\begin{aligned} \nu &= \bar{\nu} + \delta_1 \nu + \delta\nu \\ &= \bar{\nu} + \delta_1 \nu - 2K_2 r_c (\bar{X}_3 \sin \theta - \bar{X}_4 \cos \theta) + \nu_0 , \end{aligned} \tag{5.36}$$

$$\begin{aligned} \bar{\Omega}_M &= \bar{\Omega}_{MS} + \delta\Omega_M \\ &= \bar{\Omega}_{MS} - 2K_3 r_c (\bar{X}_3 \sin \theta - \bar{X}_4 \cos \theta) + \Omega_{M0} , \end{aligned} \tag{5.37a}$$

and

$$\begin{aligned} \omega_M &= \bar{\omega}_{MS} + \delta\omega_M \\ &= \bar{\omega}_{MS} - 2K_4 r_c (\bar{X}_3 \sin \theta - \bar{X}_4 \cos \theta) + \omega_{M0} . \end{aligned} \tag{5.37b}$$

In order to introduce the time into the present problem, we rewrite (4.125) as

$$\frac{\overline{dt}}{d\theta} = \frac{\bar{X}_6^3}{\mu_E}\Big\{(1 + \bar{X}_1^2 + \bar{X}_2^2)^{3/2} - \omega_4 + \frac{3}{2}\frac{K_1^2}{(1 - e_S^2)^{3/2}}\sin^2\lambda\cos 2\bar{\alpha} +$$
$$+ \frac{3}{2}\frac{\kappa K_1^2}{(1 - e_S^2)^{3/2}}[\sin^2\lambda\cos 2(\bar{\nu} + \bar{\omega}_M + \bar{\Omega}_M) +$$
$$+ \sin 2\lambda \sin I(\cos\bar{\Omega}_M - \cos(2\bar{\nu} + 2\bar{\omega}_M + \bar{\Omega}_M))] \times$$
$$\times \Big(1 + \frac{9}{4}(\bar{X}_1^2 + \bar{X}_2^2)\Big)\Big\}, \tag{5.38a}$$

where (5.20) and (5.21) are used for $\bar{X}_1$ and $\bar{X}_2$, respectively, and the corresponding expression of (4.105a) on p. 120 for $\bar{X}_6^3$ is

$$\bar{X}_6^3 = \bar{X}_{60}^3\Big\{1 + 3\frac{\bar{X}_{60}^6}{\bar{\bar{X}}_{60}^6}(\tfrac{1}{2}\varepsilon a_0(1 + \kappa)\sin 2\lambda(\cos\Psi - \cos\Psi_0) +$$
$$+ \frac{9}{4}\frac{\bar{X}_{60}^6}{\bar{\bar{X}}_{60}^6}[\tfrac{1}{2}\varepsilon a_0(1 + \kappa)\sin 2\lambda]^2(\cos 2\Psi - \cos 2\Psi_0) + F - F_0 +$$
$$+ \frac{3}{8}\varepsilon r_c^2\chi b_0\Bigg[\frac{\omega_{31}\cos\lambda - \omega_3}{K_1 + \omega_3}[\cos(\bar{\alpha} + \xi) - \cos(\bar{\alpha}_0 + \xi_0)] +$$
$$+ \frac{\omega_{31}\cos\lambda + \omega_3}{K_1 - \omega_3}[\cos(\bar{\alpha} - \xi) - \cos(\bar{\alpha}_0 - \xi_0)]\Bigg]\Big\}$$

when (5.24) is employed.

The integration of (5.38a) with the corresponding known values of $\bar{X}_1$, $\bar{X}_2$, $\bar{X}_6^3$, $\bar{\alpha}$, $\bar{\nu}$, $\bar{\Omega}_M$, and $\bar{\omega}_M$ results in the first-order approximation for $\bar{t}$:

$$\bar{t} - \bar{t}_0 = \frac{\bar{X}_{60}^3}{\mu_E^2}(1 - 2\omega_4)[f_1(\theta, \Psi) + \varepsilon f_2(\theta, \Psi, \xi, \bar{\alpha}, \bar{\nu}, \bar{\Omega}_M, \bar{\omega}_M)]\Bigg|_{\theta_0}^{\theta}, \tag{5.38b}$$

where $f_1(\theta, \Psi)$ is shown in (4.127) and

$$f_2(\theta, \Psi, \xi, \bar{\alpha}, \bar{\nu}, \bar{\Omega}_M, \bar{\omega}_M) = \text{Terms in (4.128)} + {}^{D_{40}}_{D_{41}}\sin(\bar{\alpha} \pm \xi) +$$
$$+ {}^{D_{42}}_{D_{43}}\sin(\bar{\alpha} + \xi \pm \Psi) +$$
$$+ {}^{D_{44}}_{D_{45}}\sin(\bar{\alpha} - \xi \pm \Psi) +$$
$$+ {}^{D_{46}}_{D_{47}}\sin(\bar{\alpha} + \xi \pm 2\Psi) +$$
$$+ {}^{D_{48}}_{D_{49}}\sin(\bar{\alpha} - \xi \pm 2\psi). \tag{5.39}$$

The constants D_{40}-D_{49} are presented in Appendix D.

The improved first-order approximation for t is

$$t = \bar{t} + \frac{\bar{X}_6^3}{\mu_E^2}(1 + \bar{X}_1^2 + \bar{X}_2^2)^2\Bigg[\frac{1}{(1 + \bar{X}_1^2 + \bar{X}_2^2)^{1/2}}\Big\{\tan^{-1}\Big[(1 + \bar{X}_1^2 + \bar{X}_2^2)^{1/2} \times$$
$$\times \frac{\bar{X}_2\sin\theta + \bar{X}_1\cos\theta}{\bar{X}_2\cos\theta - \bar{X}_1\sin\theta}\Big] - \Big(\theta + \tan^{-1}\Big(\frac{\bar{X}_1}{\bar{X}_2}\Big)\Big)\Big\} +$$
$$+ \frac{2\bar{X}_6^2}{\mu_E}\frac{\bar{X}_4\cos\theta - \bar{X}_3\sin\theta + (\bar{X}_1\bar{X}_4 - \bar{X}_2\bar{X}_3)(\bar{X}_2\sin\theta + \bar{X}_1\cos\theta}{[1 + (\bar{X}_1\cos\theta + \bar{X}_2\sin\theta)^2]^{1/2}}\Bigg] \tag{5.40}$$

with all $\bar{X}$'s, $\bar{\alpha}$, $\bar{\nu}$, $\bar{\Omega}_M$, and $\bar{\omega}_M$ known from (5.20)-(5.27).

5.5. Recapitulation

Since θ and $X_1, \ldots, X_4, X_6, \alpha, \nu, \Omega_M$, and ω_M are related by (5.29)-(5.37), wherein $\bar{X}_1, \ldots, \bar{X}_4, \bar{X}_6, \bar{\alpha}, \bar{\nu}, \bar{\Omega}_M$, and $\bar{\omega}_M$ are determined by (5.20)-(5.27), the three basic geometric parameters r, θ, ϕ, and the angular momentum about the polar axis of the present problem can readily be obtained through (5.4)-(5.6). Therefore, (5.4)-(5.6) are the solutions to (5.1)-(5.3). The time, t, is related to the solutions by (5.38b) and (5.40).

The reader is referred to Section 4.9. for actual trajectory calculations and the determination of constants $\bar{X}_{10}, \ldots, \bar{X}_{40}, \bar{X}_{60}, \bar{\alpha}_0, \bar{\nu}_0, \bar{\Omega}_{M0}$, and $\bar{\omega}_{M0}$.

5.6. Osculating Orbital Elements

Introducing (5.31) and (5.32) into (5.5) and comparing the resulting expressions with the osculating orbit

$$r = \frac{\ell}{1 + e\cos(\eta - \omega)} \tag{5.41}$$

defined in (2.107), we have

$$\ell = \bar{\ell} + \ell_P = \ell_s + \ell_L + \ell_P , \tag{5.42}$$

$$\begin{aligned} \ell_s &= \text{Terms in (4.136)} , \\ \ell_L &= \text{Terms in (4.137)} - \\ &\quad \frac{\bar{X}_6^6}{\mu_E}(1 + \bar{X}_1^2 + \bar{X}_2^2)(\bar{X}_3 \cos\bar{\alpha} + \bar{X}_4 \cos\lambda \sin\bar{\alpha}) , \end{aligned} \tag{5.43}$$

$$\begin{aligned} \ell_P &= \text{Terms in (4.138)} - \\ &\quad - P\frac{\bar{X}_6^6}{3\mu_E^4}[\bar{X}_3(-\cos\lambda\sin\bar{\alpha}\sin 2\theta + \cos\bar{\alpha}\cos 2\theta) + \\ &\quad + \bar{X}_4(-\cos\bar{\alpha}\sin 2\theta + \cos\lambda\sin\bar{\alpha}\cos 2\theta)] , \end{aligned} \tag{5.44}$$

and

$$e\cos(\eta - \omega) = \ell(\bar{X}_3' \cos\theta + \bar{X}_4' \sin\theta) ,$$

$$\begin{aligned} \bar{X}_3' &\equiv \bar{X}_3 + \delta\bar{X}_3' \\ &= \bar{X}_3 \text{ (in (5.22) with additional terms of (4.110) in } u_3^{(1)}) + \\ &\quad + \text{terms of } \delta\bar{X}_3' \text{ in (4.140)} - \\ &\quad - P\frac{\bar{X}_6^2}{4\mu_E^2}(\bar{X}_1 \sin\lambda\sin\bar{\alpha} + \cos\bar{\alpha}) \end{aligned} \tag{5.45}$$

and

$$\begin{aligned} \bar{X}_4' &\equiv \bar{X}_4 + \delta\bar{X}_4' \\ &= \bar{X}_4 \text{ (in (5.23) with additional terms of (4.111) in } u_4^{(1)}) + \\ &\quad \text{terms of } \delta\bar{X}_4' \text{ in (4.141)} - \\ &\quad - P\frac{\bar{X}_6^2}{4\mu_E^2}(\bar{X}_2 \sin\lambda + \cos\lambda)\sin\bar{\alpha} . \end{aligned} \tag{5.46}$$

In order to understand some of the physical aspects of the present problem from the previous analysis, only the zeroth-order solutions of the X's will be examined. These solutions are

$$\bar{X}_1 = \omega_1 a_0 \sin \Psi , \tag{5.47}$$

$$\bar{X}_2 = \omega_{11} a_0 \cos \Psi - D , \tag{5.48}$$

$$\bar{X}_3 \cong \bar{X}_3' = \omega_{31} b_0 \cos \xi - \frac{3}{4}\frac{\varepsilon\chi}{K_1} \cos \lambda (\cos \bar{\alpha} - \cos \bar{\alpha}_0) , \tag{5.49}$$

$$\bar{X}_4 \cong \bar{X}_4' = \omega_3 b_0 \sin \xi - \frac{3}{4}\frac{\varepsilon\chi}{K_1} (\sin \bar{\alpha} - \sin \bar{\alpha}_0) , \tag{5.50}$$

and

$$\bar{X}_6 = \bar{X}_{60}[1 + \frac{1}{2}(1 + \kappa)\varepsilon a_0 \sin 2\lambda \frac{\bar{X}^6_{60}}{\bar{\bar{X}}^6_{60}}(\cos \Psi - \cos \Psi_0)] , \tag{5.51}$$

where the dominant first-order solar radiation pressure terms are included in the above zeroth-order solutions because they will bring out the basic characteristics of our present problem.

For the purpose of discussing the behavior of orbital elements, we shall limit our discussion to the case of a circular geostationary orbit (r_c). This limitation will result in $b_0 = 0$. If the initial orbit is in the equatorial plane, it can be found that $a_0 = D/\omega_{11}$ and $\Psi_0 = 0$ when $\theta_0 = 0$. Under these simplified conditions, the resulting expressions will give us some meaningful interpretation of the physical phenomenon of a geostationary satellite under the combined influences of the oblate Earth, the Sun, the Moon, and solar radiation pressure.

5.6.1. Time ($\bar{t}$), Ascending Node ($\bar{\Omega}$), and Inclination ($\bar{i}$)

Upon comparing the zeroth-order solutions of the X's, which include the solar radiation pressure (shown in (5.47)-(5.51)), to the solutions without such pressure (as given in (4.142)), it can be seen that $\bar{X}_1$, $\bar{X}_2$, and $\bar{X}_6$ are identical while $\bar{X}_3$ and $\bar{X}_4$ are different. Consequently, the expressions for determining the time (4.147), the line of nodes (4.151), and the inclination of the orbital plane (4.154) remain the same. They are

$$\bar{t} = \frac{1}{\dot{\theta}_E}\Big\{\theta + \frac{3D^2}{\omega_1(1-\omega_4)}[\omega_4 + 6D_2(1 - 2\omega_4)](\sin \Psi - \omega_1\theta) + \frac{9}{4}\frac{D^4}{\omega_1(1-\omega_4)}(1 - 2\omega_4)(2\omega_1\theta - \sin 2\Psi)\Big\} , \tag{5.52}$$

$$\bar{\Omega} = \frac{\pi}{2} - \frac{\Psi}{2} , \tag{5.53}$$

and

$$\tan \bar{i} = 2D \sin \frac{\Psi}{2} , \tag{5.54}$$

where the initial condition $\bar{t}_0 = 0$, $\theta_0 = 0$ has been employed.

The predominant physical effect of the Sun's and Moon's attractions is the rotation and the regression of the orbital plane as previously discussed in Chapter 4. The solar radiation pressure has only a nominal effect on the movement of the orbital plane and the orbital period. In other words, the solar radiation pressure effect will be exhibited in the first-order solutions of the orbital period, the inclination of the orbital plane, and the ascending node.

5.6.2. Semilatus Rectum ($\bar{\ell}$) and Semimajor Axis ($\bar{a}$)

From (4.135)-(4.137) and (5.42), we have

$$\bar{\ell} = \frac{1}{\mu_E}\bar{X}_6^2(1 + \bar{X}_1^2 + \bar{X}_2^2)\left\{1 - \frac{3}{2}J_2\frac{R_E^2}{r_c^2} - \frac{1}{4}\frac{K_1^2}{(1 - e_S^2)^{3/2}} \times \right.$$
$$\left. \times\ [(1 + \kappa)(1 - 3\cos^2\lambda) + 3\sin^2\lambda\sin 2\bar{\alpha}]\right\}$$
$$= \frac{\bar{\bar{X}}_{60}^2}{\mu_E}\left[1 + \frac{\bar{X}_{60}^6}{2\bar{\bar{X}}_{60}^6}\varepsilon a_0(1 + \kappa)\sin 2\lambda(\cos\Psi - 1)\right]^2[1 + D^2\sin^2\Psi +$$
$$+ D^2(\cos\Psi - 1)^2]\left\{1 - \frac{3}{2}J_2\frac{R_E^2}{r_c^2} - \frac{1}{4}\frac{K_1^2}{(1 - e_S^2)^{3/2}} \times \right.$$
$$\left. \times\ [(1 + \kappa)(1 - 3\cos^2\lambda) + 3\sin^2\lambda\sin 2\bar{\alpha}]\right\}. \tag{5.55}$$

For a short duration, say one year, $\cos\Psi \cong 1$ and $\sin\Psi \cong 0$, and the above equation may be simplified to

$$\bar{\ell} = \frac{\bar{\bar{X}}_{60}^2}{\mu_E}\left\{1 - \frac{3}{2}J_2\frac{R_E^2}{r_c^2} - \frac{1}{4}\frac{K_1^2}{(1 - e_S^2)^{3/2}} \times \right.$$
$$\left. \times\ [(1 + \kappa)(1 - 3\cos^2\lambda) + 3\sin^2\lambda\sin 2\bar{\alpha}]\right\}. \tag{5.56}$$

This equation shows that the semilatus rectum changes from its minimum value to its maximum in a three-month cycle, as is seen from the term containing $\sin 2\bar{\alpha}$.

From the general relationship $a \equiv \ell/(1 - e^2)$, the semimajor axis is found to be

$$\bar{a} = \frac{\bar{\bar{X}}_{60}^2}{\mu_E}\left\{1 - \frac{3}{2}J_2\frac{R_E^2}{r_c^2} - \frac{1}{4}\frac{K_1^2}{(1 - e_S^2)^{3/2}}[(1 + \kappa)(1 - 3\cos^2\lambda) + \right.$$
$$\left. + 3\sin^2\lambda\sin 2\bar{\alpha}]\right\}\left\{1 + \frac{9}{16}r_c^2\left(\frac{\varepsilon\chi}{K_1}\right)^2(1 + D^2\sin^2\Psi) \times \right.$$
$$\left. \times\ [\cos^2\lambda(1 - \cos\bar{\alpha})^2 + (1 + D^2\sin^2\Psi)\sin^2\bar{\alpha}]\right\}. \tag{5.57}$$

Again, during an one-year period, $D^2\sin^2\Psi$ is of negligible importance. (5.57) denotes that the semimajor axis reaches its maximum value at the mid-year and then returns to its starting value by the end of the year.

5.6.3. Argument of Perigee ($\bar{\omega}$) and Eccentricity ($\bar{e}$)

A substitution of $\bar{X}_1, \ldots, \bar{X}_3', \bar{X}_4', \bar{X}_6$ into (4.168) and (4.169) gives

$$\tan\bar{\omega} = \frac{\bar{X}_1\bar{X}_3' + \bar{X}_2\bar{X}_4'}{\bar{X}_2\bar{X}_3' - \bar{X}_1\bar{X}_4'}\frac{1}{(1 + \bar{X}_1^2 + \bar{X}_2^2)^{1/2}}$$
$$= \frac{\sin\Psi\cos\lambda(1 - \cos\bar{\alpha}) - (\cos\Psi - 1)\sin\bar{\alpha}}{\cos\lambda(1 - \cos\bar{\alpha})(\cos\Psi - 1) + \sin\Psi\sin\bar{\alpha}} \times$$
$$\times \frac{1}{[1 + 4D^2\sin^2\frac{\Psi}{2}]^{1/2}}$$
$$\cong \cos\lambda\tan\frac{\bar{\alpha}}{2} \tag{5.58}$$

and

$$\bar{e} = \bar{\ell}[\bar{X}_3'^2 + \bar{X}_4'^2 + (\bar{X}_2\bar{X}_3' - \bar{X}_1\bar{X}_4')^2]^{1/2}$$

$$\cong \frac{\bar{\bar{X}}_{60}^2}{\mu_E}\left\{1 - \frac{3}{2}J_2\frac{R^2}{r_c^2} - \frac{1}{4}\frac{K_1^2}{(1 - e_S^2)^{3/2}}[(1 + \kappa)(1 - 3\cos^2\lambda) + \right.$$

$$\left. + 3\sin^2\lambda\sin 2\bar{\alpha}]\right\}\left(-\frac{3}{4}\frac{\varepsilon\chi}{K_1}\right)\left\{\cos^2\lambda(1 - \cos\bar{\alpha})^2 + \sin^2\bar{\alpha} + \right.$$

$$\left. + D^2[(\cos\Psi - 1)(1 - \cos\bar{\alpha})\cos\lambda + \sin\Psi\sin\bar{\alpha}]^2\right\}^{1/2}$$

$$\cong -\frac{3}{4}r_c\frac{\varepsilon\chi}{K_1}[\cos^2\lambda(1 - \cos\bar{\alpha})^2 + \sin^2\bar{\alpha}]^{1/2}, \qquad (5.59)^*$$

where $\omega_{11} \cong \omega_1$, $\bar{\alpha}_0 = 0$, and the terms containing $(\cos\Psi - 1)$ and $\sin^2\Psi$ are omitted because only a short duration of one year is under consideration.

Differentiations of (5.58) and (5.59) with respect to θ result in

$$\frac{d\bar{\omega}}{d\theta} = \frac{1}{2}\frac{1 + \tan^2\frac{\alpha}{2}}{1 + \cos^2\lambda\tan^2\frac{\bar{\alpha}}{2}}\frac{d\bar{\alpha}}{d\theta} \qquad (5.60)$$

and

$$\frac{d\bar{e}}{d\theta} = -\frac{3}{4}r_c\frac{\varepsilon\chi}{K_1}\times\frac{\cos^2\lambda(1 - \cos\bar{\alpha})\sin\bar{\alpha} + \sin\bar{\alpha}\cos\bar{\alpha}}{[\cos^2\lambda(1 - \cos\bar{\alpha})^2 + \sin^2\bar{\alpha}]^{1/2}}\times\frac{d\bar{\alpha}}{d\theta}. \qquad (5.61)$$

Equations (5.58)-(5.61) reveal that the solar radiation pressure results in an eccentricity change as well as in the rotation of the line of apsides. As is seen from (5.59), the eccentricity increases from zero, when $\bar{\alpha} = 0$, to its maximum value, $-\frac{3}{2}r_c(\varepsilon\chi/K_1)\cos\lambda$, when $\bar{\alpha} = 180°$. It then reduces to zero again when $\bar{\alpha} = 360°$. In other words, the variation of eccentricity has a cycle of one year. (5.58) shows that the total rotation of the line of apsides is approximately one half that of the Sun's rotation around the Earth. The line of apsides' rate of rotation remains practically constant, and its magnitude is also approximately one half that of the Sun's, as shown in (5.60). Further examination of (5.60) shows that the rotation rate has its minimum value, $\frac{1}{2}\,d\bar{\alpha}/d\theta$, when $\bar{\alpha} = 0$, and its maximum value, $1/(2\cos^2\lambda)\,d\bar{\alpha}/d\theta$, when $\bar{\alpha} = 180°$. On the other hand, the rate of change of the eccentricity is shown in (5.61); the maximum rate is $-\frac{3}{4}r_c(\varepsilon\chi/K_1)\times d\bar{\alpha}/d\theta$, when $\bar{\alpha} = 0°$, and the minimum rate is zero, when $\bar{\alpha} = 180°$. A picture history of the orbit changes due to solar radiation pressure over an one year period is shown in Figure 5.5. It can be seen that for an initially circular synchronous orbit, the perturbation of solar radiation pressure on the satellite will increase the orbiting velocity along one half of the orbit in which the satellite travels in the direction of the action of the radiation pressure, while in the other half of the orbit, the radiation pressure is against the orbiting velocity of the satellite. Consequently, the orbit becomes elliptic and the perigee of the orbit is established where the effect of the radiation pressure on the satellite's orbiting velocity is the greatest. This fact is depicted in the figure by noting the

---

*Because $\chi < 0$ as defined in (5.28) and $\bar{e} > 0$ for an elliptical orbit, a ' - ' sign is adopted for the value of the square root.

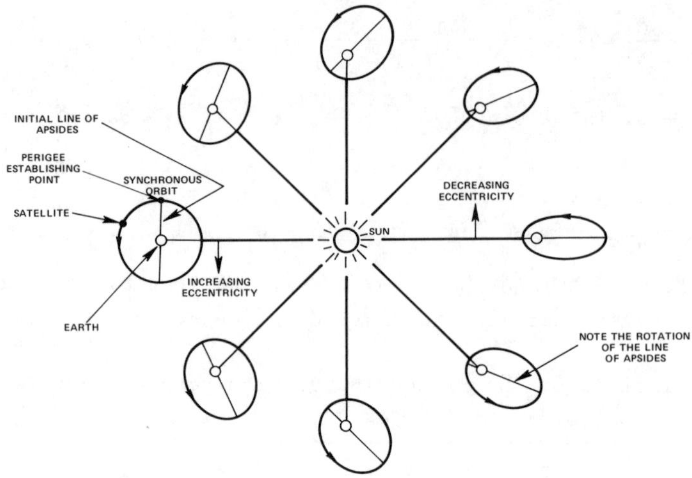

Fig. 5.5. Perturbation history of the geostationary orbit due to solar radiation pressure.

location of the 'perigee establishing point'.

5.6.4. Short-Period Terms of Orbit Elements

All short-period terms of osculating orbit elements can be obtained by applying the variational technique to the element of interest in terms of $\bar{X}$'s and δX's. These are shown in Chapter 4, so we will not repeat it here.

5.7. Canonical Method of Solution*

One of the most comprehensive solutions related to the problem of solar radiation pressure acting on geostationary satellites was presented by Voshkov'yak and Lidov (1973). Their approach may be considered as a canonical one because they developed a simplified Hamiltonian and applied it to a system of Delaunay equations. The classical method of solving the system of Delaunay equations is the von Zeipel method as was previously mentioned in Chapter 2. However in view of the simplified form of the Hamiltonian, the system of Delaunay equations is resolved into three separate independent sets of equations, and these three sets equations are not cross-coupled. Consequently, each set of the Delaunay equations can be solved by the theory in linear differential equations, and a closed-form solution is obtained.

The simplified Hamiltonian (or force) function is in a form of

---

*The material used in this section is copyrighted by the Plenum Publishing Corporation and is quoted with the permission of Plenum.

$$H(\xi, \eta, p, q, \delta, \lambda_A, \theta_g) = H_1(\xi, \eta, \theta_g) + H_2(p, q, \theta_g) + H_3(\delta, \lambda_A),$$
$$H_1 = -\tfrac{1}{2}\beta_1(\xi^2 + \eta^2),$$
$$H_2 = \tfrac{1}{2}\beta_1(p^2 + q^2) + 2\beta_2 L^*[p(-\tfrac{1}{2}\beta_3 \sin 2\lambda + I \cos 2\lambda \cos \Omega_M) - qI \cos \lambda \sin \Omega_M], \tag{5.62}$$

and

$$H_3 = \tfrac{3}{2}\delta^2 + A\delta + B(\lambda_A),$$

where a geocentric equatorial system Oxyz was taken as the basic system of coordinates with the equatorial plane of the Earth as the Oxy plane, the Ox axis is directed toward the vernal equinox, the Oz axis toward the Earth north pole, and the Oy axis completes the right-hand system of coordinates.

The independent variable for the Hamiltonian function is the angle of rotation of the Earth,

$$\theta_g = \dot{\theta}_E t, \tag{5.63}$$

where θ_g is the angular position of the Greenwich meridian in the ecliptic measured from the vernal equinox. With the usual orbital elements, a, e, i, ω, Ω, and M, as well as the values derived from them: $\tilde{\omega} \equiv \omega + \Omega$, longitude of perigee, and $T = 2\pi\, a^{3/2}/\mu_E{}^{1/2}$, orbital period, the parameters in (5.62) are defined as:

$$L^* \equiv \sqrt{\mu_E r_s}, \quad \lambda_A \equiv M + \omega + \Omega - \theta_g = M + \tilde{\omega} - \theta_g, \quad \xi \equiv (2G)^{1/2} \cos \tilde{\omega},$$
$$\eta \equiv -(2G)^{1/2} \sin \tilde{\omega}, \quad p \equiv [2(G - H)]^{1/2} \cos \Omega,$$
$$q \equiv -[2(G - H)]^{1/2} \sin \Omega, \quad \beta_1 \equiv \frac{3}{2}\left(\frac{R_E}{r_s}\right)^2 J_2, \quad \beta_2 \equiv \frac{3}{8}\left(\frac{n_M}{\dot{\theta}_E}\right)^2,$$
$$\beta_3 \equiv 1 + \left(\frac{n_S}{n_M}\right)^2,$$
$$A \equiv 2\beta_1 + (1 - \tfrac{3}{2}\sin^2 \lambda)\left[\left(\frac{n_M}{\dot{\theta}_E}\right)^2 + \left(\frac{n_S}{\dot{\theta}_E}\right)^2\right],$$

and

$$\delta \equiv \frac{L}{L^*} - 1 = \left(\frac{a}{r_s}\right)^{1/2} - 1.$$

G, H, I, L, r_s, R_E, n_S, n_M, $\dot{\theta}_E$, λ, λ_A, and Ω_M have been defined previously, and

$$B(\lambda_A) \equiv 3\left(\frac{R_E}{r_s}\right)^2 J_{22} \cos 2(\lambda_A - \lambda_{22}),$$

when only the Earth's equatorial ellipticity is taken into consideration.

In obtaining H_1, H_2, and H_3, the process of averaging over the periods of the satellite, the Sun, and the Moon was performed; consequently, the orbital elements in H have their averaged values over the respective orbital periods. The simplified form of H was derived based on the following assumptions:

(1) omission of the terms containing the eccentricity of the orbits of the satellite, the Sun, and the Moon;

(2) neglecting individual components of the disturbing function and a number of terms proportional to the inclination of the satellite orbit;

(3) approximation of $n = \dot{\theta}_E$; and

(4) dropping terms containing I^2 in H_1 and H_2 and those containing I in H_3.

As a result of the simplification of H, the Delaunay equations of the problem can be separated into three independent systems:

(1) The equations

$$\frac{d\xi}{d\theta_g} = \frac{\partial H}{\partial \eta} = \frac{\partial H_1}{\partial \eta} \quad \text{and} \quad \frac{d\eta}{d\theta_g} = -\frac{\partial H}{\partial \xi} = -\frac{\partial H_1}{\partial \xi} \tag{5.64}$$

describe the variation of the eccentricity, e, and longitude of perigee, $\tilde{\omega}$.

(2) The equations

$$\frac{dp}{d\theta_g} = \frac{\partial H}{\partial q} = \frac{\partial H_2}{\partial q} \quad \text{and} \quad \frac{dq}{d\theta_g} = -\frac{\partial H}{\partial p} = -\frac{\partial H_2}{\partial p} \tag{5.65}$$

show the change of the inclination, i, and the longitude of the ascending node, Ω.

(3) The equations

$$\frac{d\delta}{d\theta_g} = \frac{\partial H}{\partial \lambda_A} = \frac{\partial H_3}{\partial \lambda_A}, \quad \frac{d\lambda_A}{d\theta_g} = -\frac{\partial H}{\partial \delta} = -\frac{\partial H_3}{\partial \delta} \tag{5.66}$$

give the varåtions in the mean geographical longitude, λ_A, and the parameter, δ, which is related to the semimajor axis, a.

Equations (5.64) and (5.65) can be easily solved by the Lagrange method of variation of parameters in·the theory of differential equations. Their solutions are[§]

$$\xi = \gamma_2 \cos \beta_1(\theta_g - \theta_{g0}) + \tfrac{3}{2}\sqrt{L^*}\,\frac{Pr_s^2}{\mu_E}\,\frac{\beta_1 - (n_S/\dot{\theta}_E)\cos\lambda}{\beta_1^2 - (n_S/\dot{\theta}_E)^2}\cos\alpha\,,$$

$$\eta = \gamma_2 \sin \beta_1(\theta_g - \theta_{g0}) + \tfrac{3}{2}\sqrt{L^*}\,\frac{Pr_s^2}{\mu_E}\,\frac{\beta_1 \cos\lambda - (n_S/\dot{\theta}_E)}{\beta_1^2 - (n_S/\dot{\theta}_E)^2}\sin\alpha\,, \tag{5.67}$$

and

$$p = -\gamma_3 \cos \beta_1(\theta_g - \theta_{g0}) + \frac{\beta_2}{\beta_1}\sqrt{L^*}\beta_3 \sin 2\lambda - \frac{2}{(\Omega_M/\dot{\theta}_E) - \beta_1}\beta_2 I\sqrt{\Lambda^*}\cos\lambda\cos\Omega_M\,,$$

$$q = \gamma_3 \sin \beta_1(\theta_g - \theta_{g0}) - \frac{2}{(\Omega_M/\dot{\theta}_E) - \beta_1}\beta_2 I\sqrt{\Lambda^*}\cos\lambda\sin\Omega_M\,, \tag{5.68}$$

where γ_2, γ_3, and θ_{g0} are integration constants, P is the solar radiation pressure, and α is the mean longitude of the Sun.

Since H_3 is independent of θ_g, Equation (5.66) has a first integral,

$$H_3 = h = \tfrac{3}{2}\delta_0^3 + A\delta_0 + B(\lambda_{A0}) = \text{constant} \tag{5.69}$$

where δ_0 and λ_{A0} are the initial values of δ and λ_A at θ_{g0}. A substitution of (5.69) into H_3 into (5.62) gives

$$\delta = \tfrac{1}{3}[-A \pm \sqrt{G(\lambda_A)}]\,, \tag{5.70}$$

where

$$G(\lambda_A) = A^2 + 6[h - B(\lambda_A)]\,. \tag{5.71}$$

[§]The author has simplified the presentation of ξ and η in (5.67) and also modified the solutions to p and q. Vashkov'yak and Lidov's original solution to p and q is in a form of Taylor series expansion to the second-order term; readers will notice that the expressions in (5.68) are a closed-form solution to (5.65).

As (5.62), (5.66), and (5.70) depict

$$\frac{d\lambda_A}{d\theta_g} = -\frac{\partial H_3}{\partial \delta} = -3\delta + A = \pm\sqrt{G(\lambda_A)}, \tag{5.72}$$

it follows that

$$\theta_g = \pm \int_{\lambda_{A0}}^{\lambda_A} \frac{1}{\sqrt{G(\lambda_A)}}\, d\lambda_A . \tag{5.73}$$

The result of integration is an elliptical function if $B(\lambda_A)$ contains only the second sectorial harmonic (J_{22}). This expression depicts the well-known libration motion of geostationary satellites.

Because

$$\delta = \left(\frac{a}{r_s}\right)^{1/2} - 1 = \left(\frac{r_s + \Delta a}{r_s}\right)^{1/2} - 1 \cong \frac{1}{2}\frac{\Delta a}{r_s}, \tag{5.74}$$

and

$$T = 2\pi\, a^{3/2}/\mu_E^{\;1/2} = 2\pi (r_s + \Delta a)^{3/2}/\mu_E^{\;1/2}$$
$$= \frac{2\pi}{\dot{\theta}_E}\left(1 + \frac{3}{2}\frac{\Delta a}{r_s}\right), \tag{5.75}$$

combining (5.70), (5.71), and (5.75) gives

$$T = \frac{2\pi}{\dot{\theta}_E}(1 - A) \pm \frac{2\pi}{\dot{\theta}_E}\sqrt{G(\lambda_A)}$$
$$= T^* \pm \Delta T(\lambda_A) \tag{5.76}$$

and

$$T^* \equiv \frac{2\pi}{\dot{\theta}_E}(1 - A), \quad \Delta T(\lambda_A) \equiv \pm \frac{2\pi}{\dot{\theta}_E}\sqrt{G(\lambda_A)}, \tag{5.77}$$

where T^* is the constant period determined by taking into account the secular disturbances from the oblateness of the Earth, the Sun, and the Moon, and $\Delta T(\lambda_A)$ shows the deviation of the orbital period from T^*.

5.8. Comparison of the Canonical and Proposed Solutions

It is interesting to know that both the canonical and proposed solutions are derived from the same form of equations with a similar type of parameters. We note that, after utilizing the approximations of small eccentricity and near-zero inclination, the parameters used by Voshkov'yak and Lidov can be easily transformed into orbital elements, which, in turn, may be expressed in terms of our parameters by the means of (2.113), (2.117), (2.121), and (2.122). Therefore,

$$\bar{e} \cong \left(\frac{\xi^2 + \eta^2}{L^*}\right)^{1/2}, \tag{5.78}$$

$$\tan \tilde{\bar{\omega}} = -\left(\frac{\eta}{\xi}\right) = \tan(\bar{\omega} + \bar{\Omega}) = \frac{\tan\bar{\omega} + \tan\bar{\Omega}}{1 - \tan\bar{\omega}\tan\bar{\Omega}}, \tag{5.79}$$

$$\bar{i} \cong \left(\frac{p^2 + q^2}{L^*}\right)^{1/2} \cong (\bar{X}_1^2 + \bar{X}_2^2)^{1/2}, \tag{5.80}$$

and

$$\tan\bar{\Omega} = -\frac{q}{p} = -\frac{\bar{X}_1}{\bar{X}_2}. \tag{5.81}$$

Combining (5.78), (5.79), and (5.81) results in

$$\eta \cong -\sqrt{L^*}\,\bar{X}_4', \quad \xi \cong \sqrt{L^*}\,\bar{X}_3'. \tag{5.82}$$

In studying the Earth's equatorial ellipticity effect, Vashkov'yak and Lidov used λ_A and δ as their parameters, while the proposed method used Γ and Δr_c as the corresponding parameters. These two sets of parameters can be related by

$$\lambda_A - \lambda_{22} - 90° = \Gamma, \tag{5.83}$$

and

$$\delta \cong \frac{1}{2}\frac{\Delta a}{r_s} \cong \frac{1}{2}\frac{\Delta r_c}{r_c}. \tag{5.84}$$

Based on the above observations, we see the following one-to-one correspondence of the parameters used in the two methods of solution:

$$\bar{X}_1 \leftrightarrow q; \quad \bar{X}_3' \leftrightarrow \xi; \quad \Gamma \leftrightarrow \lambda_A;$$

$$\bar{X}_2 \leftrightarrow p; \quad -\bar{X}_4' \leftrightarrow \eta; \quad \Delta r_c \leftrightarrow \delta.$$

As is expected, the one-to-one correspondence of differential equations to be solved also exists:

(5.12) and (5.13) ↔ (5.65);
(5.14) and (5.15) ↔ (5.64);
(3.18) and (3.21) ↔ (5.66);

and finally their corresponding solutions are:

(5.20) and (5.21) ↔ (5.68);
(5.22) and (5.23) ↔ (5.67);
(3.35) ↔ (5.70);
(3.57) ↔ (5.73).

It should be pointed out that Vashkov'yak and Lidov's solution is only the zeroth-order solution to the problem, while the proposed solution is an improved first-order solution.

As the predominate effects of the perturbations of the oblateness of the Earth, the Sun, and the Moon on geostationary satellites are the rotation and regression of the satellite orbital plane, most investigators derived analytical expressions to compute the inclination and the longitude of ascending node in the sense of the zeroth-order solution with (Kozai, 1973) or without (Allan, 1964; Frick, 1967) the consideration of the movement of the Moon's orbit.

5.9. Summary

In addition to the attractions of the Sun, Moon, and the oblate Earth on a geostationary satellite, the introduction of solar radiation pressure acting on a satellite basically effects the eccentricity and the argument of perigee of the osculating orbit. These two orbital elements are related to our parameters, the X's, by

$$\bar{e} = \bar{\ell}[\bar{X}_3'^2 + \bar{X}_4'^2 + (\bar{X}_2\bar{X}_3' - \bar{X}_1\bar{X}_4')^2]^{1/2}, \tag{5.85}$$

$$\tan\bar{\omega} = \frac{\bar{X}_1\bar{X}_3' + \bar{X}_2\bar{X}_4'}{\bar{X}_2\bar{X}_3' - \bar{X}_1\bar{X}_4'} \times \frac{1}{(1 + \bar{X}_1^2 + \bar{X}_2^2)^{1/2}}, \tag{5.86}$$

and

$$\bar{\ell} \cong \frac{\bar{X}_6^2}{\mu_E^2}(1 + \bar{X}_1^2 + X_2^2) \ , \tag{5.87}$$

while our parameters are

$$\begin{aligned}\bar{X}_1 &\cong \omega_1 a_0 \sin\Psi + \varepsilon[\omega_1 a_0 \sin\Psi + A_3 \sin 2\bar{\alpha} + A_4 \sin\bar{\Omega}_M + \\ &\quad + A_7 \sin(\bar{\Omega}_M + \Psi) + A_8 \sin(\bar{\Omega}_M - \Psi)] \ ,\end{aligned} \tag{5.88}$$

$$\begin{aligned}\bar{X}_2 &\cong \omega_{11} a_0 \cos\Psi + \varepsilon[\omega_{11} a_0 \cos\Psi + A_{25} \cos 2\bar{\alpha} + A_{26} \cos\bar{\Omega}_M + \\ &\quad + A_{29} \cos(\bar{\Omega}_M + \Psi) + A_{30} \cos(\bar{\Omega}_M - \Psi) - G] \ ,\end{aligned} \tag{5.89}$$

$$\begin{aligned}\bar{X}_3' &\cong \omega_{31} b_0 \cos\xi + \frac{\varepsilon}{4}\Big\{B_7 \cos(\bar{\Omega}_M + \xi) + B_8 \cos(\bar{\Omega}_M - \xi) + B_9 \cos\bar{\omega}_M + \\ &\quad + \cos(\bar{\omega}_M + \bar{\Omega}_M)[B_{10} + B_{10M2} \cos(\bar{\lambda}_M - 2\bar{\lambda}_S + \bar{\omega}_M + \bar{\Omega}_M) + \\ &\quad + B_{10M3} \cos 2(\bar{\lambda}_S - \bar{\omega}_M - \bar{\Omega}_M)] + B_{11} \cos(\bar{\omega}_M + 2\bar{\Omega}_M)\Big\} + \\ &\quad + \frac{\varepsilon}{4}\chi\Big\{\frac{1}{\omega_3^2 - K_1^2}[3\omega_{31} + K_1(3\cos\lambda + D\sin\lambda)]\cos\bar{\alpha} - \\ &\quad - \frac{1}{2}\frac{a_0 \sin\lambda}{\omega_3^2 - (K_1 + \omega_1)^2}[\omega_{11}(K_1 + \omega_1) - \omega_1\omega_{31}]\cos(\bar{\alpha} + \Psi) - \\ &\quad - \frac{1}{2}\frac{a_0 \sin\lambda}{\omega_3^2 - (K_1 - \omega_1)^2}[\omega_{11}(K_1 - \omega_1) + \omega_1\omega_{31}]\cos(\bar{\alpha} - \Psi)\Big\} \ ,\end{aligned} \tag{5.90}$$

$$\begin{aligned}\bar{X}_4' &\cong \omega_3 b_0 \sin\xi + \frac{\varepsilon}{4}\Big\{B_{30} \sin(\bar{\Omega}_M + \xi) + B_{31} \sin(\bar{\Omega}_M - \xi) + \\ &\quad + B_{32} \sin\bar{\omega}_M + \sin(\bar{\omega}_M + \bar{\Omega}_M)[B_{33} + B_{33M2} \cos(\bar{\lambda}_M - 2\bar{\lambda}_S + \\ &\quad + \bar{\omega}_M + \bar{\Omega}_M) + B_{33M3} \cos 2(\bar{\lambda}_S - \bar{\omega}_M - \bar{\Omega}_M)] + \\ &\quad + \sin(\bar{\omega}_M + 2\bar{\Omega}_M)[B_{34} + B_{34M4} \cos 2(\bar{\lambda}_S - \bar{\omega}_M - \bar{\Omega}_M)]\Big\} + \\ &\quad + \frac{\varepsilon}{4}\chi\Big\{\frac{1}{\omega_3^2 - K_1^2}[3K_1 + \omega_{32}(3\cos\lambda + D\sin\lambda)]\sin\bar{\alpha} + \\ &\quad + \frac{1}{2}\frac{a_0 \sin\lambda}{\omega_3^2 - (K_1 + \omega_1)^2}[\omega_1(K_1 + \omega_1) - \omega_1\omega_{32}]\sin(\bar{\alpha} + \Psi) + \\ &\quad + \frac{1}{2}\frac{a_0 \sin\lambda}{\omega_3^2 - (K_1 - \omega_1)^2}[\omega_1(K_1 - \omega_1) + \omega_1\omega_{32}]\sin(\bar{\alpha} - \Psi)\Big\} \ ,\end{aligned} \tag{5.91}$$

and

$$\begin{aligned}\bar{X}_6 &\cong \bar{X}_{60}\Big\{1 + \frac{\bar{X}_{60}^6}{\bar{\bar{X}}_{60}^6}[\tfrac{1}{2}\varepsilon a_0(1 + \kappa)\sin 2\lambda](\cos\Psi - \cos\Psi_0) + \\ &\quad + \frac{9}{4}\left(\frac{\bar{X}_{60}^6}{\bar{\bar{X}}_{60}^6}[\tfrac{1}{2}\varepsilon a_0(1 + \kappa)\sin 2\lambda]\right)^2(\cos 2\Psi - \cos 2\Psi_0)\end{aligned} \tag{5.92}$$

where

$$\bar{X}_{60} = r_c^2\dot{\theta}_E/(1 + \bar{X}_{10}^2 + \bar{X}_{20}^2)^{1/2} \ . \tag{5.93}$$

The other orbital elements in terms of X's are

$$\tan\bar{i} = (\bar{X}_1^2 + \bar{X}_2^2)^{1/2} \tag{5.94}$$

and

$$\bar{\Omega} = \tan^{-1}\left(-\frac{\bar{X}_1}{\bar{X}_2}\right) \ . \tag{5.95}$$

As is seen from the expressions of $\bar{X}$'s in (5.88) and (5.89), $\bar{X}_1$ and $\bar{X}_2$ do not contain the solar radiation pressure factor in their dominant terms; therefore, the orbital inclination and the line of nodes are practically not effected by the solar radiation pressure. The additional χ-terms due to the solar radiation pressure in (5.90) and (5.91) result in the rotation of the line of apsides as well as the semi-annual periodic change of eccentricity.

While the above expressions are limited to the mean values, the 'true values' can be obtained by replacing $\bar{X}_n + \delta X_n$ for $\bar{X}_n$ as all δX_n's are derived in the previous sections. It is reminded that the terms in the $\bar{X}$'s in (5.88)-(5.92) are the dominant ones and the remarks on the refinement of all $\bar{X}$'s due to the Sun's perturbation on the Moon's motion are also applicable in this chapter.

Bibliography

The problem of solar radiation pressure and its effect on the motion of an Earth satellite has been documented extensively by Polyakhova (1963). This problem received wide attention after the launch of 'Echo 1' (Parkinson et al., 1960). Basically, Polyakhova collected Musen's analytical works (Musen, 1960) and applied them to the investigation of resonance cases when the motion of satellites undergoes long-period variations with large amplitudes. Musen's derivation is limited to the combined effects of an oblate Earth and solar radiation pressure on orbits in general, and the shadow effect due to solar radiation was not taken into consideration. His expression of argument of perigee becomes singular when the eccentricity of the orbit is zero. He also found that resonance will occur when the angular movement of the perigee of an orbit follows the motion of the Sun. Bryant (1961) used a spherical Earth model to study the Earth shadow effect of solar radiation pressure on an orbit. After deriving the basic equations for the problem, he employed the Kryloff and Bogoliuboff method to average these equations for further numerical integration. Kozai (1961) derived the disturbing function due to solar radiation pressure in the Lagrange's equations for the general planetary motion and obtained expressions for the changes of orbital elements over one complete orbit with the Earth shadow effect taken into consideration. Wyatt (1961) evaluated the effect of solar radiation pressure on the secular acceleration of Earth satellites by studying the change of orbital period with the consideration of the Earth shadow effect. When Allan (1962) derived the rate of change of orbital elements due to lunisolar perturbations, he also considered the solar radiation pressure. The behavior of the argument of perigee and the eccentricity for orbits in the ecliptic plane due to solar radiation alone has been well exposed by Perrine (1966) with the aid of an analog computer. Lubowe (1967) numerically integrated the equations of the problem by employing different shapes (disk, lenticular, and balloon) of a satellite. With the Earth shadow effect as his objective, Levin (1968) simplified the approach to the problem by using a spherical Earth with the ecliptic coincident with the orbital plane. For a constant rate of the Sun revolving around the Earth, he found that the perigee position precessed at half the rate of the Sun, and the Earth's shadow was shown to affect the amplitude of the perturbation on the orbital radius at perigee by an amount equal to the fraction part of the orbit in the shadow. For a distant satellite, the shadow angle (the angle subtending the shadow part of the orbit at the center of the Earth) approaches to zero; there is no appreciable correction necessary due to the shadow. Mello (1972) later approached the shadow effect problem by using the classical Delaunay variables and the von Zeipel method, and he compared the analytical approximation with the numerically integrated results for the Echo 1 case. In determining various perturbing forces acting on a satellite from the observed orbital changes, Sehnal (1969) considered the solar radiation as one of the perturbing forces.

Voshkov'yak published a series of three papers dealing with the movement of a geostationary satellite. All of them included the perturbation of solar radiation pressure. The first paper (Voshkov'yak, 1969) is essentially an application of the variation technique in mathematics to an existing orbit, and the variation terms are then computed by including various perturbing forces. In view of the nature of the approach to the problem, the result is only valid for a short time interval. The second paper (Voshkov'yak, 1972) proposed a method to predict the movement of a geostationary satellite by a semi-analytical method. This method is similar to that suggested by Kozai (1973); the secular effect is obtained by a large-step numerical integration of the osculating orbital parameters, while the short-period effect is derived by analytical means. The third paper (Voshkov'yak, 1973) is an analytical approach, which uses the angle of rotation of the Earth as the independent variable. The results are similar to those zeroth-order approximate solutions derived by Zee (1971, 1972).

In 1973, Zee extended his study on synchronous satellites by including solar radiation pressure (Zee, 1973). In addition to the four assumptions made by Zee (1972), namely

- the dominant perturbing factor of the Earth is the second harmonic (J_2);
- the Earth-Sun-Moon system lies in the plane of ecliptic;
- the center of the Earth-Moon system moves around the Sun in a circular orbit at a constant rate;
- the Earth-Moon system rotates about its center of mass at a constant angular rate;

a fifth assumption, related to the application of the solar radiation pressure on the satellite, was added. The solar panel face was assumed to be always normal to the Sun-satellite radius vector, and the Earth's shadow effect was neglected. As is seen from the assumptions, the motion of the Moon's orbit was not included. However, the principal effects of the rotation of the satellite orbital plane as well as the periodic change of eccentricity were fully revealed.

In 1985, Liu derived an improved first-order solution to the problem of the geostationary satellite by using orbital parameters similar to the p-q-h-k parameters proposed by Brouwer and Clemence (1961) for an orbit having small eccentricity and near-zero inclination. In his solution (Liu, 1985), the Sun and Moon were assumed to be moving in circular orbits, and the parallactic term in the Moon's potential was not considered. Numerically integrated values were compared with his analytical prediction for a period of 100 days. However, his comparison was based on a limited portion of his solution with terms relating to the Earth's potential and the perturbation of the Sun, while the perturbations of the Moon and solar radiation pressure were not included in his findings of comparison.

References

Allan, R.R.; 'Satellite Orbit Perturbation Due to Radiation Pressure and Luni-solar Forces', Technical Note No. Space 7, Royal Aircraft Establishment, Farnborough, England (1963).

Allan, R.R. and Cook, G.E.; 'The Long-Period Motion of the Plane of a Distant Circular Orbit', Proceeding of the Royal Society (London) A280, 97-109 (1964).

Bogoliubov, H. and Mitropolsky, J.; Asymptotic Method in the Theory of Nonlinear Oscillations, Hindustan Publishing Corporation, Delhi-6, India, 1961.

Brouwer, D. and Clemence, G.M.; Method of Celestial Mechanics, Academic Press, New York and London, 1961.

Bryant, R.W.; 'The Effect of Solar Radiation Pressure on the Motion of Artificial Satellite', Astron. J. 66, 430 (1961).

Frick, R.H.; 'Orbital Regression of Synchronous Satellites Due to the Combined Gravitational Effects of the Sun, the Moon and the Oblate Earth', R-454-NASA, RAND Corporation, Santa Monica, California (1967).

Kozai, Y.; 'Effect of Solar Radiation Pressure on the Motion of an Artificial Satellite', Special Report 56, Smithsonian Astrophysical Observatory, Cambridge, Mass. (1961).

Kozai, Y.; 'A New Method to Compute Lunisolar Perturbations in Satellite Motions', Special Report 349, Smithsonian Astrophysical Observatory, Cambrdige, Mass. (1973).

Levin, E.; 'Solar Radiation Pressure Perturbation of Earth Satellite Orbits', AIAA J. 6, 120-126 (1968).

Liu, L.; 'Orbital Variation of Synchronous Satellite and Its Calculation', Chinese Astronomy & Astrophysics 9, 27-34 (1985).

Lubowe, A.G.; 'Orbital Behavior of Large Synchronous Satellite', Astronautica Acta 13, 49-61 (1967).

Mello, S.F.; 'Analytical Studies of Earth's Shadowing Effects on Satellite Orbits', Celestial Mechanics 5, 80-101 (1972).

Musen, P.; 'The Influence of the Solar Radiation Pressure on the Motion of an Artificial Satellite', J. Geophysics Research 65, 1391-1396 (1960).

Parkinson, R.W., Jones, H.M., and Shapiro, I.I.; 'The Effect of Solar Radiation Pressure on Earth Satellite Orbit', Science 131, 920-921 (1960).

Perrine, S.; 'Solar Pressure Perturbations on the Orbit of a Flat Reflector', NASA TM X-53568, George C. Marshall Space Flight Center, Huntsville, Alabama (1966).

Polyakhova, Ye.N.; 'Solar Radiation Pressure and the Motion of Satellites', AIAA J. 1, 2893-2909 (1963).

Satellite Solar Power Station, System Engineering Report No. ASP-583-R-8, Grumman Aerospace Corporation, Bethpage, New York (1971).

Sehnal, L.; 'Radiation Pressure Effects in the Motion of Artificial Satellite', in Dynamics of Satellite, B. Morando (ed.), Springer-Verlag, Berlin, 1969.

Voshkov'yak, M.A.; 'Perturbed Motion of a Satellite during a Short Time Interval', Cosmic Research 7, 756-765 (1969).

Voshkov'yak, M.A.; 'On the Method of Approximate Computation of the Motion of a Synchronous Artificial Earth Satellite', Cosmic Research 10, 131-140 (1972).

Voshkov'yak, M.A. and Lidov, M.L.; 'An Approximate Description of the Evolution of the Orbit of a Stationary Artificial Earth Satellite', Cosmic Research 11, 305-316 (1973).

Wyatt, S.P.; 'The Effect of Radiation Pressure on the Secular Acceleration of Satellites', Special Report 60, Smithsonian Astrophysical Observatory, Cambridge, Mass. (1961).

Zee, C.H.; 'Effect of Earth Oblateness and Equator Ellipticity on a Synchronous Satellite', Astronautica Acta 16, 143-153 (1971).

Zee, C.H.; 'Effects of the Sun and the Moon on a Near-Equatorial Synchronous Satellite', Astronautica Acta 17, 891-906 (1972).

Zee, C.H.; 'Effect of the Sun, the Moon and Solar Radiation Pressure on a Near-Equatorial Synchronous Satellite', Astronautica Acta 18, 281-287 (1973).

Chapter 6

STATION-KEEPING

6.1. General

As is seen from the previous three chapters, the stationary condition relative to an Earth-equator ground station, which is expected from an equatorial, spherical-synchronous satellite, does not persist with time. The predominant perturbing effect of Earth's triaxiality (Chapter 3) gives rise to a long-period libration of the satellite about the equator's minor axis. The attractions of the Sun and Moon basically result in a rotation and regression of the satellite's orbital plane (Chapter 4). The solar radiation pressure mainly brings about a change in orbit eccentricity as well as a rotation of the line of apsides (Chapter 5). All of these perturbations, if left unattended, may lead to a significant departure from the desired geosynchronous state and a consequent failure of the satellite's mission. Nullifying the perturbative effects may be achieved through an application of corrective thrust. In this situation, the application of corrective thrust by a rocket is known as an orbit maneuver. The simplest type of orbit maneuver which may be considered is known as an impulse. It is mathematically defined as a thrust of infinite magnitude applied within an infinitesimal time interval, such that the integral of the thrust over the time interval results in a finite change in momentum. Rockets which produce a high thrust usually have a specific impulse (I_{sp}, the impulse delivered per unit weight of propellant) of 200 to 300 sec. Such rockets are frequently used for orbit maneuvers. However an orbit maneuver can also be achieved by a rocket of very low thrust, with a long thrusting time duration. Rockets which produce low thrust have a high I_{sp}, e.g., arc-jet engines ($I_{sp} \cong 1{,}000$ sec), plasma engines ($I_{sp} \cong 5{,}000$ sec), ion engines ($I_{sp} \geqq 10{,}000$ sec), etc.

In order to properly understand the requirements for station-keeping, we shall first examine the effect of an impulse and the perturbations on all orbital parameters and then present some basic analyses of the orbit maneuver. The orbit maneuvers will be considered with both high and low thrusts.

6.2. Effect of Impulsive Velocity on Orbital Parameters

For the purpose of studying the effect of impulsive velocity on orbital elements, the first-order approximate solutions of the X's (i.e. $\bar{X}$'s) will be used to depict the position of a geostationary satellite, wherein

(a) The radial position of the satellite is obtained from (3.33), (3.55), (3.56), and (3.58):

$$r = r_c + \Delta r_c - r_c^2(\bar{X}_3 \cos\theta + \bar{X}_4 \sin\theta) , \tag{6.1}$$

or

$$r = r_c - \frac{2}{3}\frac{r_c}{\dot{\theta}_E}\dot{\Gamma} - r_c^2(\bar{X}_3 \cos\theta + \bar{X}_4 \sin\theta) . \tag{6.2}$$

(b) The longitude of the satellite relative to the minor axis of the Earth's elliptical equator can be derived from (3.6a) by the approximation

$$r^2 \frac{d\theta}{dt} \cong (r_c + \Delta r_c)^2 \dot{\theta}_E , \tag{6.3}$$

which, with the help of (6.1), may be simplified as

$$[1 - 2r_c(\bar{X}_3 \cos\theta + \bar{X}_4 \sin\theta)]\, d\theta = \dot{\theta}_E\, dt . \tag{6.4}$$

This equation shows that the average rate of θ within a 2π-period is $\dot{\theta}_E$, while the angular position (or longitude, λ) of the geostationary satellite with respect to its mean daily position (Γ) within an orbit can be obtained by an integration of the second term in the brackets on the left-hand side of (6.4) with respect to θ. Therefore,

$$\lambda = \Gamma + 2r_c(\bar{X}_3 \sin\theta - \bar{X}_4 \cos\theta) . \tag{6.5}$$

(c) The latitude of the satellite is

$$\phi = \bar{X}_1 \cos\theta + \bar{X}_2 \sin\theta , \tag{6.6}$$

which is an approximation of (4.53) for the reason of a small value of ϕ.

Equations (6.2), (6.5), and (6.6) describe the position of a geostationary satellite in terms of Γ, $\dot{\Gamma}$, and $\bar{X}$'s on which the evaluation of delta-V requirements for station-keeping will be based.

The velocity vector of a satellite in an orbit is

$$\vec{V} = (V_R, V_T, V_N) , \tag{6.7}$$

where V_R is the radial component, positive outward; V_T is the transverse (normal to the radius vector $\vec{r}$) component in the orbital plane, positive in the direction of motion; and V_N, the component normal to the orbital plane, positive toward the North. These velocity components can be derived by differentiating (6.1), (6.5), and (6.6) with respect to time (t); thus

$$V_R = \dot{r} = \frac{dr}{d\theta}\frac{d\theta}{dt} = V_c r_c(\bar{X}_3 \sin\theta - \bar{X}_4 \cos\theta) , \tag{6.8}$$

$$\begin{aligned} V_T &= r_c(\dot{\lambda} + \dot{\theta}_E) \\ &= r_c(\dot{\Gamma} + \dot{\theta}_E) + 2r_c^2(\bar{X}_3 \cos\theta + \bar{X}_4 \sin\theta) , \end{aligned} \tag{6.9}$$

and

$$V_N = r_c\dot{\phi} = r_c \frac{d\phi}{d\theta}\frac{d\theta}{dt} = - V_c(\bar{X}_1 \sin\theta - \bar{X}_2 \cos\theta) , \tag{6.10}$$

where V_c is $r_c\dot{\theta}_E$, the stationary velocity, and $d\theta/dt$ is approximated by $\dot{\theta}_E$.

The application of an impulsive velocity correction on a satellite will only effect its velocity vector, while the position vector remains invariant. Hence, we apply the variation technique to both position (r, λ, ϕ) and velocity (V_R, V_T, V_N) vectors of a satellite in a geostationary orbit at $\theta = \theta_A$, and the resulting expressions are then derived from (6.2), (6.5), (6.6), and (6.8)-(6.10):

$$-\frac{2}{3}\frac{r_c}{\dot{\theta}_E}\Delta\dot{\Gamma} - r_c^2(\Delta\bar{X}_3 \cos\theta_A + \Delta\bar{X}_4 \sin\theta_A) = 0 , \tag{6.11}$$

$$\Delta\Gamma + 2r_c(\Delta\bar{X}_3 \sin\theta_A - \Delta\bar{X}_4 \cos\theta_A) = 0 , \tag{6.12}$$

$$\Delta\bar{X}_1 \cos\theta_A + \Delta\bar{X}_2 \sin\theta_A = 0 , \tag{6.13}$$

$$\Delta V_R = V_c r_c(\Delta\bar{X}_3 \sin\theta_A - \Delta\bar{X}_4 \cos\theta_A) , \tag{6.14}$$

$$\Delta V_T = r_c\Delta\dot{\Gamma} + 2r_c(\Delta\bar{X}_3 \cos\theta_A + \Delta\bar{X}_4 \sin\theta_A) , \tag{6.15}$$

and

$$\Delta V_N = - V_c\Delta\bar{X}_1 \sin\theta_A + V_c\Delta\bar{X}_2 \cos\theta_A . \tag{6.16}$$

Consequently, a solution to the above set of linear equations gives the impulsive changes of Γ, $\dot{\Gamma}$, and $\bar{X}$'s as a result of the application of $\Delta V(\Delta V_R, \Delta V_T, \Delta V_N)$ at $\theta = \theta_A$ in a geostationary orbit as

$$\Delta\Gamma = - (2/V_c)\Delta V_R , \tag{6.17}$$

$$\Delta\dot{\Gamma} = - (3\dot{\theta}_E/V_c)\Delta V_T , \tag{6.18}$$

$$\Delta\bar{X}_1 = - (\Delta V_N/V_c)\sin\theta_A , \tag{6.19}$$

$$\Delta\bar{X}_2 = (\Delta V_N/V_c)\cos\theta_A , \tag{6.20}$$

$$\Delta\bar{X}_3 = (2\Delta V_T/V_c)\cos\theta_A + (\Delta V_R/V_c)\sin\theta_A , \tag{6.21}$$

and

$$\Delta\bar{X}_4 = (2\Delta V_T/V_c)\sin\theta_A - (\Delta V_R/V_c)\cos\theta_A . \tag{6.22}$$

It should be noted that the θ_A in (6.19) and (6.20) may be different from that in (6.21) and (6.22) because the applications of ΔV_N, ΔV_T, and ΔV_R may be independent of each other.

Examination of the above set of equations gives the following facts:

(a) (6.17) and (6.18) show that Γ may be changed only by a suitable radial impulse, and $\dot{\Gamma}$ may be changed by a suitable transverse impulse. The change in Γ or $\dot{\Gamma}$ is independent of the point of application of a particular impulse.

(b) (6.19) and (6.20) depict the changes of $\bar{X}_1$ and $\bar{X}_2$ that are achieved by the normal component of a velocity increment applied at $\theta = \theta_A$.

(c) The changes in $\bar{X}_3$ and $\bar{X}_4$ can be accomplished by either a suitable transverse impulse applied at $\theta = \theta_A$, or a radial impulse of twice the magnitude applied 90° away, or a suitable impulse having both transverse and radial components.

(d) From (6.17), (6.18), (6.21), and (6.22), it is seen that any change in Γ and $\dot{\Gamma}$ will cause a change in $\bar{X}_3$ and $\bar{X}_4$, and conversely.

Equations (6.17)-(6.22) may also be viewed that the changes in Γ, $\dot{\Gamma}$, and $\bar{X}$'s on the left-hand side of these equations are resulting from the perturbations discussed in the three previous chapters, and they can be reduced or even eliminated by a suitable application of delta-V or delta-V's. Consequently, the satellite will be able to remain in a bounded region as viewed from a point on the Earth for a limited period of time. This conjecture forms our basic approach to station-keeping.

6.3. Effect of Perturbations on Orbital Parameters

As shown in (4.149) and (4.152), X_1 and X_2 are mainly related to the orbital inclination, i, and the longitude of ascending node, Ω, while the argument of perigee, ω, and the eccentricity, e, are principally associated with X_3 and X_4 as depicted in (4.168) and (4.169). The longitude, λ, of the satellite is basically governed by Γ, $\dot{\Gamma}$, and e (or $\bar{X}_3$ and $\bar{X}_4$) as shown in (6.5). For a geostationary orbit, the eccentricity is rather small and, therefore, λ is dominated by Γ and $\dot{\Gamma}$. This reminder will help us in relating our orbital parameters to the osculating orbital elements.

We shall use the first-order solutions of Γ, $\dot{\Gamma}$, and X's in our study of station-keeping because the oscillatory nature of the improved first-order solution prohibits us from establishing any meaningful station-keeping scheme. Hence, (5.20)-(5.23) will be adopted in the following station-keeping study for the effects of the combined perturbations of the Sun, Moon, Earth, and solar radiation pressure on X_1, X_2, X_3, and X_4, while (3.5), (3.36), (3.58), and (5.38) are retained for determining Γ and $\dot{\Gamma}$ in our longitudinal control of station-keeping. We shall explore further the behavior of the longitudinal drift and drift rate based on our findings developed in the previous chapters.

Since we are dealing with the first-order approximate solutions, the effect of perturbation due to the ellipticity of the equator of the Earth on i, Ω, e, and ω are negligible as can be seen from (3.67), (3.64), (3.71), and (3.72); however, its effect on Γ and $\dot{\Gamma}$ is of major importance, their values are computed from (3.5), (3.35), (3.57) and (3.58), and we shall add a subscript 'E' on Γ and $\dot{\Gamma}$ signifying the ellipticity effect of the equator of the Earth.

For station-keeping purposes, it will be seen that Γ_E is kept in the neighborhood of $\Gamma_{E\,max}$, i.e., $\Gamma_{E\,max} - \Gamma_E = \Delta\Gamma_E$ is small; therefore, substitutions of $\Gamma_0 = \Gamma_{E\,max} - \Delta\Gamma_E$ and $\Delta r_{c0} = 0$ in (3.35) and approximating $\cos 2(\Gamma_{E\,max} - \Delta\Gamma_E)$ by $\cos 2\Gamma_{E\,max} + \Delta\Gamma_E \sin 2\Gamma_{E\,max}$ result in

$$\Delta\Gamma_E = \frac{k_1}{2 \sin 2\Gamma_{E\,max}} \left(\frac{\Delta r_c}{r_c}\right)^2 . \tag{6.23}$$

The curves in Figure 3.10. can be fitted by the following equation for various $\Gamma_{E\,max}$'s:

$$\Gamma_{E\,max} - \Gamma_E = \Delta\Gamma_E = KT^2 , \tag{6.24}$$

where T is the time and K is a constant depending on $\Gamma_{E\,max}$. Table V. shows that K is in the unit of milliradian/day^2; thus, $\Delta\Gamma_E$ and T in (6.24) have their units in milliradian and day, respectively.

With (6.23) and (6.24), Equation (3.58) becomes

$$\dot{\Gamma}_E = \frac{d\Gamma_E}{dt} = -\frac{3}{2}\frac{\Delta r_c}{r_c}\dot{\theta}_E = \pm 3\dot{\theta}_E \left(\frac{K \sin 2\Gamma_{E\,max}}{2k_1}\right)^{1/2} T , \tag{6.25}$$

where the '+' sign is for '$-\Delta r_c$' and vice versa. If a geostationary satellite is assumed to have the libration along the contour shown in Figure 3.6. with Γ_E and $\dot{\Gamma}_E$ as its drift angle and drift rate, respectively, then

TABLE V

Constant K in Equation (6.24)

$\Gamma_{E\,max}$ (deg)	5	10	20	30	40	45	50	60	70	80	85
$K \frac{mrad}{day^2} \times 10^{-3}$	2.570	5.069	9.556	12.917	14.738	14.992	14.790	13.051	9.720	5.188	2.638

the corresponding values due to the Sun, Moon, and solar radiation pressure can be obtained as follows.

Since the satellite is in a geostationary orbit with a radius of r_c, the angular rate can be computed from (4.125) with $\bar{X}_1$, $\bar{X}_2$, and $\bar{X}_6$ having their expressions as depicted in (5.20), (5.21), and (5.24), respectively, and the drift rate is

$$\dot{\Gamma}_{SMR} = \overline{\frac{d\theta}{dt}} - \dot{\theta}_E , \tag{6.26}$$

where $\dot{\Gamma}_{SMR}$ is designated as the drift rate due to the Sun, Moon, and solar radiation pressure. The corresponding drift angle is obtained from the value of θ in (5.38b) and $(\bar{t} - \bar{t}_0)\dot{\theta}_E$:

$$\dot{\Gamma}_{SMR} = (\theta - \theta_0) - (\bar{t} - \bar{t}_0)\dot{\theta}_E . \tag{6.27}$$

The total drift rate ($\dot{\Gamma}_T$) and drift angle (Γ_T) are, therefore,

$$\dot{\Gamma}_T = \dot{\Gamma}_E + \dot{\Gamma}_{SMR} \tag{6.28}$$

and

$$\Gamma_T = \Gamma_E + \Gamma_{SMR} , \tag{6.29}$$

respectively.

6.4. Orbit Maneuvers

Orbit maneuvers include in-plane and out-of-plane maneuvers. The former results in changes of drift angle, drift rate, and eccentricity by applications of transverse and radial velocity impulses; the latter gives the change of the inclination of the orbital plane by means of an impulse normal to the orbital plane. Of course these impulses may be combined as components of a resultant impulse or impulses.

6.4.1. In-Plane Orbit Maneuver - Altitude Change

The in-plane orbit maneuver usually involves an altitude change and an epoch adjustment; the latter is related to the time of perigee passage of a satellite. For geostationary satellites, only the former is required because the satellites are already in their nominal desired positions. As will be seen later, maintaining the satellites within a certain range of east-west drift can be accomplished by a scheme involving an altitude change.

From an orbital maneuvering point of view, altitude change, in fact, is a transfer of a satellite from one circular orbit to another circular orbit of a different radius; the difference in radius is the change in altitude.

6.4.1.1. High Thrust Case

The minimum energy transfer of a satellite from one circular orbit to another circular orbit of different radius is known as the Hohmann transfer. This transfer requires two impulses: the first is applied at a point in the initial orbit; the second is applied 180° later and results in the establishment of the final orbit. The transfer orbit is an ellipse (as shown in Figure 6.1.).

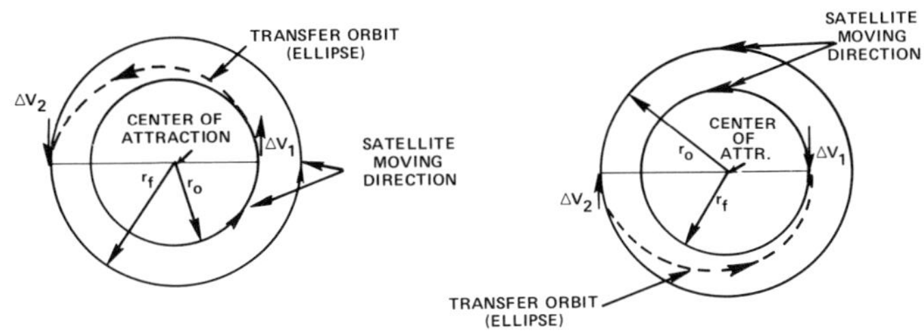

Fig. 6.1. Hohmann transfer.

The initial and final orbits have their circular orbital velocities as $(\mu_E/r_0)^{1/2}$ and $(\mu_E/r_f)^{1/2}$, respectively, and the orbital velocities of the elliptical transfer orbit at its perigee and apogee are $\{2\mu_E((1/r_0) - [1/(r_0 + r_f)])\}^{1/2}$ and $\{2\mu_E((1/r_f) - [1/(r_0 + r_f)])\}^{1/2}$, respectively. In order to transfer a satellite from the initial orbit (r_0) to the elliptical transfer orbit at its perigee (see left figure of Figure 6.1.), it is required that the satellite has an increment of orbital velocity from $(\mu_E/r_0)^{1/2}$ to $\{2\mu_E((1/r_0) - [1/(r_0 + r_f)])\}^{1/2}$. This increment is accomplished by an application of an impulse ΔV_1 in the tangential direction having a magnitude of

$$\Delta V_1 = \left[2\mu_E\left(\frac{1}{r_0} - \frac{1}{r_0 + r_f}\right)\right]^{1/2} - \left(\frac{\mu_E}{r_0}\right)^{1/2} . \quad (6.30a)$$

Similarly, after the satellite is in the elliptical transfer orbit and reaches the apogee of the transfer orbit, an impulse ΔV_2 is applied and the satellite is then in the final orbit (r_f); the magnitude of ΔV_2 is determined by the difference in the orbital velocity at the apogee of the transfer orbit and the circular orbital velocity in the final orbit, hence

$$\Delta V_2 = \left(\frac{\mu_E}{r_f}\right)^{1/2} - \left[2\mu_E\left(\frac{1}{r_f} - \frac{1}{r_0 + r_f}\right)\right]^{1/2} . \quad (6.30b)$$

Readers may use the same reasoning to derive ΔV_1 and ΔV_2 for the case in the right figure of Figure 6.1.

The amount of fuel consumed can be calculated from the general relationship

$$\Sigma\Delta V = -\bar{c} \ln \frac{M_f}{M_0} , \quad (6.31)$$

where $\bar{c}$ is the average jet velocity of the rocket, M_f is the mass of the satellite after the execution of the delta-V, and M_0 is the mass before the execution. Consequently, the mass of fuel consumed (M_p) is

$$M_p = M_f - M_0 .$$

In general, M_p is small in comparison with M_0 during the station-keeping; therefore (6.31) may be approximated as

$$\frac{\Sigma\Delta V}{\bar{c}} = \frac{M_p}{M_0} \tag{6.32}$$

when $\ln[(M_0 - M_p)/M_0]$ is expanded into a series.

Based on (6.30)-(6.32), the expression

$$\bar{c}\,\frac{M_p}{M_0} = \frac{1}{2}\,\frac{|\Delta r|}{r_0}\left(\frac{\mu_E}{r_0}\right)^{1/2} \tag{6.33}$$

can be derived by a binomial series expansion with both terms, M_p/M_0 and $\Delta r/r_0 = (r_f - r_0)/r_0$, having a small order of magnitude exhibited in our present problem of interest. Equation (6.33) gives the fuel consumption for an orbital altitude change between two circular orbits of radii r_0 and r_f. It should be noted that $\bar{c}$ is related to the specific impulse, I_{sp}, of the impulsive thrusting rocket by $I_{sp}g_e \equiv \bar{c}$ in which g_e is the gravitational acceleration at the Earth's surface.

In reality, the requirement of infinite thrust magnitude applied within an infinitesimal time interval does not exist; both the magnitude of the thrust and the thrusting time are finite. When a satellite is equipped with a constant thrust chemical propellant rocket, the difference of fuel consumptions between the impulsive and the finite thrustings is negligible (Zee, 1963a) for the Hohmann-type orbital transfer. In the Hohmann-type transfer, the transfer orbits involve circle-ellipse or ellipses having their major axes aligned, and the impulses are applied at the perigee and the apogee. The constant thrust (F) of the rocket can be computed by

$$F = \dot{m}\bar{c} , \tag{6.34}$$

where $\dot{m}$ is the constant flow rate of fuel mass, and the thrusting time, t_T, for a single ΔV is determined by

$$t_T = M_p(\text{for a single } \Delta V)/\dot{m} \tag{6.35}$$

Following Zee's derivation (Zee, 1963a), the power-flight path for the Hohmann or Hohmann-type transfer is*

$$\begin{aligned}\frac{1}{u} = {} & \frac{A^2}{2B^4}(1 - B\tau)^2[\ln(1 - B\tau)]^2 + \frac{A}{B^2}\left(\frac{7}{4}\,\frac{A}{B} \pm \frac{3}{2}C\right)(1 - B\tau)^2 - \\ & - \frac{A}{B^3}\,\frac{3}{2}\left(\frac{A}{B} \pm C\right)(1 - B\tau)^2\ln(1 - B\tau) + \frac{1}{2}(C^2 - 1)\tau^2 + \\ & + \left[D + \frac{2A}{B}\left(\frac{A}{B} \pm C\right)\right]\tau + 1 - \frac{A}{B^2}\left(\frac{7}{4}\,\frac{A}{B} \pm \frac{3}{2}C\right)\end{aligned} \tag{6.36}$$

---

*Whenever a term has two signs in front of it, the upper sign represents the positive thrust case (thrust is normal to the radius vector and in the direction of increasing θ), and the lower sign is for the negative thrust case (thrust is normal to the radius vector and in the direction of decreasing θ).

and

$$\begin{aligned}\theta = {} & \pm \frac{A^3}{12B^6}(1 - B\tau)^3[\ln(1 - B\tau)]^2 \mp \frac{A^2}{B^5}\left(\frac{7}{18}\,\frac{A}{B} \pm \frac{1}{6}C\right)(1 - B\tau)^2 \times \\ & \times \ln(1 - B\tau) \pm \frac{A^2}{B^5}\left(\frac{127}{216}\,\frac{A}{B} \pm \frac{7}{18}C\right)(1 - B\tau)^2 \pm \frac{A}{B^2}\left[\frac{1}{2B^2}(C^2 - 1) + \right. \\ & \left. + \frac{D}{B} + 1 + \frac{A^2}{4B^4} \pm \frac{AC}{2B^3}\right][\ln(1 - B\tau) - 1](1 - B\tau) \mp \frac{A}{12B}(C^2 - 1)\tau^3 \mp \\ & \mp \frac{A}{B}\left[\frac{1}{4B}(C^2 - 1) + \frac{D}{2} + \frac{A}{B^2}\left(\frac{A}{B} \pm C\right)\right]\tau^2 + C\tau \pm \frac{A^2}{B^4}\left(\frac{11}{8}\,\frac{A}{B} \pm C\right)\tau - \\ & - \frac{A}{B^2}\left[\pm \frac{73A^2}{216B^4} - \frac{AC}{9B^3} \mp \frac{1}{2B^2}(C^2 - 1) \mp \frac{D}{B} \mp 1\right], \end{aligned} \tag{6.37}$$

where the dimensionless terms are defined as

$$u \equiv r_0/r, \qquad \tau \equiv (g_0/r_0)^{1/2}t,$$

$$A \equiv \dot{m}\bar{c}/(M_0 g_0), \qquad B \equiv (\dot{m}/M_0)(r_0/g_0)^{1/2},$$

$$C \equiv V_{\theta_0}/(g_0 r_0)^{1/2}, \qquad D \equiv V_{r_0}/(g_0 r_0)^{1/2},$$

and r is the distance from the center of the Earth to the satellite at any time during the thrusting, r_0 is the distance at the beginning of the thrusting, θ is the polar angle (see Figure 6.2.), g_0 is the gravitational acceleration at r_0, V_{θ_0} and V_{r_0} are the transverse and the radial velocities, respectively, at the beginning of the thrusting, and t is the time.

For the case of the Hohmann transfer, C = 1 and D = 0.

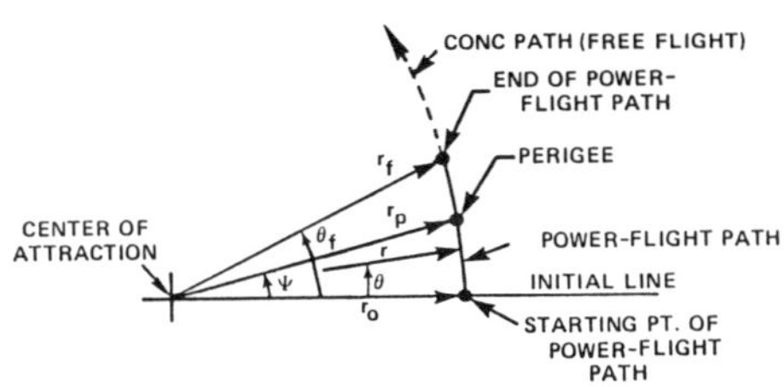

Fig. 6.2. Notations - power-flight trajectory.

Because of the finite thrusting time, the rocket must commence thrusting before the satellite reaches the point where the equivalent impulse would be applied. The angle between the radius vector at the start of thrusting and the radius vector at the impulse point is known as the lead angle (Ψ), and the time required to travel the lead angle is called the lead time (t_ℓ). As is seen in Figure 6.2., the nominal location of the impulse is predetermined at the perigee (or apogee) of the transfer orbit, which is a function of the power-flight path; thus, the lead angle can only be obtained by trial and error.

With (6.32)-(6.37), the impulse for the Hohmann or Hohmann-type transfer can be replaced by a power-flight of finite time. The analytical expressions of the trajectory may serve a useful purpose for the guidance system design for orbit maneuvers.

6.4.1.2. Low Thrust Case

The foregoing orbital transfer between two circular orbits can also be achieved by a low tangential thrust for a long duration of time. As previously mentioned, a low thrust rocket engine is inherently characterized by a high I_{sp}. In fact, by its definition, the I_{sp} itself is an indication of the efficiency of propellant consumption; thus, the propellant consumed during a low thrust orbital transfer between two orbits is comparatively small in comparison with that of a high thrust maneuver. In view of the small consumption of propellant, the relative constancy of satellite mass during a low thrust orbital transfer makes little difference in solving the problem either for a constant thrust or for a constant thrust acceleration. In the following, we shall present the solutions for both cases in a closed-form analytical expression as applied to orbit maneuvers between two circular orbits of radii r_0 (initial) and r_f (final).

6.4.1.2.1. Constant Thrust Case

For the case of a constant thrust, the thrust acceleration varies with time because of the consumption of fuel; thus, the differential equations of motion of a satellite are the nonautonomous type in nonlinear mechanics, and the stroboscopic method (Minorsky, 1962) can be employed in solving the problem. The following result is known as the improved first-order approximation of the power-flight trajectory of an Earth satellite under the influence of a low constant tangential thrust (Zee, 1963b, 1967a, 1968a):

$$\left.\begin{aligned}
\bar{u} &= \frac{1}{\bar{h}^2} + \frac{1}{K_1\bar{h}^3}\cos(\theta+\Phi_1), \qquad u = \bar{u} + \frac{9\bar{h}}{2K_1(\frac{1}{A}-\bar{\tau})}\sin(\theta+\Phi_1),\\
\bar{h} &= \left[(\bar{c}/(g_0r_0)^{1/2})\ln\left(\frac{\frac{1}{A}-\bar{\tau}}{\frac{1}{A}-\bar{\tau}_0}\right) + \frac{1}{\bar{h}_0}\right],\\
h &= \bar{h} - \frac{3\bar{h}^4}{K_1(\frac{1}{A}-\bar{\tau})}\sin(\theta+\Phi_1),
\end{aligned}\right\} \tag{6.38}$$

$$\begin{aligned}
\theta = {}& \frac{a}{\bar{h}_0^3}(\bar{\tau}-\bar{\tau}_0) - (\frac{1}{A}-\bar{\tau}_0)a^2\Big\{\frac{3}{\bar{h}_0^3}[X(\ln X-1)+1] + \\
&+ \frac{3a}{\bar{h}_0}[X(\ln X-1)^2 + (X-2)] + \\
&+ a^2[X(\ln X-1)^3 + 3X(\ln X-1) - 2(X-3)]\Big\},
\end{aligned} \tag{6.39}$$

and

$$\tau = \bar{\tau} + \frac{6\bar{h}^6}{aK_1(\frac{1}{A}-\bar{\tau})}\cos(\theta+\Phi_1), \tag{6.40}$$

where the dimensionless terms are defined as

$$u \equiv \frac{r_0}{r}, \qquad h \equiv \frac{\dot{\theta}}{u^2\dot{\theta}_0}, \qquad \tau \equiv \frac{g_0 t}{\bar{c}},$$

$$a \equiv \frac{\dot{\theta}_0\bar{c}}{g_0}, \qquad X \equiv \frac{\frac{1}{A}-\bar{\tau}}{\frac{1}{A}-\bar{\tau}_0}, \qquad \frac{1}{A} \equiv \pm\frac{M_0 g_0}{\dot{m}\bar{c}},$$

and r is the distance from the center of the Earth to the satellite at any time (t), r_0 is the distance at $t = 0$, θ is the polar angle (see Figure 6.3.), g_0 is the gravitational acceleration at r_0, $\dot{\theta}$ is the instantaneous angular velocity, $\dot{\theta}_0$ is the angular velocity at $t = 0$, $\dot{m}$ is the constant flow rate of fuel mass, $\bar{c}$ is the average jet velocity of the rocket, and M_0 is the mass of satellite at $t = 0$. $\dot{m}\bar{c}$ is the constant tangential thrust applied to the satellite, which is very small in the present case; consequently, A is very small as is seen from its definition. When $r_f > r_0$ and $1/A = +(M_0 g_0)/(\dot{m}\bar{c}) > 0$, $\dot{m}\bar{c}$ is directed to increase the tangential velocity; when $r_f < r_0$ and $1/A = -(M_0 g_0)/(\dot{m}\bar{c}) < 0$, $\dot{m}\bar{c}$ is directed to decrease the tangential velocity. Furthermore, $\bar{h}_0$, K_1, Φ_1, and τ_0 are the integration constants, which are determined from the given initial conditions t (or τ) = 0, $\theta = 0$, $u = 1$, $du/d\theta = (du/d\theta)_0$, and $h_0 = 1$. For the case of a satellite initially in a circular orbit, these constants are

$$\bar{h}_0 = 1 - 6A^2, \quad K_1 = -(A/2)[1 - (9A^2/2)] ,$$

$$\Phi_1 = \frac{\pi}{2} - (3A/2), \quad \text{and} \quad \bar{\tau}_0 = 18A^3/a.$$

Based on the first-order solutions of u and h in (6.38), the fact that $\Delta r/r_0 \ll 1$, $\Delta h/h_0 \ll 1$, $\bar{h}_0 \cong 1$, and $\bar{\tau} \cong 0$ leads to the following approximations:

$$\frac{\Delta r}{r_0} = 2\Delta h$$

and

$$-\Delta h = \frac{\bar{c}}{(g_0 r_0)^{1/2}} \ln \left[1 - \left[\pm \frac{\dot{m}t}{M_0}\right]\right] .$$

Combining these two equations results in

$$-\Delta r = 2\left(\frac{r_0}{g_0}\right)^{1/2} \bar{c} \ln \left[1 - \left[\pm \frac{\dot{m}t}{M_0}\right]\right] .$$

Since $\dot{m}t = M_p$ is the fuel spent during a change of altitude, Δr, and $M_p/M_0 \ll 1$, the expansion of $\ln [1 - (\pm \dot{m}t/M_0)]$ approximates the above equation as

$$\Delta r = \pm 2\left(\frac{r_0}{g_0}\right)^{1/2} \bar{c} \frac{\dot{m}t}{M_0} , \tag{6.41}$$

where the '+' sign is for increasing Δr and the '-' sign for decreasing Δr. Equation (6.41) can be used to calculate the fuel consumption and time required for a satellite to be transferred from a circular orbit of r_0 to another circular orbit of $r_0 \pm \Delta r$.

It should be noted that (6.38)-(6.40) are valid only if the satellite is initially in an orbit of small eccentricity.

6.4.1.2.2. Constant Thrust Acceleration Case

For the present case, the thrust acceleration remains constant during the flight, the differential equations of motion are of the autonomous type in nonlinear mechanics, and the well-known Kryloff-Bogoliuboff (1947) method is used to solve the problem. The resulting power-flight trajectory is (Zee, 1963c, 1968a)

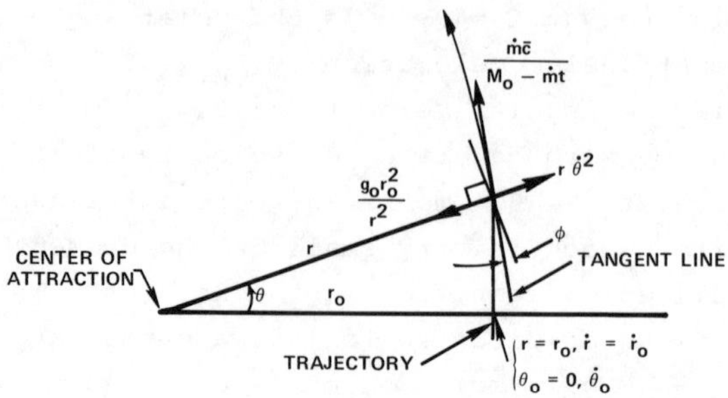

Fig. 6.3. Notations - low thrust power-flight trajectory.

$$\bar{u} = \frac{1}{\bar{h}^2} + \frac{1}{K_1\bar{h}^3}\cos(\theta + \Phi_1), \quad u = \bar{u} + \frac{9\varepsilon\bar{h}}{2K_1}\sin(\theta + \Phi_1), \tag{6.42}$$

$$\bar{h} = [\frac{1}{\bar{h}_0^4} - 4\varepsilon\theta]^{-1/4}, \quad h = \bar{h} - 3\frac{\varepsilon\bar{h}^4}{K_1}\sin(\theta + \Phi_1), \tag{6.43}$$

$$\dot{\theta}_0(\bar{t} - \bar{t}_0) = \frac{1}{\varepsilon}\left\{\frac{1}{\bar{h}_0} - [\frac{1}{\bar{h}_0^4} - 4\varepsilon\theta]^{1/4}\right\}, \tag{6.44}$$

and

$$t = \bar{t} + \frac{6\varepsilon\bar{h}^6}{K_1\omega}\cos(\theta + \Phi_1), \tag{6.45}$$

where all symbols have the same definitions as in the constant thrust case, f is the constant thrust-acceleration, and $\varepsilon = f/g_0$ ($f > 0$ when $r_f > r_0$; $f < 0$ when $r_f < r_0$). The above expressions are derived based on the assumption that the orbits are nearly circular. The integration constants $\bar{h}_0$, K_1, Φ_1, and $\bar{t}_0$ are

$$\bar{h}_0 = 1 - 6\varepsilon^2, \quad K_1 = -\frac{1}{2\varepsilon}(1 - \frac{9}{2}\varepsilon^2),$$

$$\Phi_1 = \frac{\pi}{2} - \frac{3}{2}\varepsilon, \quad \text{and} \quad \bar{t}_0 = 18\varepsilon^3/\dot{\theta}_0$$

for a satellite being initially in a circular orbit.

The fuel consumption (M_p) for the power-flight can be computed from the expressions

$$\varepsilon g_0 M = -\bar{c}\frac{dM}{dt} \tag{6.46}$$

and

$$\ln\left(\frac{M_0}{M}\right) = \frac{\varepsilon g_0}{\bar{c}}(t - t_0) = \frac{\Delta V}{\bar{c}}, \tag{6.47a}$$

where M is the mass of the satellite at any time (t) during the power-flight, and ΔV is the corresponding velocity increment. Equation (6.47a) can be further approximated as

$$\frac{M_p}{M_0} = \frac{\varepsilon g_0}{\bar{c}}(t - t_0) \tag{6.47b}$$

because $M_p/M_0 << 1$ as $\ln(M_0/M) = \ln[M_0/(M_0 - M_p)] = \ln[1 - (M_p/M_0)]^{-1}$ is expanded into a series.

If the approximation, $t \cong \bar{t}$, is made, then combining (6.43), (6.44), and (6.47) results in

$$\frac{M_p}{M_0} = \frac{\varepsilon g_0}{\bar{c}}(t - t_0) = \frac{\varepsilon g_0}{\bar{c}} \times \frac{1}{\varepsilon\dot{\theta}_0}\left(\frac{1}{\bar{h}_0} - \frac{1}{\bar{h}}\right) \cong \frac{g_0}{\bar{c}\dot{\theta}_0} \times \frac{\Delta\bar{h}}{\bar{h}_0^2} \tag{6.48}$$

because $\Delta\bar{h}/\bar{h}_0 << 1$.

The first-order solution of u in (6.42) gives

$$1 - \frac{\Delta r}{r_0} \cong \frac{1}{\bar{h}_0^2} - \frac{2}{\bar{h}_0^3}\,\Delta\bar{h}\,,$$

which, in turn, becomes

$$\Delta r = 2r_0\Delta\bar{h} \tag{6.49}$$

as $\bar{h}_0 \cong 1$.

Combining (6.48) and (6.49) leads to

$$\frac{M_p}{M_0} = \frac{g_0\Delta r_0}{2\bar{c}\dot{\theta}_0 r_0} = \frac{\varepsilon g_0}{\bar{c}}(t - t_0)\;, \tag{6.50}$$

which relates the fuel consumption, altitude change (Δr), and time required for the change. It should be noted that the sign of Δr is reflected in the value of ε.

6.4.2. In-Plane Orbit Maneuver - Drift Angle, Drift Rate, and Eccentricity Changes

Another type of in-plane orbit maneuver may be applied to the changes of the drift angle and the drift rate of a geostationary satellite with respect to a fixed point on the equator of the Earth. Unfortunately, these changes will invariably be associated with a change of the orbit eccentricity as is seen from (6.17), (6.18), (6.21), and (6.22). Equations (6.21) and (6.22) depict that the change in eccentricity is more economical by an application of a transverse impulse as compared with a radial one; however, an application of the radial impulse sometimes may be justified because of the requirement of a shift in drift angle.

In general, the interdependence of changes in eccentricity, the drift angle, and the drift rate necessitates the applications of both transverse and radial impulses; this requirement sometimes may be fulfilled by a single impulse applied at a suitable point in the orbit in a desirable direction.

When the drift angle and the drift rate are within the required bound, only the eccentricity needs to be changed. This change can be accomplished by a Hohmann-type transfer with two impulses. As is seen later, these two impulses are equal and have opposite signs. Consequently, the change in the drift rate due to the first impulse is nullified by the second one of the opposite sign.

6.4.2.1. High Thrust Case - Two-Impulse Hohmann-Type Transfer

Nullifying the eccentricity of an orbit by means of the two Hohmann-type transfer impulses is shown in Figure 6.4.

If the initial orbit has an eccentricity, e, and a semimajor axis, a, and it is desired to circularize the orbit having a radius equal to 'a', then there are two possible sets of impulses to accomplish the Hohmann-type transfer; the first set is

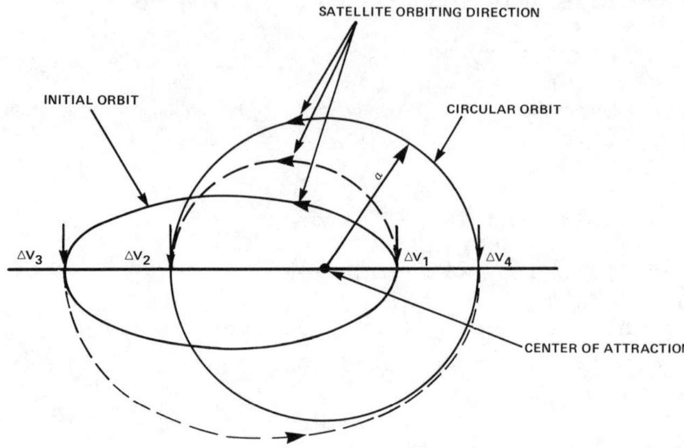

Fig. 6.4. Two-impulse Hohmann-type transfer.

$$\left.\begin{aligned}\Delta V_1 &= \left(\frac{\mu_E}{a}\right)^{1/2}\left[-\left(\frac{1+e}{1-e}\right)^{1/2} + \left(\frac{2}{(2-e)(1-e)}\right)^{1/2}\right] \cong -\frac{1}{4}\left(\frac{\mu_E}{a}\right)^{1/2} e \\ \text{and} \quad \Delta V_2 &= \left(\frac{\mu_E}{a}\right)^{1/2}\left[1 - \left(\frac{2(1-e)}{2-e}\right)^{1/2}\right] \cong \frac{1}{4}\left(\frac{\mu_E}{a}\right)^{1/2} e\,,\end{aligned}\right\} \tag{6.51}$$

when the first impulse is applied at the perigee of the initial orbit, and the second set is

$$\left.\begin{aligned}\Delta V_3 &= \left(\frac{\mu_E}{a}\right)^{1/2}\left[\left(\frac{2}{(2+e)(1+e)}\right)^{1/2} - \left(\frac{1-e}{1+e}\right)^{1/2}\right] \cong \frac{1}{4}\left(\frac{\mu_E}{a}\right)^{1/2} e \\ \text{and} \quad \Delta V_4 &= \left(\frac{\mu_E}{a}\right)^{1/2}\left[\left(\frac{2(1+e)}{2+e}\right)^{1/2} - 1\right] \cong -\frac{1}{4}\left(\frac{\mu_E}{a}\right)^{1/2} e\,,\end{aligned}\right\} \tag{6.52}$$

when the first impulse is applied at the apogee of the orbit. It can be seen that the application of ΔV_1 and ΔV_2 (or ΔV_3 and ΔV_4) will not change the drift rate because they are equal and opposite in sign. If impulses are to be replaced by a finite thrust, it is suggested that the same procedures described in Section 6.4.1.1. be followed to compute the power-flight trajectory.

The fuel consumption for either set of impulses is

$$\frac{1}{2}\left(\frac{\mu_E}{a}\right)^{1/2} e = -\bar{c}\,\ln\left(\frac{M_0 - M_p}{M_0}\right) = -\bar{c}\,\ln\left(1 - \frac{M_p}{M_0}\right) \cong \bar{c}\,\frac{M_p}{M_0} \tag{6.53}$$

as is seen from (6.51), (6.52), and (6.32), and from the fact that $M_p/M_0 << 1$. This equation shows that the fuel consumption to circularize an elliptical orbit is linearly proportional to the original eccentricity of the orbit.

As mentioned in Section 6.4.1.1., the difference of fuel consumption between the impulsive and finite thrustings in the present case is negligible because the orbital maneuvering is of the Hohmann-type.

6.4.2.2. High Thrust Case - Single Impulse

We shall examine the effects of a single impulse in the orbital plane on various orbital elements. In particular we are interested in the impulse either in the transverse or in the radial direction. These two impulses may be combined into a single impulse and vice versa. The following table summarizes the effects of these impulses on orbital elements:

TABLE VI. Effects of impulse on change of orbital elements.

Change of orbital element		Transverse impulse	Radial impulse
Major Axis	(Δa)	$2\dfrac{\Delta V_T}{V_c}a[1 + e\cos(\eta - \omega)]$	$2\dfrac{\Delta V_r}{V_c}ae\sin(\eta - \omega)$
Eccentricity	(Δe)	$\dfrac{\Delta V_T}{V_c}\cos(\eta - \omega)[1 - e\cos(\eta - \omega)]$	$\dfrac{\Delta V_r}{V_c}\sin(\eta - \omega)$
Mean motion	(Δn)	$-\dfrac{3}{a}[1 + e\cos(\eta - \omega)]\Delta V_T$	$-\dfrac{3e}{a}\sin(\eta - \omega)\Delta V_r$
Perigee	$(\Delta\omega)$	$2\dfrac{\Delta V_T}{V_c}\dfrac{1}{e}\sin(\eta - \omega)$	$-\dfrac{\Delta V_r}{V_c}\dfrac{1}{e}\cos(\eta - \omega)$

The expressions in this table are derived based on the general results of the effects of impulse on orbital elements (Ehricke, 1959) with approximations applicable to the geostationary satellite orbit, such as $\sqrt{\mu_E/a} \cong V_c$, $e^2 \cong 0$, etc. ΔV_T and ΔV_r are considered to have positive values when their application will increase the positive velocity of V_T and V_r, respectively. It can be seen that the application of any impulse to change one orbital element will invariably change the other orbital elements as well. For station-keeping, the interrelationships of changes in the orbital elements as a result of the application of an impulse should be studied before a corrective measure is adopted.

The fuel consumption for the corrective impulse can be computed from (6.32) as

$$(\Delta V_T^2 + \Delta V_r^2)^{1/2}/\bar{c} = \Delta V/\bar{c} = M_p/M_0 , \tag{6.54}$$

where ΔV_T and ΔV_r are the required corrective impulses, and the orientation of ΔV is determined from the directions of ΔV_T and ΔV_r.

When an impulse, which has the desired transverse and radial components, is replaced by a fixed-orientation finite thrust ($F = \dot{m}\bar{c}$) along with the corresponding finite thrusting time ($t_T = M_p/\dot{m}$), the power-flight trajectory is (Zee, 1963d)

$$\left.\begin{aligned} \rho &= a_0 + a_1\tau + a_2\tau^2 + a_3\tau^3 + a_4\tau^4 + a_5\tau^5 + a_6\tau^6 + a_7\tau^7 + R_{n1} \\ &= \rho_T + R_{n1} , \\ \dot{\rho} &= a_1 + 2a_2\tau + 3a_3\tau^2 + 4a_4\tau^3 + 5a_5\tau^4 + 6a_6\tau^5 + 7a_7\tau^6 + R_{n2} \\ &= \dot{\rho}_T + R_{n2} , \\ \ddot{\rho} &= 2a_2 + 6a_3\tau + 12a_4\tau^2 + 20a_5\tau^4 + 30a_6\tau^4 + 42a_7\tau^5 + R_{n3} \\ &= \ddot{\rho}_T + R_{n3} \end{aligned}\right\} \tag{6.55}$$

 184

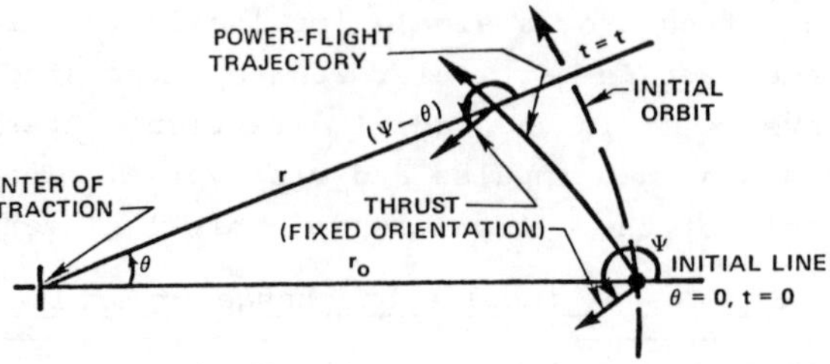

Fig. 6.5. Notations - fixed-orientation high thrust power-flight trajectory.

$$\left.\begin{aligned}\theta &= b_0 + b_1\tau + b_2\tau^2 + b_3\tau^3 + b_4\tau^4 + b_5\tau^5 + b_6\tau^6 + b_7\tau^7 + R_{n4}\\ &= \theta_T + R_{n4},\\ \dot\theta &= b_1 + 2b_2\tau + 3b_3\tau^2 + 4b_4\tau^3 + 5b_5\tau^4 + 6b_6\tau^5 + 7b_7\tau^6 + R_{n5}\\ &= \dot\theta_T + R_{n5},\end{aligned}\right.$$

and

$$\begin{aligned}\ddot\theta &= 2b_2 + 6b_3\tau + 12b_4\tau^2 + 20a_5\tau^3 + 30a_6\tau^4 + 42a_7\tau^5 + R_{n6}\\ &= \ddot\theta_T + R_{n6},\end{aligned}$$

where the dimensionless terms are defined as (6.56)

$$\rho \equiv \frac{r}{r_0}, \qquad \tau \equiv \left(\frac{g_0}{r_0}\right)^{1/2} t,$$

$$A \equiv \frac{\dot m \bar c}{M_0 g_0}, \qquad B \equiv \frac{\dot m}{M_0}\left(\frac{r_0}{g_0}\right)^{1/2},$$

$$N \equiv \sin\Psi, \qquad M \equiv \cos\Psi,$$

$$C = \left(\frac{d\rho}{d\tau}\right)_{\tau=0}, \text{ and } D = \left(\frac{d\theta}{d\tau}\right)_{\tau=0}.$$

ρ and θ are the polar coordinates, g_0 is the gravitational acceleration at r_0, and Ψ is the fixed orientation of the thrust direction with respect to the radius vector, $\vec r_0$.

The series coefficients are

$$a_0 = 1, \quad a_1 = C, \quad b_0 = 0, \quad b_1 = D,$$

$$2a_2 = AM + b_1^2 - 1,$$

$$6a_3 = A(BM + Nb_1) + a_1(b_1^2 + 2) + 4b_1b_2,$$

$$\begin{aligned}12a_4 &= A(B^2M + BNb_1 + Nb_2 - \tfrac{1}{2}Mb_1^2) + a_2(b_1^2 + 2) + 4a_1b_1b_2 +\\ &\quad + 4b_2^2 + 6b_1b_3 - 3a_1^2,\end{aligned}$$

$$\begin{aligned}20a_5 &= A(B^3M + B^2Nb_1 + BNb_2 - \tfrac{1}{2}MBb_1^2 + Nb_3 - Mb_1b_2 - \tfrac{1}{6}Nb_1^3) +\\ &\quad + a_3(b_1^2 + 2) + 4a_2b_1b_2 + a_1(4b_2^2 + 6b_1b_3) +\\ &\quad + (8b_1b_4 + 12b_2b_3) - 6a_1a_2 + 4a_1^2,\end{aligned}$$

$$\begin{aligned}30a_6 &= A[B^4M + B^3Nb_1 + B^2Nb_2 - \tfrac{1}{2}MB^2b_1^2 + B(Nb_3 - Mb_1b_2 - \tfrac{1}{6}Nb_1^3) + \\ &+ Nb_4 - \tfrac{1}{2}M(b_2^2 + 2b_1b_3) - \tfrac{1}{2}Nb_1^2b_2] + a_4(b_1^2 + 2) + 4a_3b_1b_2 + \\ &+ a_2(4b_2^2 + 6b_1b_3) + a_1(8b_1b_4 + 12b_2b_3) + (9b_3^2 + 16b_2b_4 + \\ &+ 10b_1b_5) + 12a_1^2a_2 - 5a_1^4 - 6a_1a_3 - 3a_2^2,\end{aligned}$$

$$\begin{aligned}42a_7 &= A[B^5M + B^4Nb_1 + B^3(Nb_2 - \tfrac{1}{2}Mb_1^2) + B^2(Nb_3 - Mb_1b_3 - \tfrac{1}{6}Nb_1^3) + \\ &+ B(Nb_4 - \tfrac{1}{2}Mb_2^2 - Mb_1b_3 - \tfrac{1}{2}Nb_1^2b_2) + Nb_5 - M(b_1b_4 + b_2b_3) - \\ &- \tfrac{1}{2}N(b_1^2b_3 + b_1b_2^2)] + a_5(b_1^2 + 2) + 4a_4b_1b_2 + a_3(4b_2^2 + \\ &+ 6b_1b_3) + a_2(8b_1b_4 + 12b_2b_3) + a_1(9b_3^2 + 16b_2b_4 + \\ &+ 10b_1b_5) + (12b_1b_6 + 20b_2b_5 + 24b_3b_4) - 20a_1^3a_2 + 6a_1^5 + \\ &+ 12a_1^2a_3 + 12a_1a_2^2 - 6a_1a_4 - 6a_2a_3,\end{aligned}$$

and

$$2b_2 + 2a_1b_1 = AN,$$

$$3b_3 + 4b_1b_2 + b_1(a_1^2 + 2a_2) = \tfrac{1}{2}A[N(B + a_1) - Mb_1],$$

$$\begin{aligned}&4b_4 + 6a_1b_3 + 2b_2(a_1^2 + 2a_2) + b_1(2a_3 + 2a_1a_2) \\ &\quad = \tfrac{1}{3}A[N(B^2 + a_1B + a_2) - Mb_2 - \tfrac{1}{2}Nb_1^2 - Mb_1(B + a_1)],\end{aligned}$$

$$\begin{aligned}&5b_5 + 8a_1b_4 + 3b_3(a_1^2 + 2a_2) + 4b_2(a_3 + a_1a_2) + b_1(a_2^2 + 2a_4 + 2a_1a_3) \\ &\quad = \tfrac{1}{4}A[N(B^3 + a_1B^2 + a_2B + a_3) - Mb_1(B^2 + a_1B + a_2) + \\ &\quad + \tfrac{1}{6}Mb_1^3 - (Mb_2 + \tfrac{1}{2}Nb_1^2)(B + a_1) - Mb_3 - Nb_1b_2],\end{aligned}$$

$$\begin{aligned}&6b_6 + 10a_1b_5 + 4b_4(a_1^2 + 2a_2) + 6b_3(a_3 + a_1a_2) + 2b_2(a_2^2 + 2a_4 + \\ &\quad + 2a_1a_3) + 2b_1(a_5 + a_1a_4 + a_2a_3) \\ &\quad = \tfrac{1}{5}A[N(B^4 + a_1B^3 + a_2B^2 + a_3B + a_4) - Mb_1(B^3 + a_1B^2 + \\ &\quad + a_2B + a_3) - (Mb_2 + \tfrac{1}{2}Nb_1^2)(B^2 + a_1B + a_2) + (\tfrac{1}{6}Mb_1^3 - Mb_3 - \\ &\quad - Nb_1b_2)(B + a_1) - Mb_4 - \tfrac{1}{2}N(b_2^2 + 2b_1b_3) + \tfrac{1}{2}Mb_1^2b_2],\end{aligned}$$

$$\begin{aligned}&7b_7 + 12a_1b_6 + 5b_5(a_1^2 + 2a_2) + 8b_4(a_3 + a_1a_2) + 3b_3(a_2^2 + 2a_4 + \\ &\quad 2a_1a_3) + 4b_2(a_5 + a_1a_4 + a_2a_3) + b_1(a_3^2 + 2a_1a_5 + 2a_2a_4 + 2a_6) \\ &\quad = \tfrac{1}{6}A\Big\{N(B^5 + a_1B^4 + a_2B^3 + a_3B^2 + a_4B + a_5) - \\ &\quad - Mb_1(B^4 + a_1B^3 + a_2B^2 + a_3B + a_4) - (Mb_2 + \tfrac{1}{2}Nb_1^2)(B^3 + \\ &\quad + a_1B^2 + a_2B + a_3) + (\tfrac{1}{6}Mb_1^3 - Mb_3 - Nb_1b_3)(B^2 + a_1B + a_2) + \\ &\quad + [\tfrac{1}{2}Mb_1^2b_2 - Mb_4 - \tfrac{1}{2}N(b_2^2 + 2b_1b_3)](B + a_1) - Mb_5 - \\ &\quad - N(b_1b_4 + b_2b_3) + \tfrac{1}{2}M(b_1^2b_3 + b_1b_2^2)\Big\}.\end{aligned}$$

It can be seen that all a's and b's are in terms of the preceding ones and consequently, can be determined one after another.

The remainders of the series, i.e., the R_n's, are related by

$$R_{n1} = \tfrac{1}{8}\tau R_{n2} = \tfrac{1}{56}\tau^2 R_{n3}$$

and

$$R_{n4} = \tfrac{1}{8}\tau R_{n5} = \tfrac{1}{56}\tau^2 R_{n6},$$

which, in turn, are computed from

$$\ddot{\rho}_T + R_{n3} = \frac{A}{1 - B\tau}[M(\cos\theta_T - R_{n4}\sin\theta_T) + N(\sin\theta_T + R_{n4} \times$$
$$\times \cos\theta_T)] + \rho_T\dot{\theta}_T^2 + \dot{\theta}_T^2 R_{n1} + 2\rho_T\dot{\theta}_T R_{n5} - \frac{1}{\rho_T^2} + \frac{2R_{n1}}{\rho_T^3} \tag{6.57}$$

and

$$\rho_T^2\ddot{\theta}_T + 2\rho_T\ddot{\theta}_T R_{n1} + \rho_T^2 R_{n6} + 2(\rho_T\dot{\rho}_T\dot{\theta}_T + \dot{\theta}_T\dot{\rho}_T R_{n1} + \rho_T\dot{\theta}_T R_{n2} + \rho_T\dot{\rho}_T R_{n5})$$
$$= \frac{A}{1 - B\tau}\rho_T[N(\cos\theta_T - R_{n4}\sin\theta_T) - M(\sin\theta_T +$$
$$+ R_{n4}\cos\theta_T)] + \frac{A}{1 - B\tau}R_{n1}(N\cos\theta_T - M\sin\theta_T) . \tag{6.58}$$

The accuracy of any truncated power series decreases as the magnitude of its independent variable increases; thus, for any pre-assigned accuracy, the maximum value of the independent variable can be determined. For the present problem, the independent variable is τ. After computing ρ, θ, $d\rho/d\tau$, and $d\theta/d\tau$ at the maximum τ, a new series can be developed by using the previous 'maximum τ' as the original point. This procedure will be repeated again and again until the required total τ is reached.

6.4.2.3. Low Thrust Case - Constant Thrust Acceleration

The changes of orbital elements can also be accomplished by a low thrust having a fixed orientation (Ψ) with respect to the radius vector of a geostationary satellite. The resulting trajectory is presented in Appendix E. We shall utilize the analytical approximate solution in Appendix E to examine the changes on orbital elements. As the expression of the low thrust orbit is

$$u = \frac{C}{h^2} + S\cos(\theta + \Phi) , \tag{6.59}$$

where the dimensionless terms are defined as

$$u \equiv \frac{r_0}{r}, \quad h \equiv \frac{\dot{\theta}}{u^2\dot{\theta}_0}, \quad C \equiv \frac{g_0}{r_0\dot{\theta}_0^2}, \quad \sin\Psi \equiv M, \quad \cos\Psi \equiv N, \; \varepsilon \equiv \frac{f}{g_0} ,$$

r and θ are the polar coordinates of the satellite, r_0 is the orbital distance at $t = 0$, g_0 is the gravitational acceleration at r_0, $\dot{\theta}$ is the angular velocity at any time t, $\dot{\theta}_0$ is the corresponding value at $t = 0$, and f is the constant thrust acceleration.

A comparison of (6.59) with the general orbit equation

$$r = \frac{a(1 - e^2)}{1 + e\cos(\eta - \omega)}$$

gives

$$a(1 - e^2) = r_0h^2, \quad e = Sh^2, \quad \theta = \eta, \quad \text{and} \quad \Phi = -\omega .$$

The variations of the orbital elements in relation to that of the dimensionless parameters of the low thrust trajectory are:

$$2r_0h\Delta h = \Delta a(1 - e^2) - 2ae\Delta e \cong \Delta a , \tag{6.60}$$

$$h^2\Delta S + 2hS\Delta h = \Delta e , \tag{6.61}$$

and

$$\Delta\Phi = - \Delta\omega . \tag{6.62}$$

If the first-order approximate solution of the low thrust trajectory is adopted for estimating the change of orbital elements, then $\bar{h}_0 = 1$, $\bar{S}_0 = e_0$, and $C \cong 1$. The dimensionless parameters may then be written in their incremental forms as

$$\Delta\bar{h} = \varepsilon M(\theta - \theta_0) \ , \tag{6.63}$$

$$\Delta\bar{S} = \frac{7}{2}e_0\Delta\bar{h} = \frac{7}{2}\varepsilon e_0 M(\theta - \theta_0) \ , \tag{6.64}$$

and

$$\Delta\Phi = -\frac{3}{2}\frac{N}{M}\Delta\bar{h} = -\frac{3}{2}\varepsilon N(\theta - \theta_0) \tag{6.65}$$

from the corresponding expressions in Appendix E by means of a series expansion in small parameters. The corresponding incremental forms of orbital elements are

$$\Delta a = 2\varepsilon M r_0(\theta - \theta_0) \ , \tag{6.66}$$

$$\Delta e = \frac{11}{2}\varepsilon e M(\theta - \theta_0) \ , \tag{6.67}$$

$$\Delta\omega = +\frac{3}{2}\varepsilon N(\theta - \theta_0) \ , \tag{6.68}$$

and

$$\Delta n = -3\varepsilon M\dot{\theta}_E(\theta - \theta_0) \tag{6.69}$$

because $n \equiv \mu_E^{1/2}/a^{3/2}$ and $n \cong \dot{\theta}_E$.

The time, t, required for the satellite to traverse $\theta - \theta_0$ may be approximated by $(\theta - \theta_0)/\dot{\theta}_E$ and the corresponding fuel consumption $M_p = \varepsilon g_0 M_0(t - t_0)/\bar{c}$ as is seen in (6.47b). $\Delta\omega$ and Δn are equivalent to $\Delta\Gamma$ and $\Delta\dot{\Gamma}$, respectively, in our rotating Earth longitude reference system shown in Figure 3.1. Consequently, the fuel consumption for the change of orbital elements shown in (6.66)-(6.69) are:

$$M_p(\text{for } \Delta a) = \frac{1}{2}\frac{V_c}{\bar{c}}\frac{\Delta a}{r_c}M_0 \ , \tag{6.66a}$$

$$M_p(\text{for } \Delta e) = \frac{2}{11}\frac{V_c}{\bar{c}}\frac{\Delta e}{e}\frac{1}{M}M_0 \ , \tag{6.67a}$$

$$M_p(\text{for } \Delta\omega \text{ or } \Delta\Gamma) = \frac{2}{3}\frac{V_c}{\bar{c}}\Delta\Gamma\frac{1}{N}M_0 \ , \tag{6.68a}$$

and

$$M_p(\text{for } \Delta n \text{ or } \Delta\dot{\Gamma}) = \frac{1}{3}\frac{r_c}{\bar{c}}\Delta\dot{\Gamma}\frac{1}{M}M_0 \ , \tag{6.69a}$$

where $g_0 \cong V_c^2/r_c$ is used, and $M \equiv \sin\Psi$ and $N \equiv \cos\Psi$.

6.4.3. Out-of-Plane Orbit Maneuver

An out-of-plane orbit maneuver always involves a rotation of the orbital plane and generally consumes more fuel than does an in-plane maneuver. For geostationary satellites, the inclination of the orbital plane needs correction when it deviates by more than a very small angle from the Earth's equatorial plane. If uncorrected, the angular deviation results in a 'figure 8' track of the satellite on the Earth's surface.

6.4.3.1. High Thrust Case

The inclination of a circular orbit can be adjusted by an impulse (ΔV), which is applied to the satellite as shown in Figure 6.6. This impulse is expressed as

$$\Delta V = 2V_c \sin \frac{\Delta i}{2} , \tag{6.70}$$

where V_c is the circular orbital velocity of the satellite and Δi is the change of the inclination.

Because Δi is rather small in the present case, sin $\Delta i/2$ in (6.70) can be approximated by $\frac{1}{2}\Delta i$; consequently, (6.70) reduces to

$$\Delta V = V_c \Delta i , \tag{6.71}$$

which, in turn, gives

$$M_p = V_c M_0 \Delta i/\bar{c} \tag{6.72}$$

when (6.32) is employed.

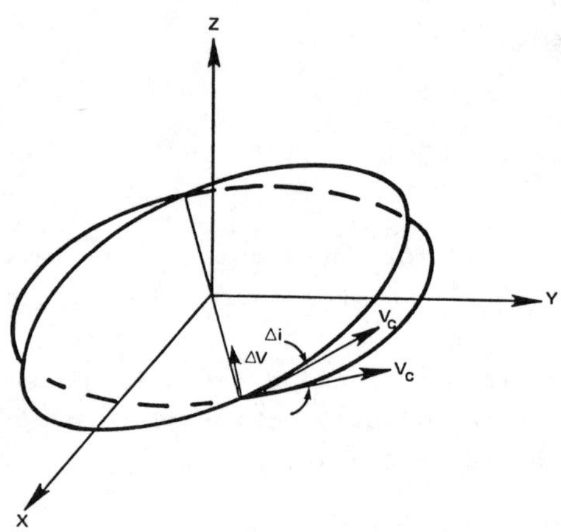

Fig. 6.6. Impulse for orbital plane inclination change.

When the impulse is replaced by a fixed-orientation finite thrust ($F = \dot{m}\bar{c}$) for the corresponding thrusting time ($t_T = M_p/\dot{m}$), the power-flight trajectory is (Zee, 1963e)

$$\left.\begin{aligned} X &= \sum_{n=0}^{7} a_n \tau^n + Rnox = X_T + Rnox , \\ Y &= \sum_{n=0}^{7} b_n \tau^n + Rnoy = Y_T + Rnoy , \\ Z &= \sum_{n=0}^{7} c_n \tau^n + Rnox = Z_T + Rnoz , \end{aligned}\right\} \tag{6.73}$$

$$\left.\begin{aligned} \dot{X} &= \sum_{n=0}^{7} na_n \tau^{n-1} + \frac{8}{\tau} Rnox = \dot{X}_T + \frac{8}{\tau} Rnox , \\ \dot{Y} &= \sum_{n=0}^{7} nb_n \tau^{n-1} + \frac{8}{\tau} Rnoy = \dot{Y}_T + \frac{8}{\tau} Rnoy , \\ \dot{Z} &= \sum_{n=0}^{7} nc_n \tau^{n-1} + \frac{8}{\tau} Rnoz = \dot{Z}_T + \frac{8}{\tau} Rnoz , \end{aligned}\right\} \tag{6.74}$$

$$\left.\begin{aligned}
\ddot{X} &= \sum_{n=0}^{7} n(n-1)\, a_n \tau^{n-2} + \frac{56}{\tau^2}\,\text{Rnox} = \ddot{X}_T + \frac{56}{\tau^2}\,\text{Rnox}\,, \\
\ddot{Y} &= \sum_{n=0}^{7} n(n-1)\, b_n \tau^{n-2} + \frac{56}{\tau^2}\,\text{Rnoy} = \ddot{Y}_T + \frac{56}{\tau^2}\,\text{Rnoy}\,, \\
\text{and}\quad \ddot{Z} &= \sum_{n=0}^{7} n(n-1)\, c_n \tau^{n-2} + \frac{56}{\tau^2}\,\text{Rnoz} = \ddot{Z}_T + \frac{56}{\tau^2}\,\text{Rnoz}\,.
\end{aligned}\right\} \tag{6.75}$$

The series coefficients are obtained from the following set of equations:

$$a_0 = X_0, \quad a_1 = \dot{X}_0$$

$$a_2 = \tfrac{1}{2}(A\ell - 4X_0 + 3X_0^3 + 3X_0Y_0^2)$$

$$b_0 = Y_0, \quad b_1 = \dot{Y}_0$$

$$b_2 = \tfrac{1}{2}(Am - 4Y_0 + 3X_0^2Y_0 + 3Y_0^3)$$

$$c_0 = 0, \quad c_1 = 0, \quad c_2 = \tfrac{1}{2}An$$

$$6a_3 = A\ell B - 4a_1 + 6X_0a_0a_1 + 3Y_0(a_0b_1 + a_1b_0)$$

$$12a_4 = A\ell B^2 - 4a_2 + 3X_0(a_1^2 + 2a_0a_2) + 3Y_0(a_0b_2 + a_1b_1 + a_2b_0)$$

$$20a_5 = A\ell B^3 - 4a_3 + 6X_0(a_0a_3 + a_1a_2) + 3Y_0(a_0b_3 + a_1b_2 + a_2b_1 + a_3b_0)$$

$$30a_6 = A\ell B^4 - 4a_4 + 3X_0(2a_0a_4 + 2a_1a_3 + a_2^2) + 3Y_0(a_0b_4 + a_1b_3 + a_2b_2 + a_3b_1 + a_4b_0)$$

$$42a_7 = A\ell B^5 - 4a_5 + 6X_0(a_0a_5 + a_1a_4 + a_2a_3) + 3Y_0(a_0b_5 + a_1b_4 + a_2b_3 + a_3b_2 + a_4b_1 + a_5b_0)$$

$$6b_3 = AmB - 4b_1 + 3X_0(a_0b_1 + a_1b_0) + 6Y_0b_0b_1$$

$$12b_4 = AmB^2 - 4b_2 + 3X_0(a_0b_2 + a_1b_1 + a_2b_0) + 3Y_0(b_1^2 + 2b_0b_2)$$

$$20b_5 = AmB^3 - 4b_3 + 3X_0(a_0b_3 + a_1b_2 + a_2b_1 + a_3b_0) + 6Y_0(b_0b_3 + b_1b_2)$$

$$30b_6 = AmB^4 - 4b_4 + 3X_0(a_0b_4 + a_1b_3 + a_2b_2 + a_3b_1 + a_4b_0) + 3Y_0(2b_0b_4 + 2b_1b_3 + b_2^2)$$

$$42b_7 = AmB^5 - 4b_5 + 3X_0(a_0b_5 + a_1b_4 + a_2b_3 + a_3b_2 + a_4b_1 + a_5b_0) + 6Y_0(b_0b_5 + b_1b_4 + b_2b_3)$$

$$6c_3 = AnB$$

$$12c_4 = AnB^2 - 4c_2 + 3X_0a_0c_2 + 3Y_0b_0c_2$$

$$20c_5 = AnB^3 - 4c_3 + 3X_0(a_0c_3 + a_1c_2) + 3Y_0(b_0c_3 + b_1c_2)$$

$$30c_6 = AnB^4 - 4c_4 + 3X_0(a_0c_4 + a_1c_3 + a_2c_2) + 3Y_0(b_0c_4 + b_1c_3 + b_2c_2)$$

$$42c_7 = AnB^5 - 4c_5 + 3X_0(a_0c_5 + a_1c_4 + a_2c_3 + a_3c_2) + 3Y_0(b_0c_5 + b_1c_4 + b_2c_3 + b_3c_2)$$

with the dimensionless terms defined as

$$X \equiv x/r_0, \qquad Y \equiv y/r_0, \qquad Z \equiv z/r_0\ ,$$
$$\dot{X} \equiv \dot{x}//g_0 r_0)^{1/2}, \quad \dot{Y} \equiv \dot{y}/(g_0 r_0)^{1/2}, \quad \dot{Z} \equiv \dot{z}/(g_0 r_0)^{1/2}\ ,$$
$$A \equiv \frac{\dot{m}\bar{c}}{M_0 g_0}, \qquad B \equiv \frac{\dot{m}}{M_0}\left(\frac{r_0}{g_0}\right)^{1/2}, \quad \text{and} \quad \tau \equiv \left(\frac{g_0}{r_0}\right)^{1/2} t.$$

The initial orbit is in the x-y plane, and the respective direction cosines of the thrust with respect to x, y, and z axis are ℓ, m, and n; x, y, and z are rectangular coordinates of the satellite, and $\dot{x}$, $\dot{y}$, and $\dot{z}$ are the corresponding velocities, wherein a subscript '0' represents the value at t = 0; r_0 is the radius of the circular orbit, and g_0 is the gravitational acceleration of the Earth at r_0. Rnox, Rnoy and Rnoz are the remainders (or errors) to be determined. Equations (6.73)-(6.75) are obtained by a Taylor series expansion in the differential equations of motion. Since the duration of thrusting is short, this approach is an effective method of solution. Because a series solution is invariably associated with truncation errors, an estimate of these errors constitutes a partial requirement of the solution. In the present case, they can be estimated by the following expression:

$$\begin{aligned} X_T^2 + 2X_T\text{Rnox} &= \\ &= \left(\frac{A\ell}{1 - B\tau} - \ddot{X}_T\right)^2 (X_T^2 + Y_T^2 + Z_T^2)^3 + 6X_T\left(\frac{A\ell}{1 - B\tau} - \ddot{X}_T\right)^2 \times \\ &\quad \times [X_T^4 + (Y_T^2 + Z_T^2)(2X_T^2 + Y_T^2 + Z_T^2)\text{Rnox}] + 6Y_T\left(\frac{A\ell}{1 - B\tau} - \ddot{X}_T\right)^2 \times \\ &\quad \times [Y_T^4 + (X_T^2 + Z_T^2)(X_T^2 + 2Y_T^2 + Z_T^2)\text{Rnoy}] + \\ &\quad + 6Z_T\left(\frac{A\ell}{1 - B\tau} - \ddot{X}_T\right)^2 [Z_T^4 + (X_T^2 + Y_T^2)(X_T^2 + Y_T^2 + 2Z_T^2)\text{Rnoz}] + \\ &\quad + (X_T^2 + Y_T^2 + Z_T^2)^3 \left|\ddot{X}_T - \frac{A\ell}{1 - B\tau}\right| \frac{112}{\tau^2}\text{Rnox}\ . \end{aligned} \tag{6.76}$$

Two similar equations for estimating errors in Y and Z may be written out from (6.76) by interchanging X and Y, and X and Z, respectively, together with their corresponding direction cosines and remainders.

It is generally known that the accuracy of the previous truncated power series decreases as the magnitude of τ increases. Thus, for any preassigned accuracy, the maximum value of τ could be determined. After computing the desired value of the maximum τ, a new series is developed by using the coordinates and velocities at the previous 'maximum τ' as the new initial conditions. The procedure can be repeated again and again until the required total τ is reached.

For a small orbital inclination adjustment of a geostationary satellite orbit, it is believed that one series expansion is sufficient for the computation of the power-flight. The calculation of truncation error only gives an added sense of accuracy.

6.4.3.2. Low Thrust Case

We shall adopt the work of Illarionov and Shkadov (1962) to study the change of inclination of an orbital plane for the low thrust case.

As shown in Figure 6.7., a satellite is initially travelling around the Earth in a circular orbit located in the x-y plane (or the equatorial plane of the Earth), and the position of the satellite is designated by the spherical coordinates r, θ, and ϕ. The satellite is now under the action of a lateral force, F, which always lies in a plane normal to the radius vector, $\vec{r}$, and is continuously directed normal to the velocity vector, $\vec{V}$. The motion of the satellite is characterized by possessing a constant velocity $V = V_0 = (g_0 r_0)^{1/2}$ and a constant altitude $r = r_0$, where g_0 is the gravitational acceleration at r_0. The resulting orbit is a small circle of the terrestrial sphere and is tangent to the original orbit at the zero time. If the lateral force, F, is discontinued at a certain point, C, along the small circle, the orbit will again become central and inclined relative to the unperturbed orbit by some angle, Ψ, as shown in Figure 6.8.

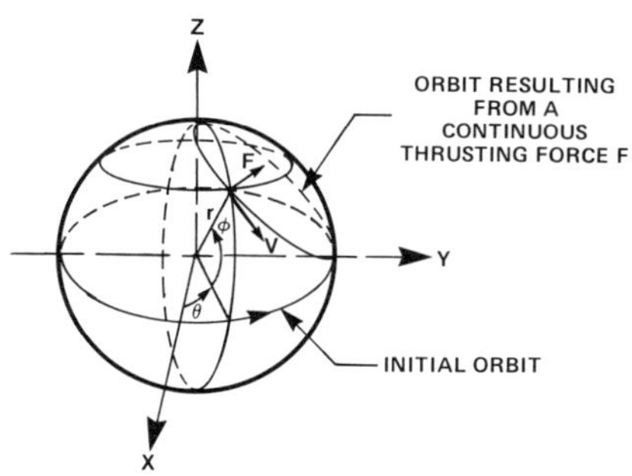

Fig. 6.7. Notations - rotation of orbital plane.

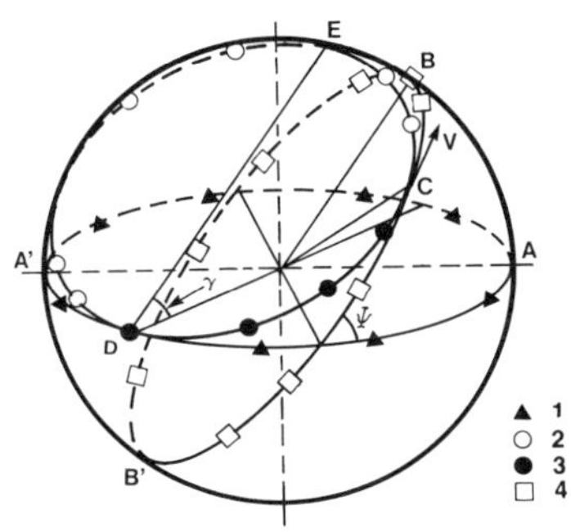

Fig. 6.8. Intermediate orbit during orbital plane rotation maneuvers.

In Figure 6.8., 1 is the original orbit, 2 is the orbit with the presence of a constant lateral force, F, 3 is a portion of power-flight, 4 is the orbit after discontinuation of the thrusting, and AB and A'B' are the maximum elevations of the satellite. With these observations in Figure 6.8., it can be seen that the satellite returns periodically once every revolution to the original point, and the alternative increment and decrement of the angle, γ, takes place during the odd and even half periods, respectively. In order to increase the total rotational angle of the orbital plane, it is necessary to reverse the direction of thrusting at the half-period point. Thus, after an integral number, n, of half-periods, the rotational angle (Ψ_n) can be calculated by

$$\Psi_n = n \sin^{-1}\left(\frac{2\varepsilon}{1 + \varepsilon^2}\right) \tag{6.77}$$

where ε is the thrust, F, to the weight of the satellite, $M_0 g_0$, ratio at the altitude of r_0, g_0 is the gravitational acceleration at r_0, and M_0 is the satellite mass at t = 0. The total travelling angle (θ_r) is equal to $n\pi$.

The time required, t_n, is

$$t_n = \frac{n\pi}{\omega_0(1 + \varepsilon^2)^{1/2}} , \tag{6.78}$$

where $\omega_0 = (g_0/r_0)^{1/2}$.

If the required rotational angle of an orbital plane is Ψ_a, then n is determined by letting

$$\Psi_a - \Psi_n < \sin^{-1}\left(\frac{2\varepsilon}{1 + \varepsilon^2}\right). \tag{6.79}$$

The duration of the power-flight and the position of the satellite (r_0, θ_a, ϕ_a) at the end of the power-flight can be calculated by the following expressions:

$$\cos 2\Psi_a = 2[\cos \theta_n + \varepsilon(\sin \theta_n - \sin \theta_a)] , \tag{6.80}$$

$$\theta_a = \sin^{-1}\left[\frac{\varepsilon(\cos \theta_n + \varepsilon \sin \theta_n) + (\sin \theta_n - \varepsilon \cos \theta_n)\cos \Phi}{1 + \varepsilon^2}\right], \tag{6.81}$$

$$\phi_a = \tan^{-1}\left[\frac{(1 + \varepsilon^2)^{1/2} \sin \Phi}{(\cos \theta_n + \varepsilon \sin \theta_n)\cos \Phi - \varepsilon(\sin \theta_n - \varepsilon \cos \theta_n)}\right], \tag{6.82}$$

$$\Phi = (1 + \varepsilon^2)^{1/2} \omega_0 \Delta t , \tag{6.83}$$

and

$$t_a = t_n + \Delta t . \tag{6.84}$$

The fuel consumption is determined by (6.47b) when $t - t_0$ is replaced by t_a (total time of the power-flight maneuver). However, if the rotation (Δi, in radians) of the orbital plane is equal to Ψ_n, then combining (6.47), (6.71), and (6.72) leads to

$$\Delta V = \frac{\pi}{2} V_0 \Delta i , \tag{6.85}$$

and

$$\frac{M_p}{M_0} = \frac{\pi}{2} \frac{V_0}{\bar{c}} \Delta i , \tag{6.86}$$

where ε^2 is omitted and $\sin[2\varepsilon/(1 + \varepsilon^2)]$ is approximated by 2ε. Equation (6.85) shows that the utilization of a small thrust rocket to rotate the orbital plane needs only $2/\pi$ of the ΔV required for the impulsive thrust maneuver. With the low thrust rocket having larger I_{sp} ($\equiv \bar{c}/g_e$), i.e., larger $\bar{c}$, the fuel savings are clearly shown when (6.72) and (6.86) are examined.

The orientation of an orbital plane includes the inclination and the longitude of the line of nodes. During the long duration of thrusting in changing the inclination, the line of nodes also regresses as discussed in Sections 4.10.2. and 4.10.6. In order to have the desired orientation of the orbital plane at the end of the thrusting maneuver, the desired inclination is governed by the duration of the thrust maneuver, while the desired longitude of the line of nodes is controlled by the initial position of the satellite at the beginning of the thrusting maneuver. In Figure 6.8., point D is the initial position of the satellite. When the satellite reaches point E, having the desired maximum elevation (the equivalent desired inclination) in the small power-flight orbit '3', the thrusting is discontinued; the discontinuation of thrusting at point E results in an Earth-center orbit, which has its line of nodes normal to the projection of line DE in the equatorial plane. Therefore, an addition of 90° to the longitude of point D gives the desired longitude of the line of nodes.

6.5. Station-Keeping Procedures

Since a geostationary satellite experiences perturbing forces from a number of sources, it is practically impossible to keep the satellite stationary with respect to an Earth observer. Station-keeping usually involves setting up a window of deviation limits relative to a desired geostationary satellite position with respect to a point on the equator of the Earth. If the limits are exceeded, corrective delta-V applications are performed so that the satellite is brought back to the window of the limits.

Our improved first-order solution of various orbital parameters includes short-period oscillatory (hourly) terms and long-period averaged terms. In general, it is not possible to correct the short-period oscillatory terms because, from a practical point of view, the correction cycle is in terms of days or weeks. This fact justifies our previous discussions on the effects of impulsive velocity (Section 6.2.) and perturbations (Section 6.3.) on orbital parameters by adopting the first-order approximate solution as a base for the study of station-keeping.

We shall recapitulate the relationships between our orbital parameters and the osculating orbital elements for the purpose of carrying out our station-keeping study.

The smallness of i, X_1, and X_2 simplifies (4.154) as

$$\bar{i} = (\bar{X}_1^2 + \bar{X}_2^2)^{1/2} , \tag{6.87}$$

and taking a variation of (6.87) gives

$$\Delta\bar{i} = (\bar{X}_1\Delta\bar{X}_1 + \bar{X}_2\Delta\bar{X}_2)/(\bar{X}_1^2 + \bar{X}_2^2)^{1/2} . \tag{6.88}$$

Following the same reasoning, (4.169) may be approximated as

$$\bar{e} = (\bar{X}_3^2 + \bar{X}_4^2)^{1/2}\,\bar{\ell} \tag{6.89}$$

and

$$\Delta\bar{e} = (\bar{X}_3\Delta\bar{X}_3 + \bar{X}_4\Delta\bar{X}_4)\bar{\ell}/(\bar{X}_3^2 + \bar{X}_4^2)^{1/2} , \tag{6.90}$$

where $(\bar{X}_3^2 + \bar{X}_4^2)^{1/2}\delta\bar{\ell}$ is neglected because it is a higher order term, and $\bar{X}_3'$ and $\bar{X}_4'$ are approximated by $\bar{X}_3$ and $\bar{X}_4$, respectively.

The drift angle and the drift rate have the same designations in both parameter systems.

The following sections will discuss possible station-keeping procedures for correcting various perturbed effects by means of high thrust impulse (delta-V application); however, some of these delta-V applications can also be accomplished by low thrust orbital maneuvers as discussed in Section 6.4., but the additional perturbing effects on the satellite during the period of low thrusting should also be kept in mind.

6.5.1. Correction of Latitudinal Variation

The attractions of the Sun and the Moon on a geostationary satellite result in the rotation and regression of the satellite orbital plane. The rotation of the orbital plane is shown in Figures 4.2. and 4.3. The rotational rate of a geostationary orbital plane is 0.853 degrees per year, which, according to (6.70), requires a corrective delta-V of 157 fps per year. The space power-flight trajectory for adjusting the inclination of the orbital plane can be calculated either by a high thrust (Section 6.4.3.1.) or by a low thrust (Section 6.4.3.2.).

If, for example, a ±2° north-south deviation from the equatorial plane of the Earth can be tolerated, then, it is possible, by selecting an initial orbit orientation of $i_0 = 2°$ and $\bar{\Omega}_0 = 285.42°$, to ensure that during a 4.55-years mission period, the inclination will not exceed the initial value, and therefore no north-south delta-V would be required. This fact can be seen clearly in Figure 4.4. Other initial inclinations (i_0) with corresponding Ω_0 may also be selected for a desired duration of tolerance.

This property of orbital plane rotation can be utilized in station-keeping by an application of ΔV_N or a low thrust maneuver as described in Section 6.4.3.2. so that the geostationary satellite orbital plane will have a small inclination with its ascending node in the neighborhood of 270°.

In general, if a correction cycle of a duration of 2T is planned to adjust the inclination of an orbital plane having an initial orientation i_0 and Ω_0, then after the application of ΔV_N, or the low thrust maneuver, the orbital plane is expected to rotate from $\bar{i}_{(-T)}$ to $\bar{i}_{(+T)}$ during the correction cycle. $\bar{i}_{(-T)}$ and $\bar{i}_{(+T)}$ are the averaged inclinations at t = -T and t = +T, respectively, as their values can be computed from the first-order solutions of $\bar{X}_1$ and $\bar{X}_2$ shown in (6.87). A detailed description of the procedures follows.

6.5.1.1. Impulsive Thrust Correction

With known i_0 and Ω_0, X_{10} and X_{20} are computed from (4.149) and (4.154), and $\bar{X}_{10}$ and $\bar{X}_{20}$ are obtained from (5.29) and (5.30). The first-order perturbation solutions, $\bar{X}_1$ and $\bar{X}_2$, shown in (5.20) and (5.21) will give $\bar{X}_{1(-T)}$, $\bar{X}_{1(+T)}$, $\bar{X}_{2(-T)}$, and $\bar{X}_{2(+T)}$ at the time '-T' and '+T' from the time of the ΔV_N application. Therefore, the application of ΔV_N results in

$$\Delta\bar{X}_{1(-T)} = \bar{X}_{10} - \bar{X}_{1(-T)} = -(\Delta V_N/V_c)\sin\theta_A \tag{6.91}$$

and

$$\Delta\bar{X}_{2(-T)} = \bar{X}_{20} - \bar{X}_{2(-T)} = (\Delta V_N/V_c)\cos\theta_A \tag{6.92}$$

as is seen from (6.19) and (6.20). The solutions to the above two equations give the magnitude (ΔV_N) and the point of application (θ_A) of delta-V, and the inclination ($\bar{i}_0$) is now adjusted to $\bar{i}_{(-T)} = \bar{i}_0 + \Delta\bar{i}_{(-T)} = \bar{i}_0 + (\bar{X}_{10}\Delta\bar{X}_{1(-T)} + \bar{X}_{20}\Delta\bar{X}_{2(-T)})/\bar{i}_0$. This adjusted inclination will become $\bar{i}_{(+T)} = \bar{i}_0 + \Delta\bar{i}_{(+T)} = \bar{i}_0 + [\bar{X}_{10}(\bar{X}_{1(+T)} - \bar{X}_{10}) + \bar{X}_{20}(\bar{X}_{2(+T)} - \bar{X}_{20})]/\bar{i}_0$ after a time period of 2T from the time of the ΔV_N application as a result of the perturbations.

The above procedure can be repeated with $\bar{X}_{1(+T)}$ and $\bar{X}_{2(+T)}$ as the initial conditions for the next ΔV_N computation, and the cycle of correction for i is established.

Since the above calculations use the first-order solutions, $\bar{X}_1$ and $\bar{X}_2$, to compute $\bar{i}_{(+T)} = \bar{i}_0 + \Delta\bar{i}_{(+T)}$, the actual i can be computed from

$$\bar{i}_{(+T)} + \delta i = \bar{i}_{(+T)} + (X_{1(T)}\delta X_{1(T)} + X_{2(T)}\delta X_{2(T)})/\bar{i}_{(+T)}, \tag{6.93}$$

where $\delta X_{1(T)}$ and $\delta X_{2(T)}$ are obtained from (5.29) and (5.30), respectively.

For the special case of maneuvering a geostationary satellite to have an orbit with its ascending node in the neighborhood of 270° as previously mentioned, it is required that the orbital plane be adjusted to have its mean orientation ($\bar{i}_{00}$, $\bar{\Omega}_{00}$) at t = 0, and it will have a zero inclination when t = T.

Therefore, we have

$$\bar{X}_{1(T)} = 0 \tag{6.94}$$

and

$$\bar{X}_{2(T)} = 0\ . \tag{6.95}$$

From these two equations, a_0 and Ψ_0 can be deduced, and, in turn, $\bar{X}_{100}$ and $\bar{X}_{200}$, which correspond to $\bar{i}_{100}$ and $\bar{\Omega}_{100}$, are obtained. The station-keeping requires

$$\Delta\bar{X}_1 = \bar{X}_{100} - \bar{X}_{10} = -(\Delta V_N/V_c)\sin\theta_A \tag{6.96}$$

and

$$\Delta\bar{X}_2 = \bar{X}_{200} - \bar{X}_{20} = (\Delta V_N/V_c)\cos\theta_A \tag{6.97}$$

to adjust the mean orientation of the orbital plane from ($\bar{i}_0$, $\bar{\Omega}_0$) to ($\bar{i}_{00}$, $\bar{\Omega}_{00}$). The procedure will be repeated with $\bar{X}_{1(2T)}$ and $\bar{X}_{2(2T)}$ as the initial conditions for the next ΔV_N computation. It should be noted that the deduction of a_0 and Ψ_0 from (6.94) and (6.95) is rather tedious. As a first approximation, (4.164) and (4.165) are employed to solve for Ψ_0 with $\Psi = 0$ and $\theta = \dot{\theta}_E T$, and a_0 is approximated by D/ω_{11}.

The above scheme of adjusting the small inclination orbital plane having its ascending node in the neighborhood of 270° is superior than the general scheme first described, because the latitudinal deviation for the former is only half that of the latter.

6.5.1.2. Low Thrust Maneuver Correction

In the case of utilizing low thrusting to adjust the inclination of the orbital plane, the perturbation of the Sun, Moon, Earth, and the solar radiation pressure on i and Ω during the period of the low thrusting maneuvering (t_a in (6.84)) should be estimated as $\Delta i_{(t_a)}$ and $\Delta\Omega_{(t_a)}$, respectively. The adjustment of the mean orientation of the orbital plane from $(\bar{i}_0, \bar{\Omega}_0)$ to $(\bar{i}_0 + \Delta\bar{i}_{(-T)}, \bar{\Omega}_0 + \Delta\bar{\Omega}_{(-T)})$ requires a low thrust maneuver to have $\Delta i = \Delta i_{(t_a)} + (\bar{i}_0 + \Delta\bar{i}_{(-T)})$ and $\Delta\bar{\Omega} = \Delta\Omega_{(t_a)} + (\bar{\Omega}_0 + \Delta\bar{\Omega}_{(-T)})$. With these two values, the initial conditions of the low thrust maneuver are established. The estimations of $\Delta i_{(t_a)}$, $\Delta i_{(-T)}$, etc. from the X's follow the same procedures described in the last section.

6.5.2. Correction of Longitudinal Variation

Earth's triaxiality affects a geostationary satellite by producing a long-period libration about the Earth's equator minor axis. The action of solar radiation pressure on the satellite results in an eccentricity change and a rotation of the line of apsides of the satellite orbit. The attractions of the Sun and Moon also induce longitudinal drift and drift rate. All these perturbations result in an oscillatory east-west motion of the satellite relative to a fixed station on the equator of the Earth, and corrections are required to minimize the longitudinal variations. We shall discuss the procedures of corrections for a large range of variation due to the libration alone and a small range of deviation encompassing the effects of all perturbations.

6.5.2.1. Corrections as a Result of Libration - Large Range Longitudinal Variation

The long-period longitudinal variation of a geostationary satellite with respect to a fixed point on the Earth is predominantly characterized by the satellite's long-period libration about the Earth's equator minor axis. Figure 6.9. shows the typical libration contours. Each contour is associated with a Γ_{max}. Since a geostationary satellite orbit has a small eccentricity, these contours may be considered as the paths of the mean daily longitudinal movement of the satellite. For an allowable range of limits of longitudinal deviations ($\pm \frac{1}{2}\Delta\Gamma$) at any desired longitude on the Earth, a contour 'A' can be chosen as a reference frame for the longitudinal station-keeping. When the satellite reaches point 'a', an orbital maneuver is required to transfer the satellite to point 'b'; this orbital transfer can be accomplished either by a high thrust Hohmann transfer (Section 6.4.1.1.) or by a low thrust spiral descending (Sections 6.4.1.2.1. and 6.4.1.2.2.). The satellite is then travelling along the contour 'A' back to the point 'a' again; the time required for the satellite travelling from points 'b' back to 'a' establishes the interval of correction cycle. This scheme forms our basic philosophy in correcting the longitudinal variation.

The contours shown in Figure 6.9. are not only symmetric about the Earth's equator minor axis and the major axis, but are also symmetric about the synchronous circular orbit (r_c).

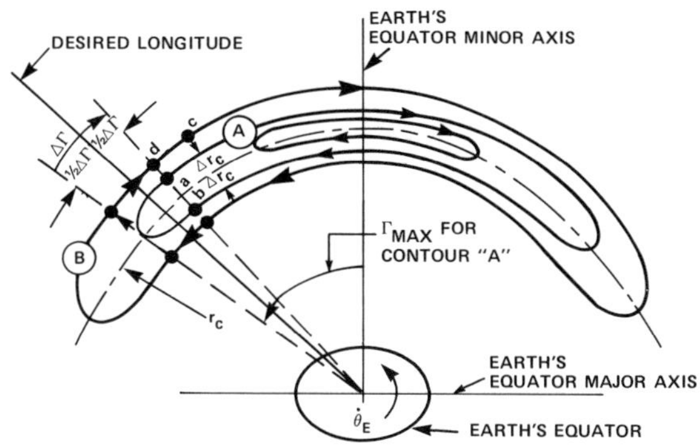

Fig. 6.9. Libration contours of geostationary satellites.

6.5.2.1.1. Impulsive Thrust Case

The orbital transfer from point 'a' to 'b' constitutes an altitude change in the in-plane orbit maneuver. We have previously studied this case in Section 6.4.1.1. Therefore, we can write (6.30) as

$$\Delta V_{1T} = \left[2\,\mu_E\left(\frac{1}{r_c - \Delta r_c} - \frac{1}{2r_c}\right)\right]^{1/2} - \left(\frac{\mu_E}{r_c - \Delta r_c}\right)^{1/2}$$

$$\cong \frac{1}{2}\,\frac{\Delta r_c}{r_c}\,V_c \tag{6.98}$$

and

$$\Delta V_{2T} = \left(\frac{\mu_E}{r_c + \Delta r_c}\right)^{1/2} - \left[2\,\mu_E\left(\frac{1}{r_c + \Delta r_c} - \frac{1}{2r_c}\right)\right]^{1/2}$$

$$\cong \frac{1}{2}\,\frac{\Delta r_c}{r_c}\,V_c\ , \tag{6.99}$$

where Δr_c is the radial distance between the points of orbital transfer ('a' or 'b') and the synchronous radius (r_c). In obtaining the above approximations, the terms containing $(\Delta r_c/r_c)^2$, etc. are omitted. The total delta-V required for the orbital transfer from point 'a' to point 'b' on contour 'A' is $(\Delta r_c/r_c)V_c$, and the corresponding fuel consumption is $(\Delta r_c/r_c)V_cM_0$ as is seen from (6.33). In general, this is true for any orbital transfer between two points, which are the intersections of a radius and a libration contour. The total delta-V for station-keeping with different ranges of limits of longitudinal deviations ($\pm\,\frac{1}{2}\Delta\Gamma$) for various contours ($\Gamma_{max}$'s) may then be calculated from the known value of Δr_c. Figures 3.9. and 3.10. furnish the necessary information for computing the total delta-V calculation and the time interval between successive delta-V applications. For example, if the limit of deviation is 1° ($= \frac{1}{2}\Delta\Gamma$) and the satellite is on the $\Gamma_{max} = 20°$ contour, then Figure 3.9. shows that $\Delta\Gamma = 2°$ and $\Delta r_c = 16{,}500$ ft. (5,030 m.). Hence $\Delta V = 1.2$ ft/sec (0.37 m/sec), and the time interval between the delta-V application is about 2 × 60 = 120 days as it can be obtained from Figure 3.10.

6.5.2.1.2. Low Thrust Case

For the case of low thrust orbital transfer from point 'a' to point 'b', the fuel consumption and the time required for the transfer are approximated by:

(a) Constant thrust case:

$$M_p = \dot{m}t \cong 4\Delta r_c \dot{\theta}_E M_0/\bar{c} \tag{6.100}$$

and

$$t = M_p/\dot{m} \; . \tag{6.101}$$

(b) Constant thrust acceleration case:

$$M_p = 4\Delta r_c \dot{\theta}_E M_0/\bar{c} \tag{6.102}$$

and

$$t = 4\Delta r_c/\varepsilon V_c \; . \tag{6.103}$$

The above set of equations can be directly derived from Sections 6.4.1.2.1. and 6.4.1.2.2. with the approximations $r \cong r_c$ and $\dot{\theta}_0 \cong \dot{\theta}_E$.

The above discussion on the correction of longitudinal variation is limited to the effect of the long-period libration alone; however, additional longitudinal drift and drift rate due to the attractions of the Sun and the Moon together with the change of the eccentricity of a geostationary orbit due to solar radiation pressure further complicate the procedures of correction. Fortunately, these additional longitudinal drifts and drift rates are rather small in comparison with those due to the ellipticity of the equator of the Earth. However, we shall discuss the correction of longitudinal variations encompassing all longitudinal drifts and drift rates in Section 6.5.2.2.1.

6.5.2.2. General Correction Procedures - Small Range Longitudinal Variation

It is generally required that the range of deviations, $\pm\,\Delta\Gamma$, is on the order of milliradians as was pointed out in Chapter 1. The station-keeping of a geostationary satellite within a small range of longitudinal variation can again be accomplished by either impulsive thrust or by a low thrusting maneuver.

6.5.2.2.1. Impulsive Thrust Case

If a geostationary satellite is initially at point 'c' on contour 'B', it is required to bring the satellite to point 'b' on contour 'A' for the station-keeping in the neighborhood of a desired longitude as shown in Figure 6.9. The following procedures will be involved:

- Assume that the orbit is already circularized by means of a two-impulse Hohmann-type transfer described in Section 6.4.2.1. Two radial impulses of equal magnitude, applied 180° apart on the orbit, will bring the satellite from point 'c' to point 'd' without changing the eccentricity as is seen from (6.17), (6.21), and (6.22);
- Transferring from point 'd' to point 'b' requires an altitude change as described in Section 6.4.1.1.

After the satellite reaches point 'b', it will travel nominally along contour 'A' within a period of time, 2T, to a point 'a' in the neighborhood of point 'a' because all perturbations are taken into consideration. From that point, we shall perform our station-keeping procedures to bring the satellite back to point 'b' as follows.

With a preassigned $\Delta\Gamma$, the drift angle and the drift rate at point 'b' are set as

$$\Gamma_b = \Gamma_{Emax} - KT^2 + \Gamma_{SMR(-T)} \tag{6.104}$$

and

$$\dot{\Gamma}_b = 3\dot{\theta}_E T \left(\frac{K \sin 2\Gamma_{Emax}}{k_1}\right)^{1/2} + \dot{\Gamma}_{SMR(-T)} \tag{6.105}$$

respectively, where (6.23)-(6.29) have been employed, and $\Gamma_{SMR(-T)}$ and $\dot{\Gamma}_{SMR(-T)}$ depict their respective value at $t = -T$.

When a geostationary satellite orbit has an eccentricity, contour 'A' may be viewed as a path of the mean daily longitudinal movement of the satellite. A reduction of the hourly longitudinal variation due to the eccentricity is required. If $\bar{e}_0$ is the initial mean eccentricity, the change of eccentricity during a time interval of (-T) is

$$\Delta\bar{e}_{(-T)} = (\bar{X}_{30}\Delta\bar{X}_{3(-T)} + \bar{X}_{40}\Delta\bar{X}_{4(-T)})\bar{\ell}/(\bar{X}_{30}^2 + \bar{X}_{40}^2)^{1/2} \tag{6.106}$$

as is seen from (6.90), where

$$\Delta\bar{X}_{3(-T)} = \bar{X}_{3(-T)} - \bar{X}_{30}$$

and

$$\Delta\bar{X}_{4(-T)} = \bar{X}_{4(-T)} - \bar{X}_{40} \,. \tag{6.107}$$

By the same procedure, we obtain $\Delta\bar{e}_{(+T)}$. Our criterion of limiting the longitudinal variation requires that the eccentricity varies from $\bar{e}_0 + \Delta\bar{e}_{(-T)}$ to $\bar{e}_0 + \bar{e}_{(+T)}$ within a time interval of 2T; therefore,

$$\bar{e}_b = \bar{e}_0 + \Delta\bar{e}_{(-T)} \tag{6.108}$$

is the eccentricity to be set at point 'b'.

When the set values of Γ_b, $\dot{\Gamma}_b$, and $\bar{e}_b$ in (6.104), (6.105), and (6.108), respectively, and their initial values of $\Gamma_{a'}$, $\dot{\Gamma}_{a'}$, and $\bar{e}_0$ at point 'a' are known, the delta-V's required to transfer the satellite from point 'a' in the neighborhood of point 'a' to point 'b' can be determined from (6.17), (6.18), (6.21), and (6.22). They are

$$\Delta\Gamma_D = \Gamma_b - \Gamma_{a'} = -(2/V_c)\Delta V_{RD} \,, \tag{6.109}$$

$$\Delta\dot{\Gamma}_D = \dot{\Gamma}_b - \dot{\Gamma}_{a'} = -(3\dot{\theta}_E/V_c)\Delta V_{TD} \,, \tag{6.110}$$

$$\Delta\bar{X}_3 = \Delta\bar{X}_{3(-T)} - \bar{X}_{30} = (2\Delta V_{Te}/V_c)\cos\theta_A + (\Delta V_{Re}/V_c)\sin\theta_A, \tag{6.111}$$

and

$$\Delta\bar{X}_4 = \Delta\bar{X}_{4(-T)} - \bar{X}_{40} = (2\Delta V_{Te}/V_c)\sin\theta_A - (\Delta V_{Re}/V_c)\cos\theta_A, \tag{6.112}$$

where subscripts 'D' and 'e' are added to ΔV_R and ΔV_T to differentiate the ΔV requirements for the corrections of the drift angle and the drift rate from that of the eccentricity.

As mentioned before, the changes of $\bar{X}_3$ and $\bar{X}_4$ in (6.111) and (6.112) can be accomplished either by a single transverse impulse (with subscript S designating 'single')

$$\Delta V_{TeS} \equiv \tfrac{1}{2}V_c(\Delta\bar{X}_3^2 + \Delta\bar{X}_4^2)^{1/2} \,, \tag{6.113}$$

or by a single radial impulse (with subscript S designating 'single')

$$\Delta V_{ReS} \equiv V_c(\Delta\bar{X}_3^2 + \Delta\bar{X}_4^2)^{1/2} \,, \tag{6.114}$$

or a combination of ΔV_{Te} and ΔV_{Re}.

Possible combinations of $|\Delta V_{RD}|$, $|\Delta V_{TD}|$, ΔV_{TeS}, and ΔV_{ReS} result in the following nine cases:

	$\Delta V_{ReS} > \|\Delta V_{RD}\|$	$\Delta V_{ReS} = \|\Delta V_{RD}\|$	$\Delta V_{ReS} < \|\Delta V_{RD}\|$
$V_{TeS} > \|\Delta V_{TD}\|$	Case I	Case II	Case III
$V_{TeS} = \|\Delta V_{TD}\|$	Case IV	Case V	Case VI
$V_{TeS} < \|\Delta V_{TD}\|$	Case VII	Case VIII	Case IX

In applying various delta-V corrections, the following facts must be recognized as is seen from (6.17)-(6.22):

- $\Delta\Gamma_D$ or $\Delta\dot{\Gamma}_D$ is independent of the point of application of the particular impulse.
- Two transverse (or radial) corrections of equal magnitude, but opposite sign, applied 180° apart will cause equal changes in $\Delta\bar{X}_3$ and $\Delta\bar{X}_4$, but cause no net change in $\Delta\dot{\Gamma}_D$ (or $\Delta\Gamma_D$).
- Two transverse (or radial) corrections of equal magnitude and same sign, applied 180° apart will cause equal changes in $\Delta\dot{\Gamma}_D$ (or $\Delta\Gamma_D$), but cause no net changes in $\Delta\dot{\Gamma}_D$ (or $\Delta\Gamma_D$).

With these facts, we shall discuss the schemes of various delta-V corrections for the nine cases:

Case I

Correct $\Delta\bar{X}_3$ and $\Delta\bar{X}_4$ partially by ΔV_{TD} and the remainders by ΔV_{Re}, if:

- $\Delta V_{Re} > |\Delta V_{RD}|$, use ΔV_{RD} in conjunction with ΔV_{TD} to correct $\Delta\bar{X}_3$ and $\Delta\bar{X}_4$ with the remainders of $\Delta\bar{X}_3$ and $\Delta\bar{X}_4$ being corrected by two radial impulses of $\frac{1}{2}(\Delta V_{Re} - |\Delta V_{RD}|)$, but opposite sign, applied 180° apart.
- $\Delta V_{Re} = |\Delta V_{RD}|$, use ΔV_{RD} in conjunction with ΔV_{DT} to correct $\Delta\bar{X}_3$ and $\Delta\bar{X}_4$.
- $\Delta V_{Re} < |\Delta V_{RD}|$, use ΔV_{Re} in conjunction with ΔV_{TD} to correct $\Delta\bar{X}_3$ and $\Delta\bar{X}_4$ with the remainder of $\Delta\Gamma_D$ being corrected by two radial impulses of $\frac{1}{2}(|\Delta V_{RD}| - \Delta V_{Re})$ of the same sign supplied 180° apart.

Case II

Correct $\Delta\bar{X}_3$ and $\Delta\bar{X}_4$ partially by ΔV_{TD} and the remainder by ΔV_{Re}, if:

- $\Delta V_{Re} = |\Delta V_{RD}|$, use ΔV_{RD} to correct $\Delta\bar{X}_3$ and $\Delta\bar{X}_4$ with ΔV_{DT}.
- $\Delta V_{Re} < |\Delta V_{RD}|$, use ΔV_{Re} to correct $\Delta\bar{X}_3$ and $\Delta\bar{X}_4$ with ΔV_{DT} and the remainder of $\Delta\Gamma_D$ being corrected by two radial impulses of $\frac{1}{2}(|\Delta V_{RD}| - \Delta V_{Re})$ of the same sign applied 180° apart.

Case III

Correct $\Delta\bar{X}_3$ and $\Delta\bar{X}_4$ partially by ΔV_{TD} and the remainders by ΔV_{Re}, which is less than $|\Delta V_{RD}|$. The remaining $\Delta\Gamma_D$ will be corrected by two radial impulses of $\frac{1}{2}(|\Delta V_{RD}| - \Delta V_{Re})$ of the same sign applied 180° apart.

Cases IV, V, VI

Correct $\Delta\bar{X}_3$ and $\Delta\bar{X}_4$ entirely by ΔV_{TD} and two radial impulses of ΔV_{RD} of the same sign applied 180° apart to correct $\Delta\Gamma_D$.

Cases VII, VIII, IX

Correct $\Delta\bar{X}_3$ and $\Delta\bar{X}_4$ entirely by ΔV_{TeS}. To correct $\Delta\Gamma_D$ and the remainder of $\Delta\dot{\Gamma}_D$, use two sets (transverse and radial) of impulses, $\frac{1}{2}(|\Delta V_{TD}| - \Delta V_{TeS})$ and $\frac{1}{2}\Delta V_{RD}$, with both sets having the same sign applied 180° apart.

In the various schemes discussed above, the required radial and transverse impulses are applied separately for the ease of discussion; however, in practice these impulses may be combined as an single impulse. Since the signs of ΔV_{RD} and ΔV_{TD} are determined by $\Delta\Gamma_D$ and $\Delta\dot{\Gamma}_D$, respectively, the signs of ΔV_{Re} and ΔV_{Te} should conform with that of the ΔV_{RD} and ΔV_{TD}, respectively; consequently, the point of application (θ_A) of ΔV_{Re} and ΔV_{Te} will be chosen to satisfy the requirement of $\Delta\bar{X}_3$ and $\Delta\bar{X}_4$. In the case of two radial impulses of equal magnitude, but opposite sign, applied 180° apart, the first radial impulse should reduce the value of $(\Delta\bar{X}_3^2 + \Delta\bar{X}_4^2)^{1/2}$, while the second one will restore its reduction. This arrangement is very important to reduce the boundness of the longitudinal variation during the station-keeping.

By having the desired stationary longitude at the center of the band

$$\begin{aligned}\Delta\Gamma &= \Gamma_{Emax} - \frac{1}{2}(\Gamma_D + \Gamma_{a'}) \\ &= KT^2 - \frac{1}{2}(\Gamma_{SMR(-T)} - \Gamma_{SMR(+T)}) \div \Big\{[\bar{X}_{30} + \frac{1}{2}(|\Delta\bar{X}_{3(-T)}| + \\ &\quad + |\Delta\bar{X}_{3(+T)}|)]^2 + [\bar{X}_{40} + \frac{1}{2}(|\Delta\bar{X}_{4(-T)}| + |\Delta\bar{X}_{4(+T)}|)]^2\Big\}^{1/2}\bar{\ell} + \\ &\quad + |\Delta\lambda|_{max} ,\end{aligned} \tag{6.115}$$

the maximum longitudinal deviation in a time interval of 2T is $\frac{1}{2}\Delta\Gamma$, where $|\Delta\lambda|_{max}$ is the maximum contribution of the short-period (hourly) oscillatory perturbation terms as they are computed from δX's and δt terms in the improved first-order solution.

As the libration contours in Figure 6.9. are symmetric about the Earth's equator minor axis and are also symmetric about the major axis, the previous discussion of station-keeping procedures limited to a geostationary satellite situated in a single quadrant can also be applied to the cases when the satellite is located in the other three quadrants.

6.5.2.2.2. Low Thrust Maneuver Case

We shall follow the same general scheme to perform the station-keeping by utilizing the impulsive thrust, which, in the present case, will be replaced by a low thrust maneuver.

In the case of the low thrust maneuver, transfer from point 'c' to 'd' along contour 'B' can be accomplished by thrusting in the positive radial direction as indicated by (6.68):

$$\Delta\omega = +\frac{3}{2}\varepsilon N(\theta - \theta_0) = \overset{\frown}{cd} , \tag{6.116}$$

where $\Delta\omega$ is the angular distance between 'c' and 'd' with due consideration of the changes in the X's during the angular displacement of $\theta - \theta_0$ of the satellite.

The transfer from point 'd' to 'b' is an altitude change, which can be calculated from (6.66):

$$- \Delta a = 2\varepsilon M r_0(\theta - \theta_0) = \overline{db} \ , \tag{6.117}$$

where $r_0 = r_c + \Delta r_c + \overline{ad}$, and the thrusting direction is opposite to the direction of the satellite revolving movement around the Earth. Again, the changes in the $\bar{X}$'s during the thrusting period should be taken into consideration.

It should be noted that the changes in $\Delta\omega$ and Δa in (6.116) and (6.117) can be accomplished by a single thrusting maneuver with its thrusting direction of

$$\tan \Psi = \frac{\sin \Psi}{\cos \Psi} = \frac{M}{N} = \frac{3}{4} \frac{\Delta a}{r_0 \Delta\omega} \tag{6.118}$$

during the angular displacement, $\theta - \theta_0$, of the satellite.

After the completion of the above maneuver, the satellite is on contour 'A', and the next altitude change will be from point 'a' to 'b' with $- \Delta a = \overline{ab}$. Thus, a complete cycle of station-keeping is established for the duration of the satellite travelling from point 'b' to 'a' along contour 'A'.

As is seen from (6.67a), the change of eccentricity by use of a low thrust is rather uneconomical. Fortunately, most of the geostationary satellites are already in a nominally circular orbit.

6.6. Summary

In the design of a geostationary satellite, the requirement of station-keeping provides two basic inputs: size of the fuel tank and the mode of guidance and control. The former determines the life of the satellite while the latter formulates its guidance and control system. As a guide for satellite system engineers to estimate the fuel consumption in various orbital maneuvers in station-keeping, the following table will serve the purpose of sizing up the fuel tank for a specified span of satellite life.

The changes of orbital elements during the station-keeping correction cycle due to the perturbations of the Sun, Moon, Earth, and the solar radiation pressure can be estimated from the first-order solutions of the X's presented in the previous three chapters; consequently, the amounts of correction of various orbital elements are established, and the corresponding fuel requirements are estimated.

Based upon the fuel consumption and the behavior of orbital elements during the correction cycle, satellite system engineers will be able to formulate a guidance and control scheme for the station-keeping mission. With a specified satellite life span, the number of correction cycles can be determined, which, in turn, gives an indication of the size of the fuel tank.

There are now more than 150 active and nonactive satellites in the geostationary orbit. As the life of most communications satellites is designed for seven to ten years, the geostationary orbit will certainly be unduly crowded in the near future. Therefore, satellite design engineers should also consider the retirement of the satellites by moving them out of the geostationary orbit through orbital transfer. For example, after more

TABLE VII. Fuel consumption (M_p) for changes of orbital elements.

	Impulsive thrust maneuver		Low thrust maneuver	
Orbital Plane Change (Δi)	$V_c \Delta i M_0/\bar{c}$	(6.72)	$\pi V_c \Delta i M_0/(2\bar{c})$	(6.86)
Altitude Change (Δr)	$\frac{1}{2}\frac{V_c}{\bar{c}}\frac{\lvert\Delta r\rvert}{r_c} M_0$ * (Hohmann-type Transfer)	(6.33)	$\frac{1}{2}\frac{V_c}{\bar{c}}\frac{\lvert\Delta r\rvert}{r_c} M_0$ *	(6.41, 6.50) (6.66a)
Eccentricity Change (Δe)	$\frac{1}{2}\frac{V_c}{\bar{c}}\Delta e M_0$ (Hohmann-type Transfer)	(6.53)	$\frac{2}{11}\frac{V_c}{\bar{c}}\frac{1}{M}\frac{\Delta e}{e} M_0$	(6.67a)**
Drift Angle Change (ΔΓ)	$\frac{1}{2}\frac{V_c}{\bar{c}}\Delta\Gamma M_0$ (ΔV_R impulse)	(6.17, 6.32)	$\frac{2}{3}\frac{V_c}{\bar{c}}\frac{1}{N}\Delta\Gamma M_0$	(6.68a)**
Drift Rate Change ($\Delta\dot{\Gamma}$)	$\frac{1}{3}\frac{V_c}{\bar{c}}\frac{1}{\dot{\theta}_E}\Delta\dot{\Gamma} M_0$ (ΔV_T impulse)	(6.18, 6.32)	$\frac{1}{3}\frac{r_c}{\bar{c}}\frac{1}{M}\Delta\dot{\Gamma} M_0$	(6.69a)**

than twelve years in space carrying international and domestic telecommunications, INTELSAT IV (F-2) was raised 400 km above the geostationary orbit by using the remaining propellant on the satellite in May 1983. In other words, in sizing up the fuel tank, the space for the 'retirement' fuel should also be allotted.

In concluding this chapter, it is interesting to mention that one of the publicized orbital maneuvers involved altitude change and circularizing, which we have discussed in the previous sections. The first Tracking and Data Relay Satellite of the NASA was launched from the Space Shuttle in April 1983; the malfunction of the second burn of the initial upper stage booster put the space craft into an elliptical orbit with a perigee 8,500 mi. short of the synchronous altitude. It was then moved into a circular geostationary orbit using its tiny altitude control thrusters. Over a period of 58 days, it took two of those 1-lb. thrusters approximately 44 hours of firing time and over 39 discrete operations to circularize the orbit. Readers will find no trouble in understanding that those discrete thrusts were applied in the vicinity of the perigees and apogees of the original and of the transfer orbits as we have previously discussed in high-thrust orbital maneuvers involving altitude change and circularizing orbit.

---

* Since the orbital maneuver is performed in the neighborhood of r_c, the V_0 and r_0 can be approximated by V_c and r_c, respectively.

** $M \equiv \sin\Psi$, $N \equiv \cos\Psi$, where Ψ is the angle measured from $+\vec{r}$ to the thrusting vector (see Appendix E).

Bibliography

Orbital maneuvers invariably require a thrust induced by rocket engines. The simplest method of investigating orbital maneuvers is by means of impulses (Ehricke, 1959), which constitute a convenient tool to study various missions and to compute fuel consumption. In practice, of course, true impulses do not exist, making the use of high thrust over a small but finite thrusting time interval the best achievable approximation. Over short thrusting periods, trajectories of high thrust power-flight can be mathematically represented by a series solution which permits estimates of the accuracy of the predicted values (Zee, 1964a, 1973). Some of the most common and practical cases of the thrusting program have been studied by the author: constant tangential thrust (Zee, 1965a); constant transverse (normal to the radius vector) thrust (Zee, 1963f); oriented constant thrust (Zee, 1963d, e); and vertical constant thrust (Zee, 1963g). As a result of these developments, the effect of finite thrusting time in orbital maneuvers and on mission analyses can be studied. In fact, the special case of the Hohmann-type transfers has been well investigated (Zee, 1963a). Owing to the nature of the solutions presented, the corresponding cases for constant thrust acceleration are also obtained.

A high thrust rocket engine has a low specific impulse (200-300 sec), and the specific impulse itself is an indication of the efficiency of fuel utilization. This fact naturally leads to the adoption of low thrust rocket engines, which have a specific impulse of one or two orders higher than that of the high thrust engines. Of course, as far as mission analysis is concerned, the economical use of fuel for low thrust engines is at the expense of mission time.

In 1953, Tsien dealt with the problem of a satellite escaping from a planet by means of a low thrust directed either in the radial or in the transverse (normal to the radius vector) direction. Later, Lawden (1958) proved that a tangential thrust is the near optimal for a satellite escaping from the Earth so far as the expenditure of fuel is concerned. As a matter of practical importance, the most useful case of a low thrust orbital maneuver is the tangential thrusting, whose purpose is to achieve altitude change. Subsequently, many papers appeared in the sixties (Moeckel, 1960; Melbourne, 1961; Zee, 1963b, c, 1967a, 1968a; King-Hele, 1963; Cohen, 1965; Evtushenko, 1967; Brofman, 1967; Eckstein and Shi, 1967). Other orbital element changes, such as eccentricity (see Appendix E), and inclination (Illarionov and Shkadov, 1962) can also be achieved by a low thrust. For satellites near the Earth, the existence of air dominates the satellites environment (Brouwer and Hori, 1961; Karranberg et al., 1962; Billik, 1962a, b; Lecar and Morrison, 1962; Kork, 1962; Zee, 1964c; Lane, 1965; Richard, 1967; Lane and Cranford, 1969), and the application of a tangential thrust provides a means for sustaining the desired satellite altitude (Zee, 1964b, 1965b).

The Earth's oblateness alone makes for the regression of the line of nodes and the rotation of the line of apsides of a geocentric orbit. Because of the magnitude of the low thrust, the oblateness and the air drag can be considered as having the same order of magnitude. A combination of the oblateness factor with either a low thrust (Zee, 1968b) or air drag (Zee, 1971) results in space trajectories of practical importance for near-Earth mission planning and guidance analysis.

For manned-satellite missions, orbital rendezvous serves the objectives of rescue, life-support supplies, equipment replacement, etc. Literature dealing with the orbital rendezvous adopts a target-centered, moving-coordinate system to find the relative motion of a satellite with respect to a target. The solution of the relative motion is obtained by linearizing the differential equations of motion. When the target is in a circular orbit, the expressions for the relative motion are rather simple (Clohessy and Wiltshire, 1960). However, if the target is in an orbit of small eccentricity, the solution is quite complicated (deVries, 1963; Anthony and Saski, 1965). For a satellite and its target both in orbits of small eccentricity, an explicit finite-thrust guidance law for orbital rendezvous can be formulated as a perturbation of a two-impulse rendezvous scheme (Zee, 1967b). This law employs a fictitious satellite traversing in a circular orbit which is close to both the satellite and the target during the entire period of the rendezvous maneuver. The use of the fictitious satellite as the center of the moving coordinate system simplifies the solution considerably; consequently, the solutions are in a manageable form, which may be easily adopted for a satellite-borne computer.

For a geostationary satellite, station-keeping is required to maintain its orbital position relative to a fixed Earth station within a certain limit. In carrying out the station-keeping, both body dynamics and particle dynamics are involved. The former deals with the stabilization of the satellite in the orbit, while the latter is related to the orbit maneuver of the satellite. Since a communications geostationary satellite requires its antenna having pointing accuracy towards its Earth receiving station, stabilization of the satellite's body axis is important. Neufeld and Anzel (1966) studied the possible stabilization and control techniques: spin stabilization, gravity-gradient stabilization, and three-axis stabilization. Among these three, spin stabilization is the most promising one. However, any misalignment of the spin axis needs to be corrected by a delta-V application. For the case of orbit maneuver, Balsam and Anzel (1969) proposed a detailed scheme to nullify the perturbed effects on a geostationary orbit by an application of delta-V's in various directions. The accuracy of the angular position of the satellite with respect to a fixed point on the Earth can be achieved on the order of milliradians. Ostrander (1969) also suggested a similar scheme to counter the perturbation of the Earth's triaxiality effect, however his discussions were limited to the generality related to station-keeping. Kamel et al. (1973a) devised schemes to facilitate east-west station-keeping by deriving a linearized solution for longitude and semimajor axis deviations in the neighborhood of a prespecified station as a result of the perturbations of the Sun, Moon, and Earth. To utilize the property of a geostationary satellite orbit plane drifting toward the equatorial plane of the Earth, Kamel et al. (1973b) obtained the optimized locations of the initial node line such that the orbit plane will remain in a preassigned inclination band for a long time for the purpose of north-south station-keeping. Both of the above investigations do not include the solar radiation pressure.

References

Anthony, M.L. and Saski, F.T.; 'The Rendezvous Problem for Nearly Circular Orbits', AIAA Paper 65-32 (1965).

Balsam, R.E. and Anzel, B.M.; 'A Simplified Approach for Correction of Perturbations on a Stationary Orbit', J. of Spacecraft and Rockets 6, 805-811 (1969).

Billik, B.; 'Survey of Current Literature on Satellite Lifetimes', ARS J. 32, 1641-1650 (1962a).

Billik, B.; 'Satellite Lifetimes', ARS J. 32, 1926 (1962b).

Brofman, W.; 'Approximate Analytical Solution for Satellite Orbits Subjected to Small Thrust or Drag', AIAA J. 5, 1121-1128 (1967).

Brouwer, D. and Hori, G-I; 'Theoretical Evaluation of Atmospheric Drag Effects in the Motion of an Artificial Satellite', Astron. J. 66, 193-225 (1961).

Clohessy, W.H. and Wiltshire, R.S.; 'Terminal Guidance System for Satellite Rendezvous', J. Aerospace Sci. 27, 653-658 (1960).

Cohen, M.J.; 'Low-Thrust Spiral Trajectory of a Satellite of Variable Mass', AIAA J. 3, 1946-1949 (1965).

deVries, J.P.; 'Elliptic Elements in Terms of Small Increment of Position and Velocity Components', AIAA J. 1, 2626-2629 (1963).

Eckstein, M.C. and Shi, Y.Y.; 'Low-Thrust Elliptic Spiral Trajectories of a Satellite of Variable Mass', AIAA J. 5, 1491-1494 (1967).

Ehricke, K.A.; Space Flight, Vol. I, Environment and Celestial Mechanics, D. Van Nostrand Co. Inc., Princeton, N.J. (1959).

Evtushenko, I.G.; 'Influence of Tangential Acceleration on the Motion of a Satellite', J. of Applied Mathematics and Mechanics 30, 710-716 (1967).

Illarionov, V.F. and Shkadov, L.M.; 'Rotation of the Plane of a Circular Satellite Orbit', PMM 26, 15-21 (1962).

Kamel, A., Ekman, D., and Tibbitts, R.; 'East-west Station-keeping of Nearly Synchronous Satellites Due to Earth's Triaxiality and Luni-Solar Effects', Celestial Mechanics 8, 129-148 (1973).

Kamel, A. and Tibbitts, R.; 'Some Useful Results on Initial Node Locations for Near-Equatorial Circular Satellite Orbits', Celestial Mechanics 8, 45-73 (1973).

Karranberg, H.K., Levin, E., and Lewis, D.H.; 'Variation of Satellite Position with Uncertainties in the Mean Atmospheric Density', ARS J. 32, 576-582 (1962).

King-Hele, D.G.; 'The Enlargement of Elliptic Satellite Orbits by Continuous Microthrust', XIV International Astronautical Congress Preprint 69 (1963).

Kork, J.; 'Satellite Lifetimes in Elliptic Orbits', J. Aerospace Sci. **29**, 1273-1290 (1962).

Kryloff, N. and Bogoliuboff, N.; Introduction to Non-linear Mechanics, Princeton University Press, Princeton N.J. (1947).

Lane, M.H.; 'The Development of an Artificial Satellite Theory Using a Power-Law Atmospheric Density Representation', AIAA Paper No. 65-35 (1965).

Lane, M.H. and Cranford, K.H.; 'An Improved Analytical Drag Theory for the Artificial Satellite Problem', AIAA Paper No. 69-925 (1969).

Lawden, D.F.; 'Optimal Escape from a Circular Orbit', Astronautica Acta **4**, 218-233 (1958).

Lecar, M. and Morrison, R.; 'A Method for Determining Atmospheric Density for Short Arcs of an Almost Circular Satellite Orbit', J. Astronautical Science **IX**, 82-85 (1962).

Melbourne, W.G.; 'Interplanetary Trajectories and Payload Capabilities of Advanced Propulsion Vehicles', Tech. Report No. 32-28, Jet Propulsion Laboratory, California Institute of Technology, Pasadena, California (1961).

Moeckel, W.E.; 'Trajectories with Constant Tangential Thrust in Central Gravitational Fields', NASA Tech. Report R-53 (1960).

Minorsky, N.; Nonlinear Oscillations, D. Van Nostrand Co. Inc., Princeton, N.J. (1962).

Neufield, M.J. and Anzel, B.M.; 'Synchronous Satellite Station Keeping', in Communications Satellite System Technology, R.B. Marsten (ed.), AIAA Progress in Astronautics & Aeronautics **19**, 323-346 (1966).

Ostrander, N.C.; 'Longitude Station Keeping of Nearly Geostationary Satellites', RAND Corporation, RM-6166-PR (1969).

Richard, T.J.; 'A Graphical Method for Predicting Satellite Lifetime on a Time-Varying Atmospheric Density Model', AAS Science & Technology Series **11**, 65-78 (1967).

Tsien, H.S.; 'Take-Off from Satellite Orbits', Jet Propulsion **23**, 233-236 (1953).

Zee, C.H.; 'Effect of Finite Thrusting Time in Orbital Maneuvers', AIAA J. **1**, 60-64 (1963a).

Zee, C.H.; 'Low Constant Tangential Thrust Spiral Trajectories', AIAA J. **1**, 1581-1583 (1963b).

Zee, C.H.; 'Low Thrust Oscillatory Spiral Trajectories', Astronautica Acta **IX**, 201-207 (1963c).

Zee, C.H.; 'Powered Flight Trajectories of Rockets Under Oriented Constant Thrust', AIAA J. **1**, 602-606 (1963d).

Zee, C.H.; 'Spatial Powered Flight Trajectories of Space Vehicles Under Oriented Constant Thrust', Astronautica Acta **IX**, 107-114 (1963e).

Zee, C.H.;' Powered Flight Trajectories of Rocket Under Constant Transverse Thrust', J. of Applied Mechanics, Am Soc. Mech. Engrs., Series E **30**,

559-561 (1963f).

Zee, C.H.; 'Rocket in Drag-Free Vertical Powered Fligth Under Constant Thrust', J. of the Astronautical Sciences X, 119-121 (1963g).

Zee, C.H.; 'On Solving Second Order Nonlinear Differential Equations', Quarterly of Applied Mathematics XXII, 71-73 (1964a).

Zee, C.H.; 'Effect of Aerodynamic Drag on Low Thrust Ascending Spiral Trajectories', J. of Spacecraft and Rockets 1, 435-437 (1964b).

Zee, C.H.; 'Trajectories of Satellite Under the Influence of Air Drag', in Celestial Mechanics and Astrodynamics, V.G. Szebehely (ed.), AIAA Progress in Astronautics & Aeronautics 14, 101-112 (1964c).

Zee, C.H. and Suh, S.; 'On Guidance Errors Induced by Thrust Decay in Vertical Powered Flight of Rockets', J. of Astronautical Sciences XI, 33-36 (1964d).

Zee, C.H.; 'Powered Flight Trajectories of Rockets Under Constant Tangential Thrust', J. of the Astronautical Sciences XII, 1-6 (1965a).

Zee, C.H.; 'Low Thrust Osciallatory Ascending Spiral Trajectories as Affected by Air Drag', J. of the Franklin Institute 280, 42-51 (1965b).

Zee, C.H.; 'Comments on Approximate Analytical Solution for Satellite Orbits Subjected to Small Thrust or Drag', AIAA J. 5, 2111-2112 (1967a).

Zee, C.H.; 'Finite Thrust Explicit Guidance Law for Nearly Circular Orbital Rendezvous', J. of Spacecraft and Rockets 4, 34-38 (1967b).

Zee, C.H.; 'Low Tangential Thrust Trajectories - Improved First-Order Solution', AIAA J. 6, 1378-1379 (1968a).

Zee, C.H.; 'Trajectories of Satellite Under the Combined Influences of Low Tangential Thrust and Earth Oblateness', Presented at the 18th Congress of the International Astronautical Federation, N.Y. (1968b).

Zee, C.H.; 'Trajectories of Satellites Under the Combined Influences of Low Radial Thrust and Earth Oblateness', Astronautica Acta 14, 289-298 (1969).

Zee, C.H.; 'Trajectories of Satellites Under the Combined Influences of Earth Oblateness and Air Drag', Celestial Mechanics 3, 148-168 (1971).

Zee, C.H. and Sobierajski, F.S.; 'Computer-Oriented Taylor Series Solution to a System of Nonlinear Differential Equations', Report No. ADN 12-01-73.1, Grumman Aerospace Corporation, Bethpage, N.Y. (1973).

Chapter 7

VERIFICATION OF THE THEORY BY THE EARLY BIRD SYNCHRONOUS SATELLITE

7.1. General

The Early Bird synchronous satellite, launched on April 6, 1965, was placed into service over the Atlantic Ocean at 25° west longitude on June 28, 1965. After approximately three and half years of service, it was retired on January 20, 1969 and put in reserve. On June 29, 1969, it was reactivated and retired again on August 21, 1969. Since then, the satellite is no longer useful for communications.

The satellite has a basically cylindrical shape of diameter 71.7 cm and height 59.0 cm. Its in-orbit final mass (no more propellant) is 34.9 kg. A 60 cm long antenna projects from one of the cylinder's bases and is oriented to coincide with the cylinder axis. The satellite has a corrective-ΔV capacity of 90 mps. It spins at a rate of 120-180 rpm with the antenna along the spin axis, which is oriented normal to the orbital plane. The satellite's effective area to mass ratio is 0.0148 m^2/kgm (0.0722 ft^2/lbm). This value will be useful in computing the effect of the solar radiation force on the satellite orbit.

The last maneuver performed on the Early Bird synchronous satellite occurred on May 27, 1970. After that maneuver, the satellite had no additional propellant on board and was therefore considered as a 'dead' one.

In order to verify the theory presented in the previous chapters, we shall compare our analytically predicted orbital elements with those derived from the tracking data and the numerically integrated results of the Early Bird. Both of them were provided by the Communications Satellite Corporation.* After evaluating various constants in our orbital parameters in terms of the initial conditions of the Sun, Moon, and the Early Bird in relation to the Earth, the solutions to these parameters can be written out directly from the derived expressions described in the previous chapters. In turn, the orbit of the Early Bird at any time can be predicted. We shall carry out our verifications to use the first-order solution with a limited number of dominant terms in order to simplify our presentation.

The following expressions, which have been derived previously, are presented for the convenience of reference in the discussions followed.

---

*The author is indebted to Dr. Victor J. Slabinski of the Communications Satellite Corporation (now at INTELSAT) for his valuable help in obtaining this information.

Inclination:

$$\bar{i} = \tan^{-1}(\bar{X}_1^2 + \bar{X}_2^2)^{1/2} \tag{7.1}$$

Ascending node:

$$\bar{\Omega} = \tan^{-1}(-\bar{X}_1/\bar{X}_2) \tag{7.2}$$

Argument of perigee:

$$\bar{\omega} = \tan^{-1}\left[\frac{\bar{X}_1\bar{X}_3' + \bar{X}_2\bar{X}_4'}{\bar{X}_2\bar{X}_3' - \bar{X}_1\bar{X}_4'}\ \frac{1}{(1 + \bar{X}_1^2 + \bar{X}_2^2)^{1/2}}\right] \tag{7.3}$$

Eccentricity:

$$\bar{e} = \frac{\bar{X}_6^2}{\mu_E}(1 + \bar{X}_1^2 + \bar{X}_2^2)[\bar{X}_3'^2 + \bar{X}_4'^2 + (\bar{X}_2\bar{X}_3' - \bar{X}_1\bar{X}_4')^2]^{1/2} \tag{7.4}$$

Radial deviation (Δr_c):

$$\theta - \theta_0 = \frac{1}{3}\left\{(2k_1)^{1/2}\mathrm{sd}^{-1}\left[\left(\frac{2k_1}{1 - k_2^2}\right)^{1/2} \frac{\Delta r_c}{r_c}\left|\frac{1 - k_2}{2}\right.\right] + \right.$$

$$\left. + \sin^{-1}\left[k_1\left(\frac{\Delta r_c}{r_c}\right)^2 + k_2\right]\right\} \tag{7.5}$$

Drift angle (Γ):

$$\cos 2\Gamma_0 - \cos 2\Gamma = \frac{r_c^2}{8J_{22}R_E^2}\left[1 + 3J_2\frac{R_E^2}{r_c^2}\right]\left[\left(\frac{\Delta r_c}{r_c}\right)^2 - \left(\frac{\Delta r_{c0}}{r_c}\right)^2\right] \tag{7.6}$$

Drift rate ($\dot{\Gamma}$):

$$\dot{\Gamma} = -\frac{3}{2}\frac{\Delta r_c}{r_c}\dot{\theta}_E\ . \tag{7.7}$$

7.2. Verification by Observed Data

The available observed data of the 'dead' Early Bird Satellite from May 27, 1970 to August 15, 1973 are listed in Table VIII. The initial conditions are taken from the observed data at 1900 (Greenwich Mean Time), May 27, 1970, except for e, which is selected to be 0.0007 instead of 0.000789. Because the difference results in a topocentric look angle error of only 0.01 degree, which is the accuracy of the angle measurements, this selection seems to be allowable. At the initial epoch, the positions of the Sun and Moon, the ascending node and the argument of perigee of the Moon's orbit were 66.82° ($\bar{\alpha}_0$), 2.68° ($\bar{\nu}_0$), 337.51° ($\bar{\Omega}_{M0}$), and 341.49° ($\bar{\omega}_M$), respectively; they were obtained from the American Ephemeris and National Almanac, U.S. Government Printing Office, Washington D.C., 1970. The intensity of solar radiation pressure (P_r) was 4.5×10^{-6} N/m^2 (0.92×10^{-7} lb/ft).

By employing (7.1) and (7.2) with known Ω_0 (75.433°) and i_0 (4.857°), $\bar{X}_{10}$ and $\bar{X}_{20}$ are obtained. After examining the magnitude of all terms in $u_1^{(1)}$ and $u_2^{(1)}$, shown in (4.97) and (4.98), and the additional terms due to the long-period terms in solar perturbation on the Moon's motion, it is found that the terms containing A_3, A_4, A_7, A_8, A_{25}, A_{26}, A_{29}, and A_{30} play a dominant role. Therefore, $\bar{X}_1$ and $\bar{X}_2$ can be approximated by

TABLE VIII. Early Bird synchronous satellite osculating orbital elements (derived from observed tracking data).

Date Time (GMT)	70/05/27 1900	70/08/12 0000	70/11/06 0000	71/04/06 1500	71/08/03 0000
Eccentricity	0.000789	0.000822	0.000683	0.000476	0.000755
Inclination (deg)	4.857	5.074	5.255	5.652	5.967
Ascending node (deg)	75.443	74.861	73.592	72.038	70.236
Arg. of perigee (deg)	55.181	69.913	91.525	71.673	84.233
Mean anomaly (deg)	325.888	78.745	114.747	119.226	1.228
Period (min)	1437.184	1437.482	1437.400	1436.440	1435.698
Drift (deg/day)	-0.253	-0.327	-0.307	-0.066	0.120
Semimajor axis (in R_E)	6.614143	6.615054	6.614800	6.611855	6.609579
East longitude (deg)	286.598	263.490	235.070	203.695	204.780
Γ (deg)*	31.516	8.408	-20.012	-51.387	-50.302
r_c (ft)	69912.9	88976.2	83661.1	22034.8	-25592.2
(m)	21309.5	27120.0	25500.0	6716.2	-7800.5
Number of days from 70/05/27	0	77	163	314	433

Date Time (GMT)	71/10/15 0000	72/02/04 1000	72/07/28 0000	73/02/07 1430	73/08/15 0000
Eccentricity	0.000741	0.000503	0.000648	0.000378	0.000573
Inclination (deg)	6.115	6.422	6.842	7.331	7.729
Ascending node (deg)	69.338	68.064	65.747	63.352	61.049
Arg. of perigee (deg)	108.488	120.188	102.737	156.579	116.289
Mean anomaly (deg)	61.767	345.067	76.913	85.357	56.957
Period (min)	1435.222	1434.867	1435.323	1436.436	1437.341
Drift (deg/day)	0.239	0.328	0.214	-0.065	-0.292
Semimajor axis (in R_E)	6.608116	6.606029	6.608427	6.11842	6.614618
East longitude (deg)	216.725	249.648	299.721	310.073	271.164
Γ (deg)*	-38.357	-5.434	44.639	54.991	16.082
r_c (ft)	-56206.5	-78952.8	-49698.6	21762.8	79852.6
(m)	-17131.8	-24064.9	-15148.2	6633.3	24339.1
Number of days from 70/05/27	506	618	793	987	1176

*Earth equator minor axis located at East Longitude 75°.082 (see Section 3.2.).

$$\bar{X}_1 = \omega_1 a_0 \sin\Psi + \varepsilon[\omega_1 a_0 \sin\Psi + A_3 \sin 2\bar{\alpha} + A_4 \sin\bar{\Omega}_M + A_7 \sin(\bar{\Omega}_M + \Psi) + A_8 \sin(\bar{\Omega}_M - \Psi)] \tag{7.8}$$

and

$$\bar{X}_2 = \omega_{11} a_0 \cos\Psi + \varepsilon[\omega_{11} a_0 \cos\Psi + A_{25} \cos 2\bar{\alpha} + A_{26} \cos\bar{\Omega}_M + A_{29} \cos(\bar{\Omega}_M + \Psi) + A_{30} \cos(\bar{\Omega}_M - \Psi)], \tag{7.9}$$

where the coefficient A's in the above equations are shown in Appendix B.

A substitution of $\bar{X}_{10}$ and $\bar{X}_{20}$ together with $\bar{\alpha}_0$ and $\bar{\Omega}_{M0}$ into (7.8) and (7.9) results in $a_0 = -2510.84$ and $\Psi_0 = 38.6069°$, respectively; consequently,

$$\bar{X}_1 = -0.1271 \sin\Psi + 0.00037 \sin 2\bar{\alpha} + 0.0085 \sin\bar{\Omega}_M - 0.00171 \sin(\bar{\Omega}_M + \Psi) - 0.00053 \sin(\bar{\Omega}_M - \Psi) \tag{7.10}$$

and

$$\bar{X}_2 = -0.1307 \cos\Psi + 0.131266 - 0.0004 \cos 2\bar{\alpha} - 0.0097 \cos\bar{\Omega}_M + 0.00070 \cos(\bar{\Omega}_M + \Psi) + 0.00053 \cos(\bar{\Omega}_M - \Psi), \tag{7.11}$$

where

$$\Psi = \omega_1(\theta - \theta_0) + 38.6069°,$$

$$\bar{\Omega}_M = 337.51° + K_3(\theta - \theta_0),$$

$$\bar{\alpha} = 66.82° + K_1(\theta - \theta_0),$$

and

$$\theta_0 = 96.4430°,$$

computed from the mean anomaly at the initial epoch.

In the Sun-Moon-Earth system, the change of the semimajor axis, a, is dominated by the Earth's equatorial ellipticity term, J_{22}, while the Sun and Moon play a minor role. The J_{22}-term has practically no effect on the angular velocity of the synchronous satellite as is seen from (3.5) because the maximum value of Γ is 90° and θ increases monotonically with time. In computing the effect on the angular velocity due to the Sun and Moon, the angular momentum should be computed based on the long-term average of the synchronous radius, which is equal to r_c. Therefore,

$$\bar{X}_{60} = r_c^2\dot{\theta}_E/(1 + \bar{X}_{10}^2 + \bar{X}_{20}^2)^{1/2} = 0.99641\, r_c^2\dot{\theta}_E \tag{7.12}$$

and

$$\bar{X}_6 \cong \bar{X}_{60}[1 + 0.01589(\cos\Psi - \cos\Psi_0) - 0.00032(\cos 2\Psi - \cos 2\Psi_0)], \tag{7.13}$$

where a similar procedure in obtaining (4.105a) in Section 4.8. is employed to derive (7.13) from (5.24) by neglecting F and F_0, and the terms containing solar radiation pressure are dropped because they are all smaller than ε. It can be seen that the change of $\bar{X}_6$ during a period of three years is about $0.0042\ \bar{X}_{60}$.

After eliminating the nonessential terms in $u_3^{(1)}$ and $u_4^{(1)}$, shown in (4.103) and (4.104), and those of B_M's in Appendix C, $\bar{X}_3'$ and $\bar{X}_4'$ may be approximated by

$$\bar{X}_3' \cong \omega_{31} b_0 \cos\xi + \frac{\varepsilon}{4}[B_7 \cos(\bar{\Omega}_M + \xi) + B_8 \cos(\bar{\Omega}_M - \xi) + B_9 \cos\bar{\omega}_M + B_{10} \cos(\bar{\omega}_M + \bar{\Omega}_M) + B_{10M2} \cos(\bar{\omega}_M + \bar{\Omega}_M) \times \cos(\bar{\lambda}_M - 2\bar{\lambda}_S + \bar{\omega}_M + \bar{\Omega}_M) + B_{10M3} \cos(\bar{\omega}_M + \bar{\Omega}_M) \times$$

$$\times \cos(2\bar{\lambda}_S - 2\bar{\omega}_M - 2\bar{\Omega}_M) + B_{11} \cos(\bar{\omega}_M + 2\bar{\Omega}_M)] +$$
$$+ \text{ solar radiation pressure terms in (5.22)} \tag{7.14}$$

and

$$\bar{X}_4' \cong \omega_3 b_0 \sin \xi + \frac{\varepsilon}{4}[B_{30} \sin(\bar{\Omega}_M + \xi) + B_{31} \sin(\bar{\Omega}_M - \xi) +$$
$$+ B_{32} \sin \bar{\omega}_M + B_{33} \sin(\bar{\omega}_M + \bar{\Omega}_M) + B_{33M2} \sin(\bar{\omega}_M + \bar{\Omega}_M) \times$$
$$\times \cos(\bar{\lambda}_M - 2\bar{\lambda}_S + \bar{\omega}_M + \bar{\Omega}_M) + B_{33M3} \sin(\bar{\omega}_M + \bar{\Omega}_M) \times$$
$$\times \cos(2\bar{\lambda}_S - 2\bar{\omega}_M - 2\bar{\Omega}_M) + B_{34} \sin(\bar{\omega}_M + 2\bar{\Omega}_M) +$$
$$+ B_{34M4} \sin(\bar{\omega}_M + 2\bar{\Omega}_M)\cos(2\bar{\lambda}_S - 2\bar{\omega}_M - 2\bar{\Omega}_M)] +$$
$$+ \text{ solar radiation pressure terms in (5.23)}, \tag{7.15}$$

where the coefficient B's and B_M's are shown in Appendixes B and C.

With known ω_0 and e_0, $\bar{X}_{30}'$ and $\bar{X}_{40}'$ can be obtained from (7.3) and (7.4) because $\bar{X}_{10}$, $\bar{X}_{20}$, and $\bar{X}_{60}$ have all been solved in terms of the given initial conditions. In turn, (7.14) and (7.15) yield $b_0 = 0.8447361 \times 10^{-7}$ and $\xi_0 = 155.703°$, where a_0 in the constants B's is $- 2510.84$. Thus,

$$\bar{X}_3' \times 10^{11} = 0.3688 \cos \xi + 0.00859 \cos(\bar{\Omega}_M + \xi) -$$
$$- 0.00383 \cos(\bar{\Omega}_M - \xi) + 0.03577 \cos \bar{\omega}_M -$$
$$- 0.10716 \cos(\bar{\Omega}_M + \bar{\omega}_M) + 0.00018 \cos(\bar{\Omega}_M + \bar{\omega}_M) \times$$
$$\times \cos(\bar{\lambda}_M - 2\bar{\lambda}_S + \bar{\omega}_M + \bar{\Omega}_M) + 0.00063 \cos(\bar{\Omega}_M + \bar{\omega}_M) \times$$
$$\times \cos(2\bar{\lambda}_S - 2\bar{\omega}_M - 2\bar{\Omega}_M) + 0.02475 \cos(\bar{\omega}_M + 2\bar{\Omega}_M) +$$
$$+ 0.11940 \cos \bar{\alpha} + 0.00105 \cos(\bar{\alpha} + \Psi) + 0.00113 \cos(\bar{\alpha} - \Psi) \tag{7.16}$$

and

$$\bar{X}_4' \times 10^{11} = 0.42431 \sin \xi + 0.00361 \sin(\bar{\Omega}_M + \xi) +$$
$$+ 0.00186 \sin(\bar{\Omega}_M - \xi) + 0.00454 \sin \bar{\omega}_M -$$
$$- 0.19220 \sin(\bar{\Omega}_M + \bar{\omega}_M) + 0.00039 \sin(\bar{\Omega}_M + \bar{\omega}_M) \times$$
$$\times \cos(\bar{\lambda}_M - 2\bar{\lambda}_S + \bar{\omega}_M + \bar{\Omega}_M) - 0.00137 \sin(\bar{\Omega}_M + \bar{\omega}_M) \times$$
$$\times \cos(2\bar{\lambda}_S - 2\bar{\omega}_M - 2\bar{\Omega}_M) + 0.04245 \sin(\bar{\omega}_M + 2\bar{\Omega}_M) +$$
$$+ 0.00014 \sin(\bar{\omega}_M + 2\bar{\Omega}_M)\cos(2\bar{\lambda}_S - 2\bar{\omega}_M - 2\bar{\Omega}_M) +$$
$$+ 0.12852 \sin \bar{\alpha} - 0.00102 \sin(\bar{\alpha} + \Psi) - 0.00110 \sin(\bar{\alpha} - \Psi) , \tag{7.17}$$

where

$$\xi = \omega_3(\theta - \theta_0) + 155.703° ,$$
$$\bar{\nu} = K_2(\theta - \theta_0) + 2.68° , \qquad \bar{\lambda}_S = \bar{\alpha} ,$$
$$\bar{\omega}_M = K_4(\theta - \theta_0) + 341.49° , \qquad \bar{\lambda}_M = \bar{\nu} + \bar{\Omega}_M + \bar{\omega}_M ,$$

and θ_0, Ψ, $\bar{\Omega}_M$, and $\bar{\alpha}$ have been defined previously.

The radial deviation (Δr_c), drift angle (Γ), and drift rate ($\dot{\Gamma}$) are computed from (7.5), (7.6), and (7.7), respectively, with $J_{22} = - 1.819 \times 10^{-6}$, $\lambda_{22} = - 14.918°$, $\Delta r_{c0} = 69{,}912$ ft, and $\Gamma_0 = 31.516°$, as they are principally affected by the Earth's equatorial ellipticity.

The time, t, and θ relationship may be derived from (5.38b) as

$$\bar{t} - \bar{t}_0 = \frac{1}{\dot{\theta}_E}[0.99955(\theta - \theta_0) + 10.81740(\sin\Psi - \sin\Psi_0) - 5.50049(\sin 2\Psi - \sin 2\Psi_0) + 0.10686(\sin 3\Psi - \sin 3\Psi_0)], \quad (7.18)$$

where the terms in εf_2 are neglected.

After obtaining $\bar{X}_1$, $\bar{X}_2$, $\bar{X}_3$, $\bar{X}_4$, and $\bar{X}_6$, the orbital elements $\bar{i}$, $\bar{\Omega}$, $\bar{\omega}$, and $\bar{e}$ can be derived from (7.1)-(7.4), and they are plotted in Figures 7.1. and 7.2. with the aid of (7.18). The variations of Γ, $\dot{\Gamma}$ and Δr_c with respect to time, t, are shown in Figures 7.3.-7.5., wherein (7.18) has again been employed.

In these five figures, the derived osculating values from tracking data shown in Table VIII. are also plotted, and these values do include the short-period effects of the perturbations. It should be mentioned that every data point in Table VIII. was derived from one hundred to two hundred observations consisting of range (from radar) and angular (azimuth and elevation) measurements with the associated time at each measurement for a duration of five to six days.

Because of the finite velocity of light, the apparent position of a satellite depends on the motion both of the satellite and of the Earth during the interval of time required for light to travel from the satellite to the observer. Furthermore, the Earth and its equator are in continuous motion. Therefore, all observations are subject to the corrections for the effects of precession, nutation, aberration, and parallax. Readers are

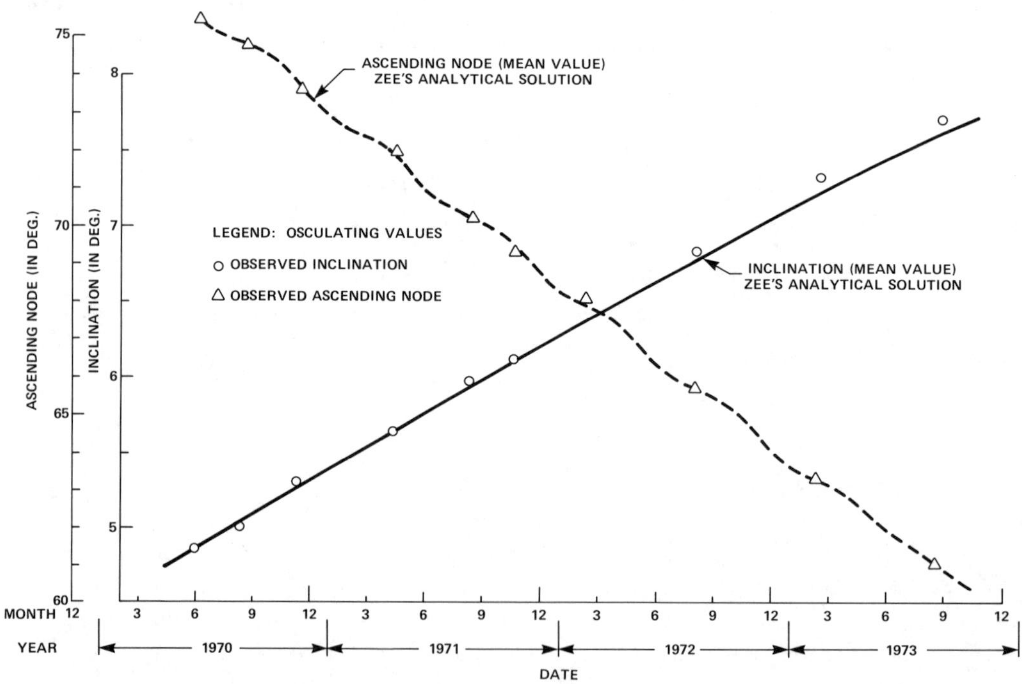

Fig. 7.1. Early Bird synchronous satellite - inclination and ascending node (observed data vs. analytical prediction).

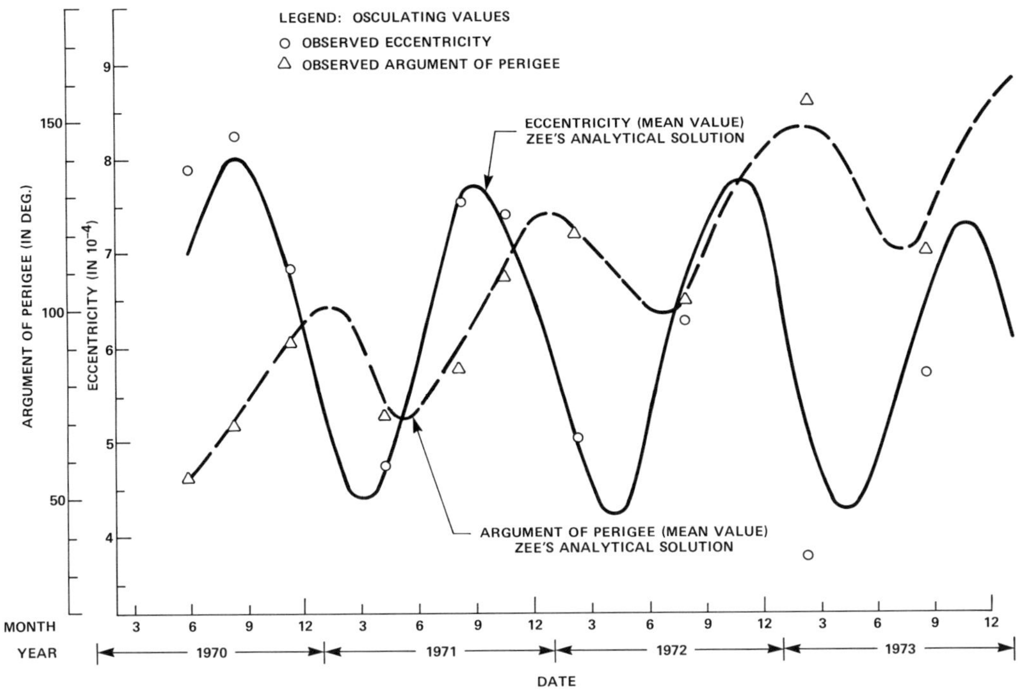

Fig. 7.2. Early Bird synchronous satellite - eccentricity and argument of perigee (observed data vs. analytical prediction).

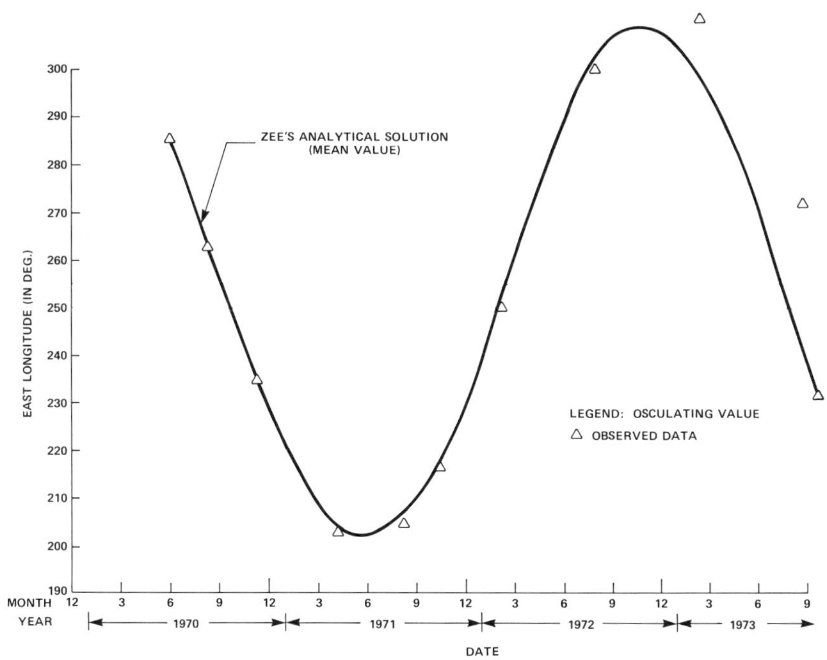

Fig. 7.3. Early Bird synchronous satellite - longitudinal drift angle (observed data vs. analytical prediction).

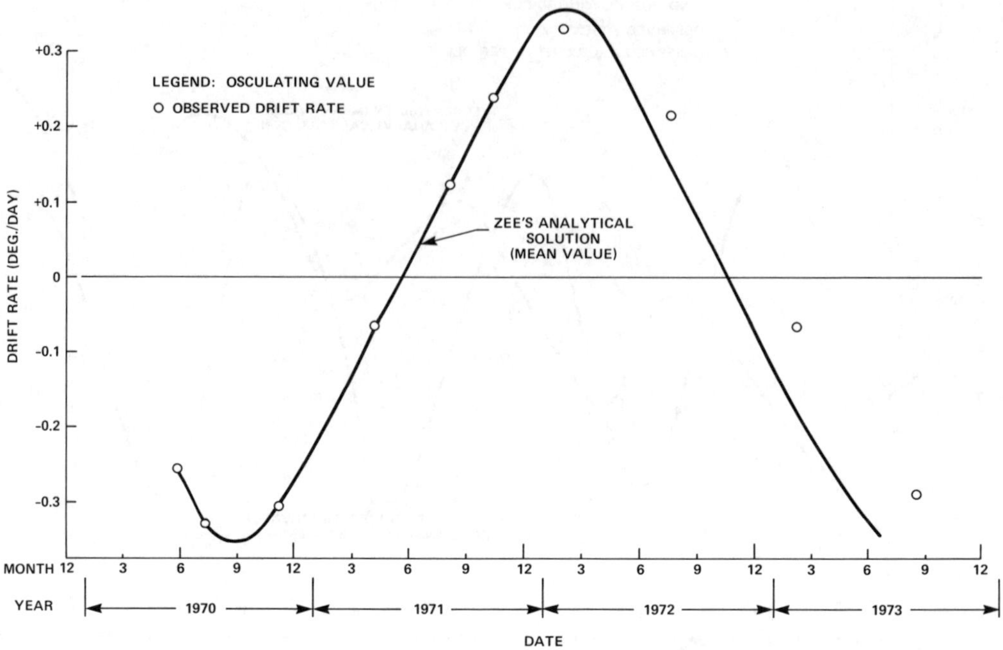

Fig. 7.4. Early Bird synchronous satellite - longitudinal drift rate (observed data vs. analytical prediction).

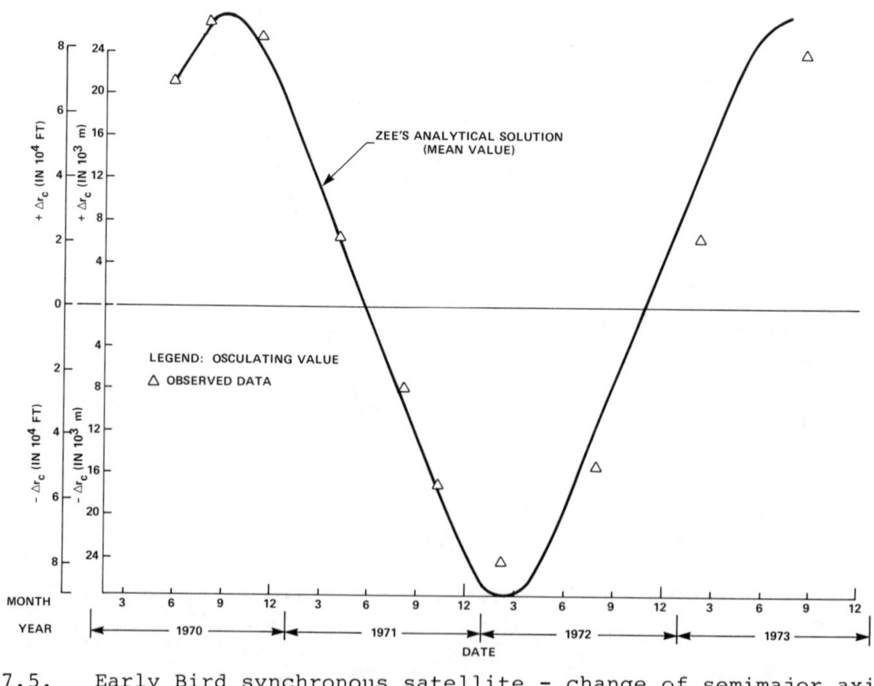

Fig. 7.5. Early Bird synchronous satellite - change of semimajor axis (observed data vs. analytical prediction).

referred to Spherical Astronomy (Smart, 1944) for a detailed treatment of these corrections.

Once all of the observed data is transformed to the Earth-center coordinates and is corrected for the effects of precession, nutation, etc., the resulting data presented at a known time sequence will be further processed in order to obtain the position of the satellite at a specific time as shown in Table VIII. This process involves a curve-fitting technique of the resulting data by means of the method of least squares.

In addition to the usual inherent errors of measurements in angles and distances, errors are also introduced in the data correction, coordinate transformation, data processing, etc. Readers are reminded that the observed data do include the effects of the attractions of the other heavenly bodies, the higher harmonics of the Earth and the Moon, etc., while our solution deals only with the perturbations of the Sun, the spherical Moon, the oblated Earth, ellipticity of the Earth's equator, and solar radiation pressure. With these observations, a comparison of the predicted mean values with those observed osculating orbital parameters data physically processed shown in Figures 7.1.-7.5. could be said to be in reasonable agreement, and this agreement certainly gives readers some confidence in our analytical theory. In the following, we shall employ the numerical integration technique to verify our theory.

7.3. Verification by Numerical Integration

The results of numerical integration of the Early Bird geostationary orbit, provided by the Communications Satellite Corporation, have a duration of three years (12/26/73 - 1/27/77). The digital computer program for the numerical integration has included the Earth's gravitational potential (see Appendix F), the perturbations of the Sun, Moon, and solar radiation. The Encke's method of orbit prediction together with the fourth-order Runge-Kutta integrator is used. The integration step size is one hour. The orbit is rectified whenever the perturbing acceleration is equal to the difference in the (μ_E/r^2)-acceleration between the true and the reference orbit, and the rectification occurs several times each day.

At the initial epoch (0 hr Greenwich Mean Time, December 26, 1973), the following orbital conditions are given for the Early Bird:

$a_0 = 6.6144590\ R_E$	East Longitude = 228.05°
$e_0 = 0.0004817$	Drift Rate $(\dot{\Gamma}_0) = -0.275°/\text{day}$
$i_0 = 8.04468°$	Mean Anomaly = 102.8031°
$\Omega_0 = 59.1705°$	Period = 1437.2883 minutes
$\omega_0 = 160.5566°$.	

In addition, the orbital conditions for the Sun and Moon at the initial epoch are

$$\bar{\alpha}_0 = 275.2363°, \qquad \bar{\Omega}_{M0} = 268.2377°,$$
$$\bar{\nu}_0 = 184.8492°, \quad \text{and} \quad \bar{\omega}_{M0} = 196.5007°.$$

The acceleration of the Early Bird due to the solar radiation pressure (P) is 0.70861×10^{-7} m/sec^2 (2.32484×10^{-7} ft/sec^2).

Following the same procedures of obtaining $\bar{X}$'s in Section 7.2., we have

$$\bar{X}_1 = -\,0.12798 \sin\Psi - 0.00173 \sin(\bar{\Omega}_M + \Psi) - 0.00052 \sin(\bar{\Omega}_M - \Psi) + 0.000830 \sin\bar{\Omega}_M + 0.00037 \sin 2\bar{\alpha}\,, \tag{7.19}$$

$$\bar{X}_2 = -\,0.13156 \cos\Psi + 0.00070 \cos(\bar{\Omega}_M + \Psi) + 0.00054 \cos(\bar{\Omega}_M - \Psi) - 0.00970 \cos\bar{\Omega}_M - 0.00040 \cos 2\bar{\alpha} + 0.13127\,, \tag{7.20}$$

$$\begin{aligned} \bar{X}_3' \times 10^{11} = {} & 0.30485 \cos\xi + 0.00710 \cos(\bar{\Omega}_M + \xi) - \\ & - 0.00316 \cos(\bar{\Omega}_M - \xi) + \\ & + 0.03577 \cos\bar{\omega}_M - 0.10716 \cos(\bar{\Omega}_M + \bar{\omega}_M) + \\ & + 0.00018 \cos(\bar{\Omega}_M + \bar{\omega}_M)\cos(\bar{\lambda}_M - 2\bar{\lambda}_S + \bar{\omega}_M + \bar{\Omega}_M) + \\ & + 0.00063 \cos(\bar{\Omega}_M + \bar{\omega}_M)\cos(2\bar{\lambda}_S - 2\bar{\omega}_M - 2\bar{\Omega}_M) + \\ & + 0.02475 \cos(\bar{\omega}_M + 2\bar{\Omega}_M) + 0.11940 \cos\bar{\alpha} + \\ & + 0.00106 \cos(\bar{\alpha} + \Psi) + 0.00133 \cos(\bar{\alpha} - \Psi)\,, \end{aligned} \tag{7.21}$$

$$\begin{aligned} \bar{X}_4' \times 10^{11} = {} & 0.35072 \sin\xi + 0.00298 \sin(\bar{\Omega}_M + \xi) + \\ & + 0.00154 \sin(\bar{\Omega}_M - \xi) + 0.00454 \sin\bar{\omega}_M - \\ & - 0.19220 \sin(\bar{\Omega}_M + \bar{\omega}_M) + \\ & + 0.00039 \sin(\bar{\Omega}_M + \bar{\omega}_M)\cos(\bar{\lambda}_M - 2\bar{\lambda}_S + \bar{\omega}_M + \bar{\Omega}_M) - \\ & - 0.00137 \sin(\bar{\Omega}_M + \bar{\omega}_M)\cos(2\bar{\lambda}_S - 2\bar{\omega}_M - 2\bar{\Omega}_M) + \\ & + 0.04245 \sin(\bar{\omega}_M + 2\bar{\Omega}_M) + \\ & + 0.00014 \sin(\bar{\omega}_M + 2\bar{\Omega}_M)\cos(2\bar{\lambda}_S - 2\bar{\omega}_M - 2\bar{\Omega}_M) + \\ & + 0.12852 \sin\bar{\alpha} - 0.00103 \sin(\bar{\alpha} + \Psi) - \\ & - 0.00111 \sin(\bar{\alpha} - \Psi)\,, \end{aligned} \tag{7.22}$$

$$\bar{X}_{60} = 0.990159\, r_c^2 \dot{\theta}_E\,, \tag{7.23}$$

$$\bar{X}_6 \cong \bar{X}_{60}[1 + 0.015404(\cos\Psi - \cos\Psi_0) - 0.0002966(\cos 2\Psi - \cos 2\Psi_0)]\,, \tag{7.24}$$

and

$$\bar{t} - \bar{t}_0 = \frac{1}{\dot{\theta}_E}[1.0005637(\theta - \theta_0) - 41.2471(\sin\Psi - \sin\Psi_0) - 5.0750(\sin 2\Psi - \sin 2\Psi_0) + 0.10608(\sin 3\Psi - \sin 3\Psi_0)]\,, \tag{7.25}$$

where

$a_0 = -\,2527.177$, $\quad b_0 = 0.6982307 \times 10^{-7}$,

$\bar{\Omega}_M = 268.2377 - K_3(\theta - \theta_0)$, $\quad \bar{\omega}_M = 196.5007° + K_4(\theta - \theta_0)$,

$\Psi_0 = 63.0328°$, $\quad \xi_0 = 166.8525°$,

$\Psi = \omega_1(\theta - \theta_0) + 63.0328°$, $\quad \xi = \omega_3(\theta - \theta_0) + 166.8525°$,

$\bar{\alpha} = K_1(\theta - \theta_0) + 275.2363°$, $\quad \bar{\nu} = K_2(\theta - \theta_0) + 184.8492°$,

$\lambda_S = \bar{\alpha}$, $\quad \bar{\lambda}_M = \bar{\nu} + \bar{\Omega}_M + \bar{\omega}_M$,

and $\theta_0 = 322.4648°$, computed from the mean anomaly at the initial epoch.

The orbital parameters Δr_c, Γ, and $\dot{\Gamma}$ are solved by (7.5), (7.6), and (7.7), respectively, with $J_{22} = -1.766 \times 10^{-6}$, $\lambda_{22} = -12.35°$, $\Gamma_0 = -29.6°$, and $\Delta r_{c0} = 22{,}677$ m (74,400 ft). It should be noted that the given Δr_{c0} is 23,348 m (76,600 ft). The adoptions of J_{22}, λ_{22}, and the adjustment of Δr_{c0} in our computation are basically for the reason that the numerical integration program has a fourth-order Earth potential model, while our analytical prediction procedures contain a triaxial Earth potential field. Furthermore, the short-period variation of the semimajor axis, a, around its long-period value due to the lunisolar perturbations is rather large as is experienced by Kamel, et al. (1973) in their numerical investigation on synchronous satellites.

Based on (7.19)-(7.25), $\bar{i}$, $\bar{\Omega}$, $\bar{e}$, and $\bar{\omega}$ can be obtained from (7.1)-(7.4), and they are plotted in Figures 7.6.-7.9. With the aid of (7.25), the plottings of Γ, $\dot{\Gamma}$, and Δr are shown in Figures 7.10.-7.12. as they are computed utilizing (7.5), (7.6), and (7.7). When the corresponding numerically integrated values of various osculating orbital parameters are plotted onto these figures, the scattering of these values about the predicted mean value curves gives an indication of good agreement of the results between our analytical prediction and the numerical integration at least for a period of a year and a half. It should be reminded that the numerically integrated osculating values do include the short-period effects of the perturbations, while the predicted mean values do not.

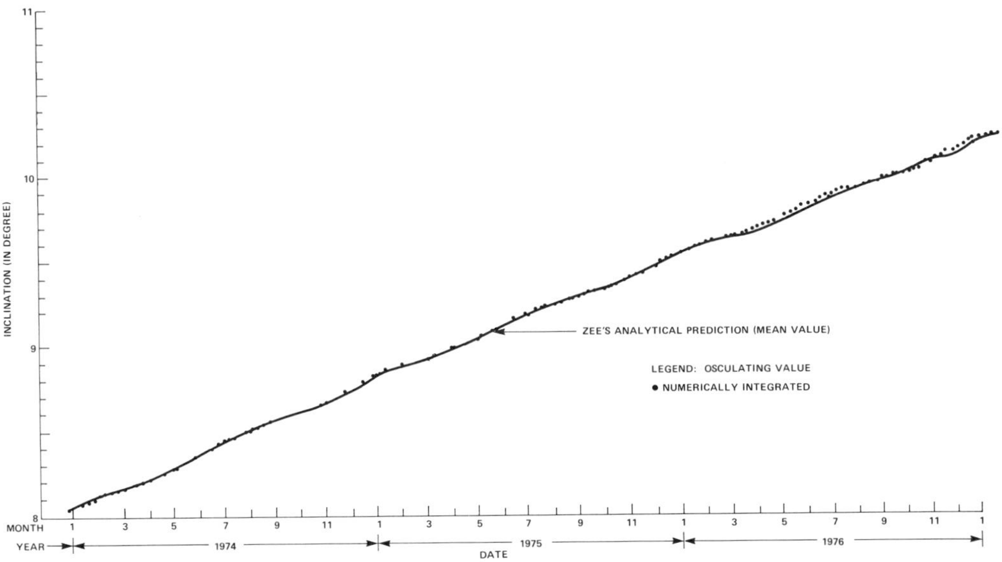

Fig. 7.6. Early Bird synchronous satellite - orbit inclination (numerical integration vs. analytical prediction).

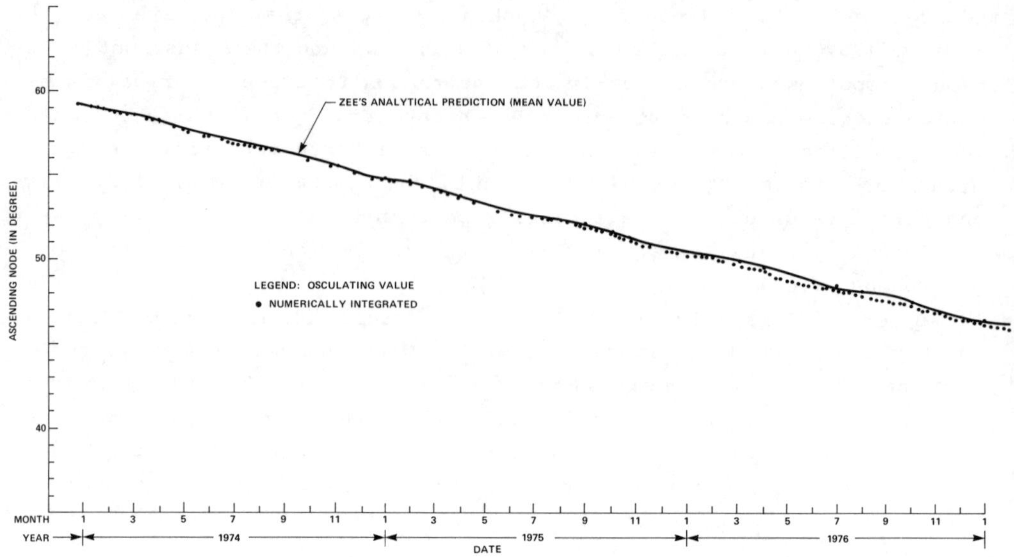

Fig. 7.7. Early Bird synchronous satellite - orbit ascending node (numerical integration vs. analytical prediction).

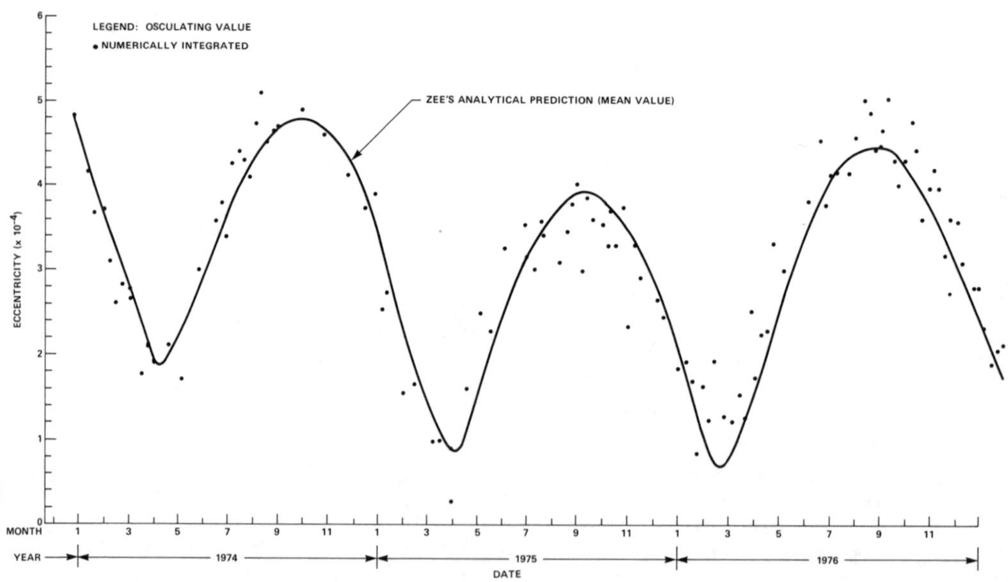

Fig. 7.8. Early Bird synchronous satellite - orbit eccentricity (numerical integration vs. analytical prediction).

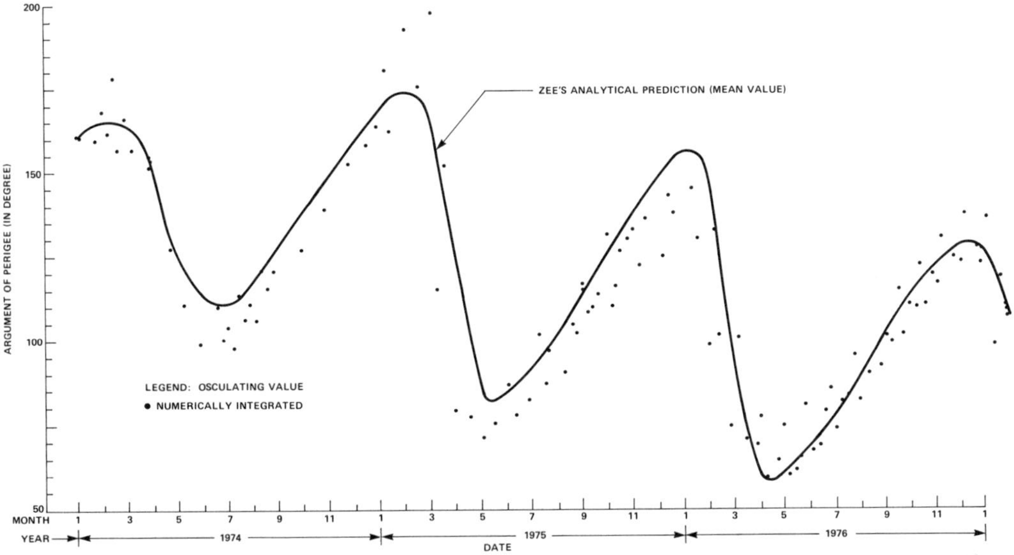

Fig. 7.9. Early Bird synchronous satellite - argument of perigee (numerical integration vs. analytical prediction).

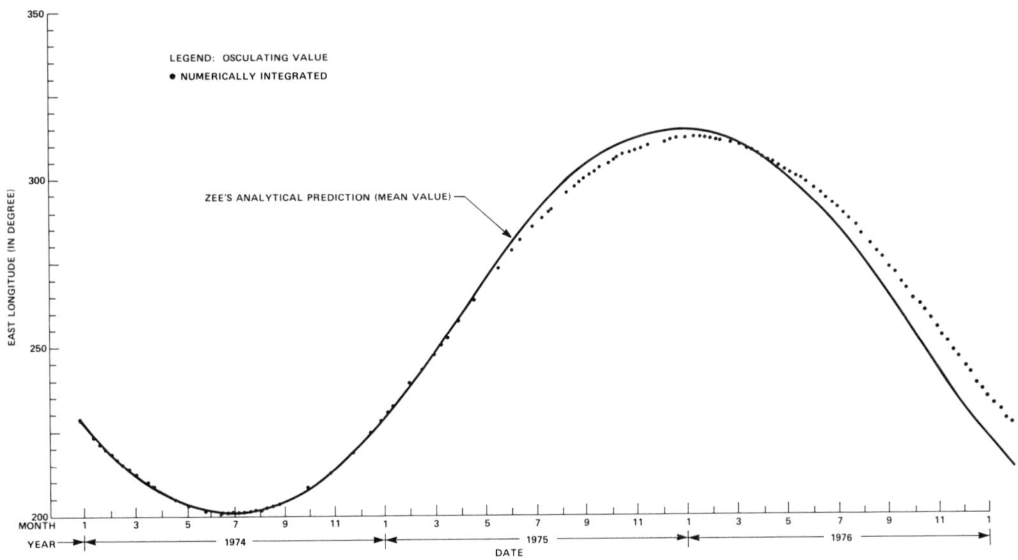

Fig. 7.10. Early Bird synchronous satellite - longitudinal drift angle (numerical integration vs. analytical prediction).

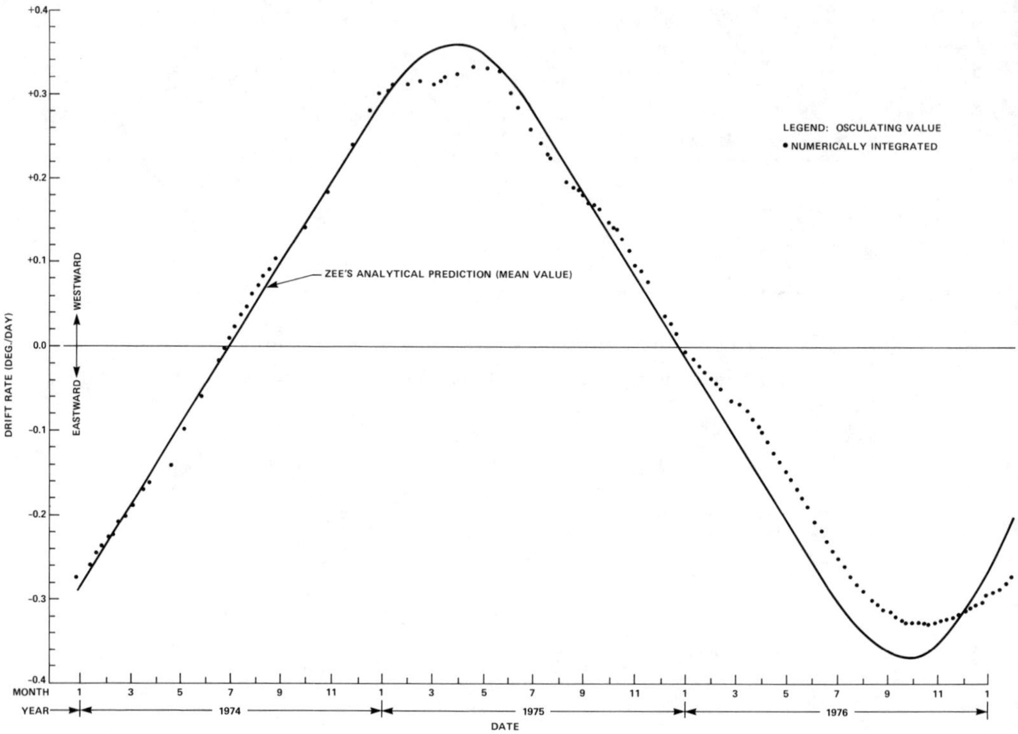

Fig. 7.11. Early Bird synchronous satellite - longitudinal drift rate (numerical integration vs. analytical prediction).

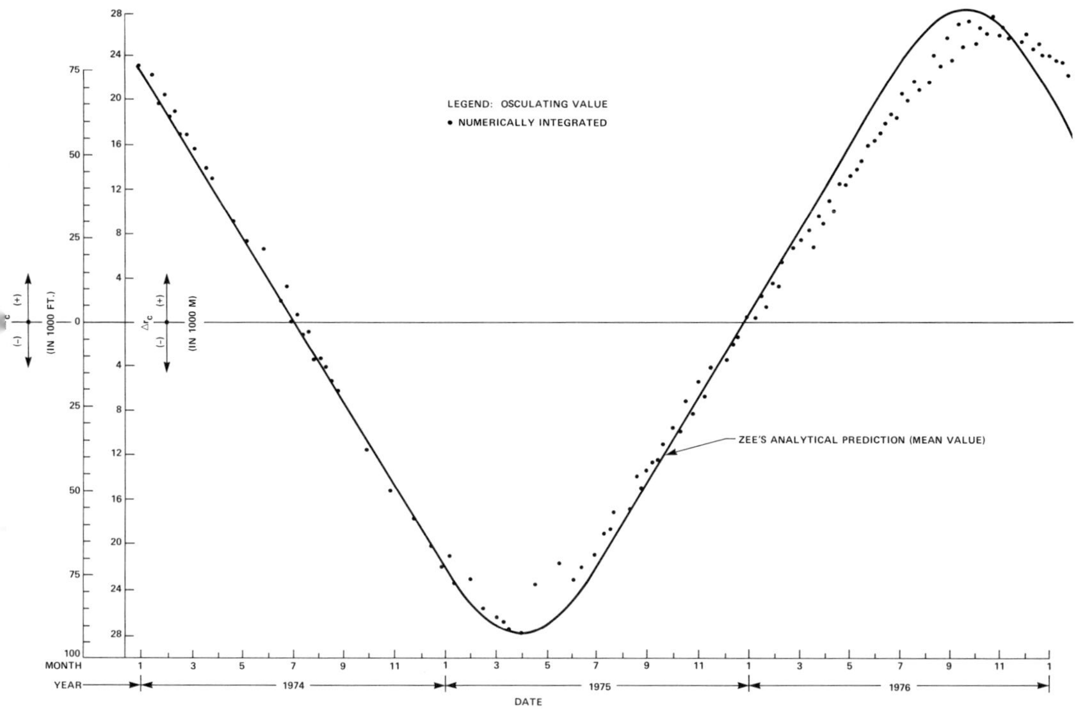

Fig. 7.12. Early Bird synchronous satellite - change of semimajor axis (numerical integration vs. analytical prediction).

7.4. Application of the Theory

After studying the separate effects of various perturbing forces on a geostationary satellite in Chapters 3, 4, and 5, it is found that the J_{22}-term in the Earth potential dominates the drift (Γ), drift rate ($\dot{\Gamma}$), and the semimajor axis (a or Δr_c), while its effects on other orbital parameters ($\bar{\Omega}$, $\bar{i}$, $\bar{\omega}$, and $\bar{e}$) are rather minute as is seen from (3.64), (3.67), (3.71), and (3.72), respectively. On the contrary, the lunisolar perturbing forces have very little effect on these three orbital parameters because they are basically derived from Δr_c, which, in turn, may be expressed in terms of X_6 as is seen from (3.37). The change of X_6 due to the lunisolar forces is depicted in (7.13) and (7.24). These equations show that a rather small reduction of X_6 occurs as the inclination of the orbital plane increases. The practically linear relationships between time, t, and θ shown in (7.18) and (7.25) due to the lunisolar perturbing forces, and those shown in (3.5) due to the Earth triaxial potential field, will further enhance our approach to solve the geostationary satellite problem by superimposing the separate solutions under the force fields considered.

Based on the above observations and the verifications of our theory by the observed data and those from numerical integration, we are therefore justified to combine our findings in Chapter 3 with those in Chapters 4 and 5 as our theory of the geostationary satellite. For an active communications geostationary satellite, it is generally required to perform maneuvers of station-keeping every few months; this short duration will make our analytical theory a valuable tool in accurate orbit predictions of a communications geostationary satellite.

Readers are reminded that our solution is referred to an inertial geocentric coordinate system. In practice, the effects of the nutation and precession of the equator and equinox to the geostationary satellites should be considered. In other words, our solution should be updated by a precession transformation (Escobal, 1965) from the initial epoch to the epoch of interest. Additionally, the dynamic effects of the nutation and precession of the equator and equinox on the motion of geostationary satellites should also be taken into consideration for precise orbit computation; however, these effects are not so large.

Also related to the attractions of the Sun and the Moon is the tidal deformation of the Earth, which, in turn, effects the motion of geostationary satellites. Readers are referred to Kozai (1973) for a detailed derivation of the disturbing function and the corresponding Lagrange's equations for various orbital elements.

References

Escobal, P.R.; Method of Orbit Determination, John Wiley & Sons, Inc., New York, 1965.

Kamel, A., Ekman, D., and Tibbits, R.; 'East-West Stationkeeping Requirements of Nearly Synchronous Satellites due to Earth's Triaxiality and Luni-Solar Effects', Celestial Mechanics **8**, 129-148 (1973).

Kozai, Y.; 'A New Method to Compute Lunisolar Perturbations in Satellite Motions', Special Report 349, Smithsonian Astrophysical Observatory, Cambridge, Mass. (1973).

Smart, M.M.; Spherical Astronomy, Cambridge University Press, London and New York, 1944.

APPENDIX A

The Fourth-Order Earth Gravitational Potential Model

The truncated fourth-order Earth gravitational potential (V_E) in the form of (2.20a) can be written as

$$
\begin{aligned}
V_E = -\frac{\mu_E}{r}\left(\frac{R_E}{r}\right)^2 [& \tfrac{1}{2}J_2(3\sin^2\phi - 1) + 3J_{21}\sin\phi\cos\phi\cos(\lambda - \lambda_{21}) + \\
& + 3J_{22}\cos^2\phi\cos 2(\lambda - \lambda_{22}) + \frac{1}{2}\left(\frac{R_E}{r}\right)J_3(5\sin^2\phi - 3)\sin\phi + \\
& + \frac{3}{2}\left(\frac{R_E}{r}\right)J_{31}(5\sin^2\phi - 1)\cos\phi\cos(\lambda - \lambda_{31}) + \\
& + 15\left(\frac{R_E}{r}\right)J_{32}\cos^2\phi\sin\phi\cos(\lambda - \lambda_{32}) + \\
& + 15\left(\frac{R_E}{r}\right)J_{33}\cos^3\phi\cos 3(\lambda - \lambda_{33}) + \\
& + \frac{1}{8}\left(\frac{R_E}{r}\right)^2 J_4(35\sin^4\phi - 30\sin^2\phi + 3) + \\
& + \frac{3}{2}\left(\frac{R_E}{r}\right)^2 J_{41}(7\sin^2\phi - 3)\cos\phi\sin\phi\cos(\lambda - \lambda_{41}) + \\
& + \frac{15}{2}\left(\frac{R_E}{r}\right)^2 J_{42}(7\sin^2\phi - 1)\cos^2\phi\cos 2(\lambda - \lambda_{42}) + \\
& + 105\left(\frac{R_E}{r}\right)^2 J_{43}\cos^3\phi\sin\phi\cos 3(\lambda - \lambda_{43}) + \\
& + 105\left(\frac{R_E}{r}\right)^2 J_{44}\cos^4\phi\cos 4(\lambda - \lambda_{44})] ,
\end{aligned} \tag{A1}
$$

where

μ_E = the gravitational constant of the Earth,
R_E = the mean equatorial radius of the Earth,
r, θ, ϕ = spherical coordinates as shown in Figure 3.1.,
λ = satellite's geographic longitude with respect to the Greenwich meridian (positive eastward),

and

J_{nm}, λ_{nm} = constants of the Earth's gravitational potential.

If we adopt the Goddard Earth Model (Gem) 6 (Smith et al., 1976), the constants in (A1) are:

$$
\begin{array}{llll}
J_2 = +\ 1082.6283 \times 10^{-6} & & \\
J_{21} = -\ 4.2915 \times 10^{-9} & & \lambda_{21} = 254.2914° \\
J_{22} = -\ 1.8083 \times 10^{-6} & & \lambda_{22} = -\ 14.8949° \\
J_3 = -\ 2.5418 \times 10^{-6} & & \\
J_{31} = -\ 2.1791 \times 10^{-6} & & \lambda_{31} = 7.0669° \\
J_{32} = -\ 3.8480 \times 10^{-7} & & \lambda_{32} = -\ 17.0348° \\
J_{33} = -\ 2.2132 \times 10^{-7} & & \lambda_{33} = 21.3183° \\
J_4 = -\ 1.6086 \times 10^{-6} & & \\
J_{41} = -\ 6.7614 \times 10^{-7} & & \lambda_{41} = 220.7042° \\
J_{42} = -\ 1.6853 \times 10^{-7} & & \lambda_{42} = 31.3316° \\
J_{43} = -\ 5.9015 \times 10^{-8} & & \lambda_{43} = -\ 4.3093° \\
J_{44} = -\ 7.3147 \times 10^{-9} & & \lambda_{44} = 29.5502°
\end{array}
$$

For a satellite situated in the Earth's potential field, the force acting on the satellite in any direction is denoted by the gradient of the potential in the corresponding direction; thus, the acting forces in the fourth-order Earth potential field in the directions in the spherical coordinates r, θ, and ϕ are:

$$
\begin{aligned}
F_r &= \frac{\partial V_E}{\partial r} \\
&= \frac{\mu_E}{r^2}\left(\frac{R_E}{r}\right)^2 [\tfrac{3}{2}J_2(3 \sin^2 \phi - 1) + 9J_{21} \sin \phi \cos \phi \cos(\lambda - \lambda_{21}) + \\
&\quad + 9J_{22} \cos^2 \phi \cos 2(\lambda - \lambda_{22}) + 2\left(\frac{R_E}{r}\right)J_3(5 \sin^2 \phi - 3)\sin \phi + \\
&\quad + 6\left(\frac{R_E}{r}\right)J_{31}(5 \sin^2 \phi - 1)\cos \phi \cos(\lambda - \lambda_{31}) + \\
&\quad + 60\left(\frac{R_E}{r}\right)J_{32} \cos^2 \phi \sin \phi \cos 2(\lambda - \lambda_{32}) + \\
&\quad + 60\left(\frac{R_E}{r}\right)J_{33} \cos^3 \cos 3(\lambda - \lambda_{33}) + \frac{5}{8}\left(\frac{R_E}{r}\right)^2 J_4(35 \sin^4 \phi - \\
&\quad - 30 \sin^2 \phi + 3) + \frac{15}{2}\left(\frac{R_E}{r}\right)^2 J_{41}(7 \sin^2 \phi - 3)\cos \phi \sin \phi \times \\
&\quad \times \cos(\lambda - \lambda_{41}) + \frac{75}{2}\left(\frac{R_E}{r}\right)^2 J_{42}(7 \sin^2 \phi - 1)\cos^2 \phi \times \\
&\quad \times \cos 2(\lambda - \lambda_{42}) + 525\left(\frac{R_E}{r}\right)^2 J_{43} \cos^3 \phi \sin \phi \cos 3(\lambda - \lambda_{43}) + \\
&\quad + 525\left(\frac{R_E}{r}\right)^2 J_{44} \cos^4 \phi \cos 4(\lambda - \lambda_{44})] , \qquad \text{(A2)}
\end{aligned}
$$

$$
\begin{aligned}
F_\theta &= \frac{1}{r \cos \phi} \frac{\partial V_E}{\partial \theta} \\
&= \frac{\mu_E}{r^2}\left(\frac{R_E}{r}\right)^2 [3J_{21} \sin \phi \cos \phi \sin(\lambda - \lambda_{21}) + \\
&\quad + 6J_{22} \cos^2 \phi \sin 2(\lambda - \lambda_{22}) + \frac{3}{2}\left(\frac{R_E}{r}\right)J_{31} \times \\
&\quad \times (5 \sin^2 \phi - 1)\sin(\lambda - \lambda_{31}) +
\end{aligned}
$$

$$+ 30\left(\frac{R_E}{r}\right)J_{32}\cos\phi\sin\phi\sin 2(\lambda - \lambda_{32}) + 45\left(\frac{R_E}{r}\right)J_{33} \times$$

$$\times \cos^2\phi\sin 3(\lambda - \lambda_{33}) + \frac{5}{2}\left(\frac{R_E}{r}\right)^2 J_{41}(7\sin^2\phi - 3) \times$$

$$\times \sin\phi\sin(\lambda - \lambda_{41}) + 15\left(\frac{R_E}{r}\right)^2 J_{42}(7\sin^2\phi - 1) \times$$

$$\times \cos\phi\sin 2(\lambda - \lambda_{42}) + 315\left(\frac{R_E}{r}\right)^2 J_{43} \times$$

$$\times \cos^2\phi\sin\phi\sin 3(\lambda - \lambda_{43}) +$$

$$+ 420\left(\frac{R_E}{r}\right)^2 J_{44}\cos^3\phi\sin 4(\lambda - \lambda_{44})] , \qquad \text{(A3)}$$

and

$$F_\phi = \frac{1}{r}\frac{\partial V_E}{\partial\phi}$$

$$= \frac{\mu_E}{r^2}\left(\frac{R_E}{r}\right)^2 [- 3J_2\sin\phi\cos\phi - 3J_{21}\cos 2\phi \times$$

$$\times \cos(\lambda - \lambda_{21}) + 6J_{22}\sin\phi\cos\phi\cos 2(\lambda - \lambda_{22}) -$$

$$- \frac{3}{2}\left(\frac{R_E}{r}\right)J_3(5\sin^2\phi - 1)\cos\phi + \frac{3}{2}\left(\frac{R_E}{r}\right)J_{31}(15\sin^2\phi - 11) \times$$

$$\times \sin\phi\cos(\lambda - \lambda_{31}) + 15\left(\frac{R_E}{r}\right)J_{32}(3\sin^2\phi - 1) \times$$

$$\times \cos\phi\cos 2(\lambda - \lambda_{32}) + 45\left(\frac{R_E}{r}\right)J_{33} \times$$

$$\times \cos^2\phi\sin\phi\cos 2(\lambda - \lambda_{33}) - \frac{5}{2}\left(\frac{R_E}{r}\right)^2 J_4(7\sin^2\phi - 3) \times$$

$$\times \sin\phi\cos\phi + \frac{5}{2}\left(\frac{R_E}{r}\right)^2 J_{41}(28\sin^4\phi - 27\sin^2\phi + 3) \times$$

$$\times \cos(\lambda - \lambda_{41}) + 30\left(\frac{R_E}{r}\right)^2 J_{42}(7\sin^2\phi - 4)\cos\phi\sin\phi \times$$

$$\times \cos 2(\lambda - \lambda_{42}) + 105\left(\frac{R_E}{r}\right)^2 J_{43}(4\sin^2\phi - 1) \times$$

$$\times \cos^2\phi\cos 3(\lambda - \lambda_{43}) +$$

$$+ 420\left(\frac{R_E}{r}\right)^2 J_{44}\cos^3\phi\sin\phi\cos 4(\lambda - \lambda_{44})] . \qquad \text{(A4)}$$

Reference

Smith, D.E., Lerch, F.J., Marsh, J.G., Wagner, C.A., Kolenkiewicz, R., and Khan, M.A.; 'Contribution to the National Geodetic Satellite Program by Goddard Space Flight Center', *J. of Geophysical Research* **81**, 1006-1026 (1976).

APPENDIX B

Constants in (4.97), (4.98), (4.103), and (4.104)

1. Expressions of constants in (4.97) and (4.98):

$$\begin{matrix} A_1 \\ A_2 \end{matrix} = \frac{1}{2}\frac{a_0}{\omega_1^2 - (2K_1 \pm \omega_1)^2}\Big\{\pm (2K_1 \pm \omega_1)(\omega_1 \cos\lambda \pm \omega_{11}\cos 2\lambda) \mp$$
$$\mp \omega_{12}[\omega_1(1 + \sin^2\lambda) \pm \omega_{11}\cos\lambda]\Big\}$$

$$A_3 = \frac{1}{\omega_1^2 - 4K_1^2}(K_1 \sin 2\lambda - \omega_{12}\sin\lambda)$$

$$A_4 = \frac{\kappa}{\omega_1^2 - K_3^2}\sin I(K_3\cos 2\lambda - \omega_{12}\cos\lambda)$$

$$A_5 = \frac{-\kappa\sin I}{\omega_1^2 - (2K_2 + 2K_4 + K_3)^2}[(2K_2 + 2K_4 + K_3)\cos 2\lambda - \omega_{12}\cos\lambda]$$

$$A_6 = \frac{\kappa}{\omega_1^2 - 4(K_2 + K_3 + K_4)^2}[(K_2 + K_3 + K_4)\sin 2\lambda - \omega_{12}\sin\lambda]$$

$$\begin{matrix} A_7 \\ A_8 \end{matrix} = \frac{1}{2}\frac{a_0\kappa\sin I}{\omega_1^2 - (K_3 \pm \omega_1)^2}[\mp(\omega_1\sin\lambda \pm 2\omega_{11}\sin 2\lambda)(K_3 \pm \omega_1) \mp$$
$$\mp \omega_{12}(\omega_1\sin 2\lambda \mp \omega_{11}\sin\lambda)]$$

$$\begin{matrix} A_9 \\ A_{10} \end{matrix} = \frac{1}{2}\frac{a_0\kappa\sin I}{\omega_1^2 - (2K_2 + K_3 + 2K_4 \pm \omega_1)^2}[(2\omega_{11}\sin 2\lambda \pm \omega_1\sin\lambda)\times$$
$$\times(2K_2 + K_3 + 2K_4 \pm \omega_1) \pm \omega_{12}(\omega_1\sin 2\lambda \mp \omega_{11}\sin\lambda)]$$

$$\begin{matrix} A_{11} \\ A_{12} \end{matrix} = \frac{1}{2}\frac{a_0\kappa}{\omega_1^2 - (2K_2 + 2K_3 + 2K_4 \pm \omega_1)^2}\Big\{(\omega_{11}\cos 2\lambda \pm \omega_1\cos\lambda)\times$$
$$\times(2K_2 + 2K_3 + 2K_4 \pm \omega_1) - \omega_{12}[\omega_{11}\cos\lambda \pm \omega_1(1 + \sin^2\lambda)]\Big\}$$

$$A_{13} = -\frac{1}{2}\frac{e_S}{\omega_1^2 - K_1^2}K_1\sin 2\lambda$$

$$A_{14} = -\frac{1}{2}\frac{e_S}{\omega_1^2 - K_1^2}[K_1 \sin 2\lambda - \omega_{12} \sin \lambda]$$

$$A_{15} = \frac{1}{2}\frac{e_S}{\omega_1^2 - 9K_1^2}[3K_1 \sin 2\lambda - \omega_{12} \sin \lambda]$$

$$A_{16} = -\frac{1}{2}\frac{\kappa e_M}{\omega_1^2 - K_2^2} K_2 \sin 2\lambda$$

$$\begin{matrix}A_{17}\\A_{18}\end{matrix} = \frac{1}{2}\frac{\kappa e_M}{\omega_1^2 - (K_2 \pm K_3)^2} \sin I[(K_2 \pm K_3)\cos 2\lambda \mp \omega_{12} \cos \lambda]$$

$$A_{19} = \frac{1}{2}\frac{\kappa e_M}{\omega_1^2 - (K_2 + 2K_4 + K_3)^2} \sin I[-(K_2 + 2K_4 + K_3)\cos 2\lambda + + \omega_{12} \cos \lambda]$$

$$A_{20} = \frac{1}{2}\frac{\kappa e_M}{\omega_1^2 - (K_2 + 2K_4 + 2K_3)^2}[\tfrac{1}{2}(K_2 + 2K_4 + 2K_3)\sin 2\lambda - - \omega_{12} \sin \lambda]$$

$$A_{21} = \frac{1}{2}\frac{\kappa e_M}{\omega_1^2 - (3K_2 + 2K_4 + K_3)^2} \sin I[-(3K_2 + 2K_4 + K_3)\cos 2\lambda + + \omega_{12} \cos \lambda]$$

$$A_{22} = \frac{1}{2}\frac{\kappa e_M}{\omega_1^2 - (3K_2 + 2K_4 + 2K_3)^2}[(3K_2 + 2K_4 + 2K_3)\sin 2\lambda - - \omega_{12} \sin \lambda]$$

$$\begin{matrix}A_{23}\\A_{24}\end{matrix} = \frac{1}{2}\frac{a_0}{\omega_1^2 - (2K_1 \pm \omega_1)^2}\Big\{\mp (2K_1 \pm \omega_1) \times \times [\omega_1(1 + \sin^2 \lambda) \pm \omega_{11} \cos \lambda] \pm \omega_{11}(\omega_1 \cos \lambda \pm \omega_{11} \cos 2\lambda)\Big\}$$

$$A_{25} = \frac{1}{\omega_1^2 - 4K_1^2}(\tfrac{1}{2}\omega_{11} \sin 2\lambda - 2K_1 \sin \lambda)$$

$$A_{26} = \frac{\kappa}{\omega_1^2 - K_3^2} \sin I(\omega_{11} \cos 2\lambda - K_3 \cos \lambda)$$

$$A_{27} = \frac{\kappa}{\omega_1^2 - (2K_2 + 2K_4 + K_3)^2}[(2K_2 + 2K_4 + K_3)\cos \lambda - - \omega_{11} \cos 2\lambda] \sin I$$

$$A_{28} = \frac{\kappa}{\omega_1^2 - 4(K_2 + K_3 + K_4)^2}[\tfrac{1}{2}\omega_{11} \sin 2\lambda - 2(K_2 + K_3 + K_4)\sin \lambda]$$

$$\begin{matrix}A_{29}\\A_{30}\end{matrix} = \frac{1}{2}\frac{a_0\kappa}{\omega_1^2 - (K_3 \pm \omega_1)^2} \sin I[(\omega_{11} \sin \lambda \mp \omega_1 \sin 2\lambda)(K_3 \pm \omega_1) - - \omega_{11}(2\omega_{11} \sin 2\lambda \pm \omega_1 \sin \lambda)]$$

$$\begin{matrix}A_{31}\\A_{32}\end{matrix} = \frac{1}{2}\,\frac{a_0\kappa\sin I}{\omega_1^2-(2K_2+2K_4+K_3\pm\omega_1)^2}[-(\mp\omega_1\sin 2\lambda+\omega_{11}\sin\lambda)\times$$
$$\times(2K_2+2K_4+K_3\pm\omega_1)\pm\omega_{11}(\pm 2\omega_{11}\sin 2\lambda+\omega_1\sin\lambda)]$$

$$\begin{matrix}A_{33}\\A_{34}\end{matrix} = \frac{1}{2}\,\frac{a_0\kappa}{\omega_1^2-(2K_2+2K_3+2K_4\pm\omega_1)^2}\Big\{-[\omega_{11}\cos\lambda\pm\omega_1(1+\sin^2\lambda)]\times$$
$$\times(2K_2+2K_3+2K_4\pm\omega_1)+\omega_{11}(\omega_{11}\cos 2\lambda\pm\omega_1\cos\lambda)\Big\}$$

$$A_{35} = -\frac{1}{2}\,\frac{e_S}{\omega_1^2-K_1^2}\,\omega_{11}\sin 2\lambda$$

$$A_{36} = -\frac{1}{2}\,\frac{e_S}{\omega_1^2-K_1^2}[K_1\sin\lambda-\omega_{11}\sin 2\lambda]$$

$$A_{37} = -\frac{1}{2}\,\frac{e_S}{\omega_1^2-9K_1^2}[3K_1\sin\lambda-\omega_{11}\sin 2\lambda]$$

$$A_{38} = -\frac{1}{2}\,\frac{\kappa e_M}{\omega_1^2-K_2^2}\,\omega_{11}\sin 2\lambda$$

$$\begin{matrix}A_{39}\\A_{40}\end{matrix} = \frac{1}{2}\,\frac{\kappa e_M}{\omega_1^2-(K_2\pm K_3)^2}\sin I[\mp(K_2\pm K_3)\cos\lambda+\omega_{11}\cos 2\lambda]$$

$$A_{41} = \frac{1}{2}\,\frac{\kappa e_M}{\omega_1^2-(K_2+2K_4+K_3)^2}\times$$
$$\times\sin I[(K_2+2K_4+K_3)\cos\lambda-\omega_{11}\cos 2\lambda]$$

$$A_{42} = \frac{1}{2}\,\frac{\kappa e_M}{\omega_1^2-(K_2+2K_4+2K_3)^2}\times$$
$$\times[-(K_2+2K_4+2K_3)\sin\lambda+\tfrac{1}{2}\omega_{11}\sin 2\lambda]$$

$$A_{43} = \frac{1}{2}\,\frac{\kappa e_M}{\omega_1^2-(3K_2+2K_4+K_3)^2}\times$$
$$\times\sin I[+(3K_2+2K_4+K_3)\cos\lambda-\omega_{11}\cos 2\lambda]$$

$$A_{44} = \frac{1}{2}\,\frac{\kappa e_M}{\omega_1^2-(3K_2+2K_4+2K_3)^2}\times$$
$$\times[-(3K_2+2K_4+2K_3)\sin\lambda+\tfrac{1}{2}\omega_{11}\sin 2\lambda]$$

2. Expressions of constants in (4.103) and (4.104):

$$\begin{matrix}B_1\\B_2\end{matrix} = \frac{2b_0}{\omega_3^2-(2K_1\pm\omega_3)^2}\Big\{[-5\omega_{31}\cos\lambda\pm\omega_3(1+4\cos^2\lambda)]\times$$
$$\times(2K_1\pm\omega_3)\mp\omega_{31}[-5\omega_3\cos\lambda\pm\omega_{31}(4+\cos^2\lambda)]\Big\}$$

$$\begin{matrix}B_3\\B_4\end{matrix} = \frac{2\kappa b_0}{\omega_3^2-(2K_2+2K_3\pm\omega_3)^2}\Big\{[-5\omega_{31}\cos\lambda\pm\omega_3(1+4\cos^2\lambda)]\times$$
$$\times(2K_2+2K_3\pm\omega_3)\mp\omega_{31}[-5\omega_3\cos\lambda\pm\omega_{31}(4+\cos^2\lambda)]\Big\}$$

$$\begin{matrix} B_5 \\ B_6 \end{matrix} = \frac{2\kappa b_0 \sin I}{\omega_3^2 - (2K_2 + K_3 \pm \omega_3)^2}[(- 5\omega_{31} \sin\lambda \pm 4\omega_3 \sin 2\lambda) \times$$

$$\times (2K_2 + K_3 \pm \omega_3) \mp \omega_{31}(- 5\omega_3 \sin\lambda \pm \omega_{31} \sin 2\lambda)]$$

$$\begin{matrix} B_7 \\ B_8 \end{matrix} = \frac{2\kappa b_0 \sin I}{\omega_3^2 - (K_3 \pm \omega_3)^2}[(5\omega_{31} \sin\lambda \mp 4\omega_3 \sin 2\lambda)(K_3 \pm \omega_3) -$$

$$- \omega_{31}(\pm 5\omega_3 \sin\lambda - \omega_{31} \sin 2\lambda)]$$

$$B_9 = \frac{\bar{\kappa} e_M}{\omega_3^2 - K_4^2} K_4(8 - \frac{15}{2} \sin^2\lambda)\sin\lambda \sin I$$

$$B_{10} = \frac{\bar{\kappa} e_M}{\omega_3^2 - (K_3 + K_4)^2} [\cos\lambda(1 - \frac{15}{4} \sin^2\lambda)(K_3 + K_4) -$$

$$- \omega_{31}(\frac{5}{4} \sin^2\lambda - 1)]$$

$$B_{11} = \frac{\bar{\kappa} e_M}{\omega_3^2 - (2K_3 + K_4)^2} \frac{1}{4} \sin I[3 \sin\lambda(2 - 5 \sin^2\lambda)(2K_3 + K_4) +$$

$$+ 5\omega_{31} \sin 2\lambda]$$

$$B_{12} = \frac{\bar{\kappa}}{\omega_3^2 - (K_2 + K_4)^2}(K_2 + K_4)(8 - \frac{15}{2} \sin^2\lambda)\sin\lambda \sin I$$

$$B_{13} = \frac{\bar{\kappa}}{\omega_3^2 - (K_2 + K_3 + K_4)^2}[\cos\lambda(1 - \frac{15}{4} \sin^2\lambda)(K_2 + K_3 + K_4) -$$

$$- \omega_{31}(\frac{5}{4} \sin^2\lambda - 1)]$$

$$B_{14} = \frac{\bar{\kappa}}{\omega_3^2 - (K_2 + 2K_3 + K_4)^2} \frac{1}{4} \sin I[3 \sin\lambda(2 - 5 \sin^2\lambda) \times$$

$$\times (K_2 + 2K_3 + K_4) + 5\omega_{31} \sin 2\lambda]$$

$$B_{15} = \frac{\bar{\kappa} e_M}{\omega_3^2 - (2K_2 + K_4)^2}(2K_2 + K_4)(8 - \frac{15}{2} \sin^2\lambda)\sin\lambda \sin I$$

$$B_{16} = \frac{\bar{\kappa} e_M}{\omega_3^2 - (2K_2 + K_3 + K_4)^2}[\cos\lambda(1 - \frac{15}{4} \sin^2\lambda)(2K_2 + K_3 + K_4) -$$

$$- \omega_{31}(\frac{5}{4} \sin^2\lambda - 1)]$$

$$B_{17} = \frac{\bar{\kappa} e_M}{\omega_3^2 - (2K_2 + 2K_3 + K_4)^2} \frac{1}{4} \sin I[3 \sin\lambda(2 - 5 \sin^2\lambda) \times$$

$$\times (2K_2 + 2K_3 + K_4) + 5\omega_{31} \sin 2\lambda]$$

$$B_{18} = \frac{\bar{\kappa} e_M}{\omega_3^2 - (2K_2 + 2K_3 + 3K_4)^2} \frac{1}{4} \sin I[- 3 \sin\lambda(2 - 5 \sin^2\lambda) \times$$

$$\times (2K_2 + 2K_3 + 3K_4) - 5\omega_{31} \sin 2\lambda]$$

$$B_{19} = \frac{\bar{\kappa} e_M}{\omega_3^2 - (2K_2 + 3K_3 + 3K_4)^2} \frac{1}{4} \sin^2 \lambda [5 \cos \lambda (2K_2 + 3K_3 + 3K_4) + + 3\omega_{31}]$$

$$B_{20} = \frac{\bar{\kappa}}{\omega_3^2 - (3K_2 + 2K_3 + 3K_4)^2} \frac{1}{4} \sin I[- 3 \sin \lambda (2 - 5 \sin^2 \lambda) \times \times (2K_2 + 2K_3 + 3K_4) - 5\omega_{31} \sin 2\lambda]$$

$$B_{21} = \frac{\bar{\kappa}}{\omega_3^2 - 9(K_2 + K_3 + K_4)^2} \frac{3}{4} \sin^2 \lambda [5 \cos \lambda (K_2 + K_3 + K_4) + \omega_{31}]$$

$$B_{22} = \frac{\bar{\kappa} e_M}{\omega_3^2 - (4K_2 + 2K_3 + 3K_4)^2} \frac{1}{4} \sin I[- 3 \sin \lambda (2 - 5 \sin^2 \lambda) \times \times (4K_2 + 2K_3 + 3K_4) - 5\omega_{31} \sin 2\lambda]$$

$$B_{23} = \frac{\bar{\kappa} e_M}{\omega_3^2 - (4K_2 + 3K_3 + 3K_4)^2} \frac{1}{4} \sin^2 \lambda [5 \cos \lambda (4K_2 + 3K_3 + 3K_4) + + 3\omega_{31}]$$

$$\begin{matrix} B_{24} \\ B_{25} \end{matrix} = \frac{2b_0}{\omega_3^2 - (2K_1 \pm \omega_3)^2} \left\{ -[\omega_{31}(4 + \cos^2 \lambda) \mp 5\omega_3 \cos \lambda](2K_1 \pm \omega_3) + + \omega_{32}[- 5 \cos \lambda \pm \omega_3 (1 + 4 \cos^2 \lambda)] \right\}$$

$$\begin{matrix} B_{26} \\ B_{27} \end{matrix} = \frac{2\kappa b_0}{\omega_3^2 - (2K_2 + 2K_3 \pm \omega_3)^2} \left\{ -[\omega_{31}(4 + \cos^2 \lambda) \mp \omega_3 \cos \lambda] \times \times (2K_2 + 2K_3 \pm \omega_3) + \omega_{32}[- 5\omega_{31} \cos \lambda \pm \omega_3 (1 + 4 \cos^2 \lambda)] \right\}$$

$$\begin{matrix} B_{28} \\ B_{29} \end{matrix} = \frac{2\kappa b_0 \sin I}{\omega_3^2 - (2K_2 + K_3 \pm \omega_3)^2} [-(\omega_{31} \sin 2\lambda \mp 5\omega_3 \sin \lambda) \times \times (2K_2 + K_3 \pm \omega_3) + \omega_{32}(\mp 5\omega_{31} \sin \lambda \pm 4\omega_3 \sin 2\lambda)]$$

$$\begin{matrix} B_{30} \\ B_{31} \end{matrix} = \frac{2\kappa b_0 \sin I}{\omega_3^2 - (K_3 \pm \omega_3)^2} [-(\pm 5\omega_3 \sin \lambda - \omega_{31} \sin 2\lambda)(K_3 \pm \omega_3) + + \omega_{32}(5\omega_{31} \sin \lambda \mp 4\omega_3 \sin 2\lambda)]$$

$$B_{32} = \frac{\bar{\kappa} e_M}{\omega_3^2 - K_4^2} \omega_{32} (8 - \frac{15}{2} \sin^2 \lambda) \sin \lambda \sin I$$

$$B_{33} = \frac{\bar{\kappa} e_M}{\omega_3^2 - (K_3 + K_4)^2} [-(\frac{5}{4} \sin^2 \lambda - 1)(K_3 + K_4) + + \omega_{32} \cos \lambda (1 - \frac{15}{4} \sin^2 \lambda)]$$

$$B_{34} = \frac{\bar{\kappa} e_M}{\omega_3^2 - (2K_3 + K_4)^2} \frac{1}{4} \sin I[5 \sin 2\lambda (2K_3 + K_4) + + 3 \sin \lambda (2 - 5 \sin^2 \lambda) \omega_{32}]$$

$$B_{35} = \frac{\bar{\kappa}}{\omega_3^2 - (K_2 + K_4)^2}\,\omega_{32}(8 - \frac{15}{2}\sin^2\lambda)\sin\lambda\sin I$$

$$B_{36} = \frac{\bar{\kappa}}{\omega_3^2 - (K_2 + K_3 + K_4)^2}[(1 - \frac{5}{4}\sin^2\lambda)(K_2 + K_3 + K_4) + {} + \cos\lambda(1 - \frac{15}{4}\sin^2\lambda)\omega_{32}]$$

$$B_{37} = \frac{\bar{\kappa}}{\omega_3^2 - (K_2 + 2K_3 + K_4)^2}\,\frac{1}{4}\sin I[5\sin 2\lambda(K_2 + 2K_3 + K_4) + {} + 3\sin\lambda(2 - 5\sin^2\lambda)\omega_{32}]$$

$$B_{38} = \frac{\bar{\kappa}e_M}{\omega_3^2 - (2K_2 + K_4)^2}\,\omega_{32}(8 - \frac{5}{2}\sin^2\lambda)\sin\lambda\sin I$$

$$B_{39} = \frac{\bar{\kappa}e_M}{\omega_3^2 - (2K_2 + K_3 + K_4)^2}[(-\frac{5}{4}\sin^2\lambda + 1)(2K_2 + K_3 + K_4) + {} + \cos\lambda(1 - \frac{15}{4}\sin^2\lambda)\omega_{32}]$$

$$B_{40} = \frac{\bar{\kappa}e_M}{\omega_3^2 - (2K_2 + 2K_3 + K_4)^2}\,\frac{1}{4}\sin I[5\sin 2\lambda(2K_2 + 2K_3 + K_4) + {} + 3\sin\lambda(2 - 5\sin^2\lambda)\omega_{32}]$$

$$B_{41} = \frac{-\bar{\kappa}e_M}{\omega_3^2 - (2K_2 + 2K_3 + 3K_4)^2}\,\frac{1}{4}\sin I[5\sin 2\lambda(2K_2 + 2K_3 + 3K_4) + {} + 3\sin\lambda(2 - 5\sin^2\lambda)\omega_{32}]$$

$$B_{42} = \frac{\bar{\kappa}e_M}{\omega_3^2 - (2K_2 + 3K_3 + 3K_4)^2}\,\frac{1}{4}\sin^2\lambda[3(2K_2 + 3K_3 + 3K_4) + {} + 5\omega_{32}\cos\lambda]$$

$$B_{43} = \frac{-\bar{\kappa}}{\omega_3^2 - (3K_2 + 2K_3 + 3K_4)^2}\,\frac{1}{4}\sin I[5\sin 2\lambda(3K_2 + 2K_3 + 3K_4) + {} + 3\sin\lambda(2 - 5\sin^2\lambda)\omega_{32}]$$

$$B_{44} = \frac{\bar{\kappa}}{\omega_3^2 - 9(K_2 + K_3 + K_4)^2}\,\frac{1}{4}\sin^2\lambda[9(K_2 + K_3 + K_4) + 5\omega_{32}\cos\lambda]$$

$$B_{45} = \frac{-\bar{\kappa}e_M}{\omega_3^2 - (4K_2 + 2K_3 + 3K_4)^2}\,\frac{1}{4}\sin I[5\sin 2\lambda(4K_2 + 2K_3 + 3K_4) + {} + 3\sin\lambda(2 - 5\sin^2\lambda)\omega_{32}]$$

$$B_{46} = \frac{\bar{\kappa}e_M}{\omega_3^2 - (4K_2 + 3K_3 + 3K_4)^2}\,\frac{1}{4}\sin^2\lambda[3(4K_2 + 3K_3 + 3K_4) + {} + 5\omega_{32}\cos\lambda]$$

where

$$\bar{\kappa} \equiv \frac{5\kappa}{(1 - e_M^2)^{3/2}}\,\frac{1}{a_M}\,.$$

APPENDIX C

Perturbed Solutions to (4.69)-(4.73) Due to $\delta_1\bar{\Omega}_M$ and $\delta_1\bar{\omega}_M$

We shall consider the case of $A_4 \sin \bar{\Omega}_M$ and $A_{26} \cos \bar{\Omega}_M$ in $u_1^{(1)}$ and $u_2^{(1)}$ as an example in obtaining our perturbed solutions due to $\delta_1\bar{\Omega}_M$ related to our solution terms, $A_4 \sin \bar{\Omega}_M$ and $A_{26} \cos \bar{\Omega}_M$. Readers will be able to obtain perturbed solutions due to $\delta_1(\ )$'s for other solution terms by following the same procedure.

For the case of $A_4 \sin \bar{\Omega}_M$ and $A_{26} \cos \bar{\Omega}_M$, the corresponding terms in (4.69) and (4.70) are

$$- \kappa \sin I \cos 2\lambda \cos \bar{\Omega}_M \equiv C \cos \bar{\Omega}_M$$

and (C1)

$$- \kappa \sin I \cos \lambda \sin \bar{\Omega}_M \equiv D \sin \bar{\Omega}_M ,$$

respectively, as they can be written out from the expressions of A_4 and A_{26}. In our method of perturbed solution, the addition of $\delta_1\bar{\Omega}_M$ to $\bar{\Omega}_M$ may be considered as an application of perturbation to (C1), which results in

$$\begin{aligned}\delta(C \cos \bar{\Omega}_M) &= - C \sin \bar{\Omega}_M \times \delta_1\bar{\Omega}_M \\ &= - \frac{3}{8}mC \sin \bar{\Omega}_M \sin(2\bar{\lambda}_S - 2\bar{\Omega}_M) \\ &= - \frac{3}{16}mC\, [\cos(\bar{\Omega}_M - 2\bar{\lambda}_S + 2\bar{\Omega}_M) - \cos(\bar{\Omega}_M + 2\bar{\lambda}_S - 2\bar{\Omega}_M)]\end{aligned} \tag{C2}$$

and

$$\begin{aligned}\delta(D \sin \bar{\Omega}_M) &= D \cos \bar{\Omega}_M \times \delta_1\bar{\Omega}_M \\ &= \frac{3}{8}mD \cos \bar{\Omega}_M \sin(2\bar{\lambda}_S - 2\bar{\Omega}_M) \\ &= \frac{3}{16}mD\, [\sin(\bar{\Omega}_M + 2\bar{\lambda}_S - 2\bar{\Omega}_M) - \sin(\bar{\Omega}_M - 2\bar{\lambda}_S + 2\bar{\Omega}_M)]\, .\end{aligned} \tag{C3}$$

The perturbation is applicable because m is small.

With the terms in (C2) and (C3) being placed in (4.69) and (4.70), respectively, our perturbed solution due to $\delta_1\bar{\Omega}_M$ in $u_1^{(1)}$ is

$$\begin{aligned}&\frac{\frac{3}{16}mC(3K_3 - 2K_1) - \frac{3}{16}mD\omega_{12}}{\omega_1^2 - (3K_3 - 2K_1)^2} \sin(3\bar{\Omega}_M - 2\lambda_S) - \\ &- \frac{\frac{3}{16}mC(2K_1 - K_3) + \frac{3}{16}mD\omega_{12}}{\omega_1^2 - (2K_1 - K_3)^2} \sin(2\bar{\lambda}_S - \bar{\Omega}_M) \\ &\cong \frac{3}{16}mC[\frac{1}{2K_1} \sin(3\bar{\Omega}_M - 2\bar{\lambda}_S) + \frac{1}{2K_1} \sin(2\bar{\lambda}_S - \bar{\Omega}_M)] \\ &= \frac{3}{16K_1}\, mC \sin \bar{\Omega}_M \cos(2\bar{\lambda}_S - 2\bar{\Omega}_M)\end{aligned}$$

$$= -\frac{3}{16K_1} m\kappa \sin I \cos 2\lambda \sin \bar{\Omega}_M \cos(2\bar{\lambda}_S - 2\bar{\Omega}_M)$$

$$= A_{4M} \sin \bar{\Omega}_M \cos(2\bar{\lambda}_S - 2\bar{\Omega}_M) , \qquad \text{(C4)}$$

where terms containing ω_1, ω_{12}, and K_3 are omitted because they are small in comparison with K_1. Similarly, we have, in $u_2^{(1)}$

$$-\frac{3}{16K_1} mD \cos \bar{\Omega}_M \cos(2\bar{\lambda}_S - 2\bar{\Omega}_M)$$

$$= \frac{3}{16K_1} m\kappa \sin I \cos \lambda \cos \bar{\Omega}_M \cos(2\bar{\lambda}_S - 2\bar{\Omega}_M)$$

$$= A_{26M} \cos \bar{\Omega}_M \cos(2\bar{\lambda}_S - 2\bar{\Omega}_M) .$$

By following the same methodology, the other perturbed solution terms in $u_1^{(1)}$, $u_2^{(1)}$, $u_3^{(1)}$, and $u_4^{(1)}$ due to $\delta_1\bar{\Omega}_M$ and $\delta_1\bar{\omega}_M$ are obtained.

Constants in (4.108), (4.109), (4.110), and (4.111):

$$A_{4M} = -\frac{3}{16K_1} m\kappa \sin I \cos 2\lambda$$

$$\begin{matrix} A_{7M} \\ A_{8M} \end{matrix} = +\frac{3}{32K_1} ma_0\kappa \sin I(\pm \omega_1 \sin \lambda + 2\omega_{11} \sin 2\lambda)$$

$$A_{26M} = +\frac{1}{16K_1} m\kappa \sin I \cos \lambda$$

$$\begin{matrix} A_{29M} \\ A_{30M} \end{matrix} = -\frac{3}{32K_1} ma_0\kappa \sin I(\mp \omega_1 \sin 2\lambda + \omega_{11} \sin \lambda)$$

$$\begin{matrix} B_{7M} \\ B_{8M} \end{matrix} = -\frac{3}{8K_1} m\kappa b_0 \sin I(5\omega_{31} \sin \lambda \mp 4\omega_3 \sin 2\lambda)$$

$$B_{9M1} = -\frac{1}{4}\frac{m^2}{3K_2 - 2K_1} \bar{\kappa} \sin \lambda \sin I(8 - \frac{15}{2} \sin^2 \lambda)$$

$$B_{9M2} = +\frac{9}{4}\frac{m^2}{K_2 - 2K_1} \bar{\kappa} \sin \lambda \sin I(8 - \frac{15}{2} \sin^2 \lambda)$$

$$B_{9M3} = -\frac{15}{16K_1} m\bar{\kappa} e_M \sin \lambda \sin I(8 - \frac{15}{2} \sin^2 \lambda)$$

$$B_{9M4} = +\frac{3}{16K_1} m\bar{\kappa} e_M \sin \lambda \sin I(8 - \frac{15}{2} \sin^2 \lambda)$$

$$B_{10M1} = +\frac{1}{4}\frac{m^2}{2K_1 - 3K_2} \bar{\kappa} \cos \lambda(1 - \frac{15}{4} \sin^2 \lambda)$$

$$B_{10M2} = +\frac{9}{4}\frac{m^2}{K_2 - 2K_1} \bar{\kappa} \cos \lambda(1 - \frac{15}{4} \sin^2 \lambda)$$

$$B_{10M3} = -\frac{15}{16K_1} m\bar{\kappa} e_M \cos \lambda(1 - \frac{15}{4} \sin^2 \lambda)$$

$$B_{11M1} = -\frac{9}{64K_1} m\bar{\kappa} e_M \sin I \sin \lambda(2 - 5 \sin^2 \lambda)$$

$$B_{11M2} = -\frac{3}{16}\frac{m^2}{3K_2 - 2K_1} \bar{\kappa} \sin I \sin \lambda(2 - 5 \sin^2 \lambda)$$

$$B_{11M3} = +\frac{27}{16}\frac{m^2}{K_2 - 2K_1} \bar{\kappa} \sin I \sin \lambda(2 - 5 \sin^2 \lambda)$$

$$B_{11M4} = - \frac{45}{64K_1} m\bar{\kappa}e_M \sin I \sin \lambda (2 - 5 \sin^2 \lambda)$$

$$\begin{matrix} B_{30M} \\ B_{31M} \end{matrix} = + \frac{3}{8K_1} m\kappa b_0 \sin I (\pm 5\omega_3 \sin \lambda - \omega_{31} \sin 2\lambda)$$

$$B_{32M} \cong 0 \quad \text{(very small coefficient, hence omitted)}$$

$$B_{33M1} = + \frac{1}{4} \frac{m^2}{3K_2 - 2K_1} \bar{\kappa} (\frac{5}{4} \sin^2 \lambda - 1)$$

$$B_{33M2} = + \frac{9}{4} \frac{m^2}{2K_1 - K_2} \bar{\kappa} (\frac{5}{4} \sin^2 \lambda - 1)$$

$$B_{33M3} = + \frac{15}{16K_1} m\bar{\kappa}e_M (\frac{5}{4} \sin^2 \lambda - 1)$$

$$B_{34M1} = - \frac{15}{64K_1} m\bar{\kappa}e_M \sin I \sin 2\lambda$$

$$B_{34M2} = - \frac{5}{16} \frac{m^2}{3K_2 - 2K_1} \bar{\kappa} \sin I \sin 2\lambda$$

$$B_{34M3} = - \frac{45}{16} \frac{m^2}{2K_1 - K_2} \bar{\kappa} \sin I \sin 2\lambda$$

$$B_{34M4} = - \frac{75}{64K_1} m\bar{\kappa}e_M \sin I \sin 2\lambda \; .$$

APPENDIX D

Expressions of Constants in (4.127), (4.128), and (5.39)

$$E_1 = \left[1 - 3\frac{\bar{X}^6_{60}}{\bar{\bar{X}}^6_{60}} a_0\omega_{12}D\cos\Psi_0\right]\left[\frac{1-\omega_4}{1-2\omega_4} + \frac{3}{4}a_0^2(\omega_1^2 + \omega_{11}^2) + \frac{3}{2}D^2\right] - $$

$$- \frac{9}{2}\frac{\bar{X}^6_{60}}{\bar{\bar{X}}^6_{60}} a_0^2\omega_1^2D^2$$

$$E_2 = \frac{3}{\omega_1}\left\{\frac{\bar{X}^6_{60}}{\bar{\bar{X}}^6_{60}} a_0\omega_{12}D\left[\frac{1-\omega_4}{1-2\omega_4} + \frac{3}{8}a_0^2(\omega_1^2 + 3\omega_{11}^2) + \frac{3}{2}D^2\right] - \right.$$

$$\left. - \omega_{11}a_0D\left[1 - 3\frac{\bar{X}^6_{60}}{\bar{\bar{X}}^6_{60}} a_0\omega_{12}D\cos\Psi_0\right]\right\}$$

$$E_3 = \frac{3}{4\omega_1}\left\{\frac{1}{2}a_0^2(\omega_{11}^2 - \omega_1^2)\left[1 - 3\frac{\bar{X}^6_{60}}{\bar{\bar{X}}^6_{60}} a_0\omega_{12}D\cos\Psi_0\right] - \right.$$

$$\left. - 3\frac{\bar{X}^6_{60}}{\bar{\bar{X}}^6_{60}} a_0^2\omega_1^2D^2\right\}$$

$$E_4 = \frac{3}{8\omega_1}\frac{\bar{X}^6_{60}}{\bar{\bar{X}}^6_{60}} a_0^3\omega_{12}(\omega_{11}^2 - \omega_1^2)D$$

$$D_0 = YY_1 + a_0^2(\omega_1^2 + \omega_{11}^2)Y_4 - 2a_0\omega_{11}DY_6 - \frac{1}{2}Y_2Y_{10}$$

$$D_1 = \frac{1}{\omega_1}[YY_3 - a_0\omega_{11}DY_4 + a_0^2(\omega_1^2 + 3\omega_{11}^2)Y_6 - \frac{1}{2}Y_3Y_{10}]$$

$$D_2 = \frac{1}{2\omega_1}[YY_2 + a_0^2(\omega_{11}^2 - \omega_1^2)Y_4 - 2a_0\omega_{11}DY_6 - Y_1Y_{10}]$$

$$D_3 = \frac{1}{3\omega_1}[a_0^2(\omega_{11}^2 - \omega_1^2)Y_6 - \frac{1}{2}Y_3Y_{10}]$$

$$D_4 = \frac{-1}{8\omega_1}Y_2Y_{10}$$

$$D_5 = \frac{1}{2K_1}\left\{\frac{D}{2K_1}Y_1\sin\lambda - \frac{1}{4}a_0Y_3\left(\frac{\omega_{11} + \omega_1\cos\lambda}{2K_1 + \omega_1} + \frac{\omega_{11} - \omega_1\cos\lambda}{2K_2 - \omega_1}\right)\times\right.$$

$$\times\sin\lambda + [(A_1 - A_2)\omega_1a_0 + (A_{23} + A_{24})\omega_{11}a_0 - DA_{25}]Y_4 +$$

$$+ 4Y_5Y_8 \sin^2 \lambda + 2[A_{25}\omega_{11}a_0 - D(A_{23} + A_{24})]Y_6 -$$
$$- 24a_0\omega_{11}DY_7 \sin^2 \lambda\Big\}$$

$$\begin{matrix} D_6 \\ D_7 \end{matrix} = \frac{1}{2K_1 \pm \omega_1}\Big\{ - \frac{1}{2}a_0Y_1 \frac{\omega_{11} \pm \omega_1 \cos \lambda}{2K_1 \pm \omega_1} \sin \lambda -$$
$$- \frac{1}{4}a_0Y_2 \frac{\omega_{11} \mp \omega_1 \cos \lambda}{2K_1 \mp \omega_1} \sin \lambda + \frac{1}{4K_1} DY_3 \sin \lambda +$$
$$+ \left[a_0(\mp A_3\omega_1 + A_{25}\omega_{11}) - D\begin{matrix} A_{23} \\ A_{24} \end{matrix}\right]Y_4 - 12a_0\omega_{11}DY_5 \sin^2 \lambda +$$
$$+ \left[\mp A_2a_0\omega_1 + a_0\omega_{11}\left(\begin{matrix} 2A_{23} \\ A_{23} \end{matrix} + \begin{matrix} A_{24} \\ 2A_{24} \end{matrix}\right) - 2DA_{25}\right]Y_6 +$$
$$+ [4Y_8 \sin^2 \lambda - 3a_0^2(\omega_1^2 - \omega_{11}^2)\sin^2 \lambda]Y_7\Big\}$$

$$\begin{matrix} D_8 \\ D_9 \end{matrix} = \frac{1}{2(K_1 \pm \omega_1)}\Big\{\frac{1}{4} \frac{D}{K_1} Y_2 \sin \lambda - \frac{1}{4}a_0Y_3 \frac{\omega_{11} \pm \omega_1 \cos \lambda}{2K_1 \pm \omega_1} \sin \lambda +$$
$$+ \left(\begin{matrix} -A_1 \\ A_2 \end{matrix}\omega_1 + \begin{matrix} A_{23} \\ A_{24} \end{matrix}\omega_{11}\right)a_0Y_4 - 3a_0^2(\omega_1^2 - \omega_{11}^2)Y_5 \sin^2 \lambda +$$
$$+ \left[a_0(\mp A_3\omega_1 + A_{25}\omega_{11}) - 2D\begin{matrix} A_{23} \\ A_{24} \end{matrix}\right]Y_6 - 12a_0\omega_{11}D\kappa Y_7 \sin^2 \lambda\Big\}$$

$$\begin{matrix} D_{10} \\ D_{11} \end{matrix} = \frac{1}{2K_1 \pm 3\omega_1}\Big\{ - \frac{1}{4}a_0Y_2 \frac{\omega_{11} \pm \omega_1 \cos \lambda}{2K_1 \pm \omega_1} \sin \lambda +$$
$$+ \left(\begin{matrix} -A_1 \\ A_2 \end{matrix}\omega_1 + \begin{matrix} A_{23} \\ A_{24} \end{matrix}\omega_{11}\right)a_0Y_6 - 3a_0^2(\omega_1^2 - \omega_{11}^2)Y_7 \sin^2 \lambda\Big\}$$

$$D_{12} = \frac{1}{K_3}\Big\{\frac{1}{K_3} \kappa DY_1 \sin I \cos \lambda +$$
$$+ \frac{1}{4}\kappa a_0Y_3 \sin I \left[\frac{\omega_1 \cos 2\lambda - \omega_{11} \cos \lambda}{K_3 - \omega_1} -\right.$$
$$\left.- \frac{\omega_1 \cos 2\lambda + \omega_{11} \cos \lambda}{K_3 + \omega_1}\right] +$$
$$+ [\omega_1a_0(A_7 - A_8) + \omega_{11}a_0(A_{29} + A_{30}) - DA_{26}]Y_4 +$$
$$+ 4\kappa Y_5Y_8 \sin I \sin 2\lambda + 2[A_{26}\omega_{11}a_0 - D(A_{29} + A_{30})]Y_6 -$$
$$- 24a_0\omega_{11}D\kappa Y_7 \sin I \sin 2\lambda\Big\}$$

$$\begin{matrix} D_{13} \\ D_{14} \end{matrix} = \frac{1}{K_3 \pm \omega_1}\Big\{\frac{1}{2} \frac{1}{K_3 \pm \omega_1} \kappa a_0Y_1 \sin I(- \omega_{11} \cos \lambda \mp \omega_1 \cos 2\lambda) \pm$$
$$\pm \frac{1}{4} \frac{1}{K_3 \mp \omega_1} \kappa a_0Y_2 \sin I(\mp \omega_{11} \cos \lambda + \omega_1 \cos 2\lambda) +$$
$$+ \frac{1}{2K_3} \kappa DY_3 \sin I \cos \lambda + \left[(\mp A_4\omega_1 + A_{26}\omega_{11})a_0 - D\begin{matrix} A_{29} \\ A_{30} \end{matrix}\right]Y_4 -$$
$$- 12a_0\omega_{11}\kappa DY_5 \sin I \sin 2\lambda + \left[a_0\left(\begin{matrix} -A_8 \\ A_7 \end{matrix}\omega_1 + \begin{matrix} 2A_{29} \\ A_{29} \end{matrix}\omega_{11} + \begin{matrix} A_{30} \\ 2A_{30} \end{matrix}\omega_{11}\right) -\right.$$
$$\left.- 2DA_{26}\right]Y_6 + \kappa[4Y_8 - 3a_0^2(\omega_1^2 - \omega_{11}^2)]Y_7 \sin I \sin 2\lambda\Big\}$$

$$\begin{aligned}
D_{15,16} &= \frac{1}{K_3 \pm 2\omega_1}\left\{\frac{1}{2K_3}\kappa D Y_2 \sin I \cos\lambda \mp\right. \\
&\mp \frac{1}{4}\frac{1}{K_3 \pm \omega_1}\kappa a_0 Y_3 \sin I(\pm\,\omega_{11}\cos\lambda + \omega_1\cos 2\lambda) + \\
&+ a_0\left(\begin{matrix}-A_7\\ A_8\end{matrix}\omega_1 + \begin{matrix}A_{29}\\ A_{30}\end{matrix}\omega_{11}\right)Y_4 - 3\kappa a_0^2(\omega_1^2 - \omega_{11}^2)Y_5 \sin I \sin 2\lambda + \\
&+ \left[a_0(\mp A_4\omega_1 + A_{26}\omega_{11}) - 2D\begin{matrix}A_{29}\\ A_{30}\end{matrix}\right]Y_6 - \\
&\left.- 12a_0\omega_{11}D\kappa Y_7 \sin I \sin 2\lambda\right\}
\end{aligned}$$

$$\begin{aligned}
D_{17,18} &= \frac{1}{K_3 \pm 3\omega_1}\left[\mp\frac{1}{4}\frac{1}{K_3 \pm \omega_1}\kappa a_0 Y_2 \sin I(\pm\,\omega_{11}\cos\lambda + \omega_1\cos 2\lambda) + \right. \\
&\left.+ a_0\left(\begin{matrix}-A_7\\ A_8\end{matrix}\omega_1 + \begin{matrix}A_{29}\\ A_{30}\end{matrix}\omega_{11}\right)Y_6 - 3\kappa a_0^2(\omega_1^2 - \omega_{11}^2)Y_7 \sin I \sin 2\lambda\right]
\end{aligned}$$

$$\begin{aligned}
D_{19} &= \frac{1}{2K_2 + 2K_4 + K_3}\left\{\frac{-1}{2K_2 + 2K_4 + K_3}\kappa D Y_1 \sin I + \right. \\
&+ \frac{1}{4}\kappa a_0\omega_{11}Y_3 \sin I\left(\frac{1}{2K_2 + 2K_4 + K_3 + \omega_1} + \frac{1}{2K_2 + 2K_4 + K_3 - \omega_1}\right) + \\
&+ [(A_9 - A_{10})\omega_1 a_0 + (A_{31} + A_{32})\omega_{11}a_0 - DA_{27}]Y_4 - \\
&- 4\kappa Y_5 Y_8 \sin I \sin 2\lambda + 2[A_{27}a_0\omega_{11} - D(A_{31} + A_{32})]Y_6 + \\
&\left.+ 24a_0\omega_{11}D\kappa Y_7 \sin I \sin 2\lambda\right\}
\end{aligned}$$

$$\begin{aligned}
D_{20} &= \frac{1}{2(K_2 + K_4 + K_3)}\left\{\frac{1}{2(K_2 + K_4 + K_3)}\kappa D Y_1 \sin\lambda + \frac{1}{8}\kappa a_0 Y_3 \times\right. \\
&\times \left[\frac{\omega_1 \sin 2\lambda - 2\omega_{11}\sin\lambda}{2K_2 + 2K_4 + 2K_3 - \omega_1} - \frac{\omega_1 \sin 2\lambda + 2\omega_{11}\sin\lambda}{2K_2 + 2K_4 + 2K_3 + \omega_1}\right] + \\
&+ [(A_{11} - A_{12})\omega_1 a_0 + (A_{33} + A_{34})\omega_{11}a_0 - DA_{28}]Y_4 + \\
&+ 4\kappa Y_5 Y_8 \sin^2\lambda + 2[A_{28}a_0\omega_{11} - D(A_{33} + A_{34})]Y_6 - \\
&\left.- 24a_0\omega_{11}D\kappa Y_7 \sin^2\lambda\right\}
\end{aligned}$$

$$\begin{aligned}
D_{21,22} &= \frac{1}{2K_2 + 2K_4 + K_3 \pm \omega_1}\left\{\frac{1}{2}\frac{1}{2K_2 + 2K_4 + K_3 \pm \omega_1}a_0\omega_{11}\kappa Y_1 \sin I + \right. \\
&+ \frac{1}{4}\frac{1}{2K_2 + 2K_4 + K_3 \mp \omega_1}a_0\omega_{11}\kappa Y_2 \sin I - \\
&- \frac{1}{2}\frac{1}{2K_2 + 2K_4 + K_3}\kappa D Y_3 \sin I + \left[(A_{27}\omega_{11} \mp A_5\omega_1)a_0 - D\begin{matrix}A_{31}\\ A_{32}\end{matrix}\right]Y_4 + \\
&+ 12a_0\omega_{11}D\kappa Y_5 \sin I \sin 2\lambda + \\
&+ \left[\begin{matrix}-A_{10}\\ A_9\end{matrix}a_0\omega_1 + \left(\begin{matrix}2A_{31}\\ A_{31}\end{matrix} + \begin{matrix}A_{32}\\ 2A_{32}\end{matrix}\right)a_0\omega_{11} - 2DA_{27}\right]Y_6 + \\
&\left.+ [3a_0^2(\omega_1^2 - \omega_{11}^2) - 4Y_8]\kappa Y_7 \sin I \sin 2\lambda\right\}
\end{aligned}$$

$$\begin{matrix} D_{23} \\ D_{24} \end{matrix} = \frac{1}{2K_2 + 2K_4 + K_3 \pm 2\omega_1}\left\{ -\frac{1}{2}\,\frac{1}{2K_2 + 2K_4 + K_3}\,\kappa D Y_2 \sin I + \right.$$
$$+ \frac{1}{4}\,\frac{1}{2K_2 + 2K_4 + K_3 \pm \omega_1}\,\kappa a_0 \omega_{11} Y_3 \sin I +$$
$$+ \left(\substack{-A_9 \\ A_{10}}\,\omega_1 + \substack{A_{31} \\ A_{32}}\,\omega_{11} \right) a_0 Y_4 + 3a_0^2(\omega_1^2 - \omega_{11}^2)\kappa Y_5 \sin I \sin 2\lambda +$$
$$\left. + \left[(A_{27}\omega_{11} \mp A_5\omega_1)a_0 - 2D\substack{A_{31} \\ A_{32}} \right] Y_6 + 12a_0\omega_{11} D\kappa Y_7 \sin I \sin 2\lambda \right\}$$

$$\begin{matrix} D_{25} \\ D_{26} \end{matrix} = \frac{1}{2K_2 + 2K_4 + K_3 \pm 3\omega_1}\left\{ \frac{1}{4}\,\frac{1}{2K_2 + 2K_4 + K_3 \pm \omega_1}\,\kappa a_0 \omega_{11} Y_2 \sin I + \right.$$
$$\left. + \left(\substack{-A_9 \\ A_{10}}\,\omega_1 + \substack{A_{31} \\ A_{32}}\,\omega_{11} \right) a_0 Y_6 + 3a_0^2(\omega_1^2 - \omega_{11}^2)\kappa Y_7 \sin I \sin 2\lambda \right\}$$

$$\begin{matrix} D_{27} \\ D_{28} \end{matrix} = \frac{1}{2K_2 + 2K_4 + 2K_3 \pm \omega_1}\left\{ \mp \frac{1}{2K_2 + 2K_4 + 2K_3 \pm \omega_1} \times \right.$$
$$\times (\omega_1 \sin 2\lambda \pm \tfrac{1}{2}\omega_{11} \sin \lambda)\kappa a_0 Y_1 \pm \frac{1}{8}\,\frac{1}{2K_2 + 2K_4 + 2K_3 \mp \omega_1} \times$$
$$\times (\omega_1 \sin 2\lambda \mp 2\omega_{11} \sin \lambda)\kappa a_0 Y_2 + \frac{1}{4}\,\frac{1}{K_2 + K_4 + K_3}\,\kappa D Y_3 \sin \lambda +$$
$$+ \left[(A_{28}\omega_{11} \mp A_6\omega_1)a_0 - D\substack{A_{33} \\ A_{34}} \right] Y_4 - 12a_0\omega_{11} D\kappa Y_5 \sin^2 \lambda +$$
$$+ \left[\left(\substack{-A_{12} \\ A_{11}}\,\omega_1 + \substack{2A_{33} \\ A_{33}}\,\omega_{11} + \substack{A_{34} \\ 2A_{34}}\,\omega_{11} \right) a_0 - 2DA_{28} \right] Y_6 +$$
$$\left. + [4Y_8 - 3a_0^2(\omega_1^2 - \omega_{11}^2]\kappa Y_7 \sin^2 \lambda \right\}$$

$$\begin{matrix} D_{29} \\ D_{30} \end{matrix} = \frac{1}{2(K_2 + K_4 + K_3 \pm \omega_1)}\left\{ \frac{1}{4}\,\frac{1}{K_2 + K_4 + K_3}\,\kappa D Y_2 \sin \lambda \mp \right.$$
$$\mp \frac{1}{8}\,\frac{1}{2K_2 + 2K_4 + 2K_3 \pm \omega_1}(\omega_1 \sin 2\lambda \pm 2\omega_{11} \sin \lambda)\kappa a_0 Y_3 +$$
$$+ \left(\substack{-A_{11} \\ A_{12}}\,\omega_1 + \substack{A_{33} \\ A_{34}}\,\omega_{11} \right) a_0 Y_4 - 3a_0^2(\omega_1^2 - \omega_{11}^2)\kappa Y_5 \sin^2 \lambda +$$
$$\left. + \left[(A_{28}\omega_{11} \mp A_6\omega_1)a_0 - 2D\substack{A_{33} \\ A_{34}} \right] Y_6 - 12a_0\omega_{11} D\kappa Y_7 \sin^2 \lambda \right\}$$

$$\begin{matrix} D_{31} \\ D_{32} \end{matrix} = \frac{1}{2K_2 + 2K_4 + 2K_3 \pm 3\omega_1}\left\{ \mp \frac{1}{8}\,\frac{1}{2K_2 + 2K_4 + 2K_3 \pm \omega_1} \times \right.$$
$$\times (\omega_1 \sin 2\lambda \pm 2\omega_{11} \sin \lambda)\kappa a_0 Y_2 + \left(\substack{-A_{11} \\ A_{12}}\,\omega_1 + \substack{A_{33} \\ A_{34}}\,\omega_{11} \right) a_0 Y_6 -$$
$$\left. - 3a_0^2(\omega_1^2 - \omega_{11}^2)\kappa Y_7 \sin^2 \lambda \right\}$$

$$D_{33} = \frac{1}{4(2K_2 + 2K_4 + K_3)}\left(\frac{1}{2K_2 + 2K_4 + K_3 - \omega_1} - \right.$$
$$\left. - \frac{1}{2K_2 + 2K_4 + K_3 + \omega_1} \right) \kappa a_0 \omega_1 Y_3 \sin I \cos 2\lambda$$

$$\begin{matrix} D_{34} \\ D_{35} \end{matrix} = \frac{1}{2K_2 + 2K_4 + K_3 \pm \omega_1}\left[\mp \frac{\frac{1}{2}}{2K_2 + 2K_4 + K_3 \pm \omega_1} \times \right.$$

$$\times \kappa a_0 \omega_1 Y_1 \sin I \cos 2\lambda \pm$$

$$\left. \pm \frac{1}{4} \frac{1}{2K_2 + 2K_4 + K_3 \mp \omega_1} \kappa a_0 \omega_1 Y_2 \sin I \cos 2\lambda \right]$$

$$\begin{matrix} D_{36} \\ D_{37} \end{matrix} = \frac{1}{(2K_2 + 2K_4 + K_3 \pm 2\omega_1)(2K_2 + 2K_4 + K_3 \pm \omega_1} (\mp \tfrac{1}{4})\kappa\, a_0 \omega_1 Y_3 \sin I \cos 2\lambda$$

$$\begin{matrix} D_{38} \\ D_{39} \end{matrix} = \frac{1}{(2K_2 + 2K_4 + K_3 \pm 3\omega_1)(2K_2 + 2K_4 + K_3 \pm \omega_1)} \times$$

$$\times (\mp \tfrac{1}{4})\kappa a_0 \omega_1 Y_2 \sin I \cos 2\lambda$$

$$\begin{matrix} D_{40} \\ D_{41} \end{matrix} = Y_1 Y_9 \frac{\omega_{31} \cos \lambda \mp \omega_3}{(K_1 \pm \omega_3)^2}$$

$$\begin{matrix} D_{42} \\ D_{43} \end{matrix} = \tfrac{1}{2} Y_3 Y_9 \frac{\omega_{31} \cos \lambda - \omega_3}{(K_1 + \omega_3)(K_1 + \omega_3 \pm \omega_1)}$$

$$\begin{matrix} D_{44} \\ D_{45} \end{matrix} = \tfrac{1}{2} Y_3 Y_9 \frac{\omega_{31} \cos \lambda + \omega_3}{(K_1 - \omega_3)(K_1 - \omega_3 \pm \omega_1)}$$

$$\begin{matrix} D_{46} \\ D_{47} \end{matrix} = \tfrac{1}{2} Y_2 Y_9 \frac{\omega_{31} \cos \lambda - \omega_3}{(K_1 + \omega_3)(K_1 + \omega_3 \pm 2\omega_1)}$$

$$\begin{matrix} D_{48} \\ D_{49} \end{matrix} = \tfrac{1}{2} Y_2 Y_9 \frac{\omega_{31} \cos \lambda + \omega_3}{(K_1 - \omega_3)(K_1 - \omega_3 \pm 2\omega_1)}$$

where

$$Y = \tfrac{1}{2} a_0 \sin \lambda \left[\frac{\omega_{11} + \omega_1 \cos \lambda}{2K_1 + \omega_1} \cos(2\bar{\alpha}_0 + \Psi_0) + \frac{\omega_{11} - \omega_1 \cos \lambda}{2K_1 - \omega_1} \times \right.$$

$$\left. \times \cos(2\bar{\alpha}_0 - \Psi_0) \right] - \frac{1}{2K_1} D \sin \lambda \cos 2\bar{\alpha}_0 -$$

$$- \kappa D \left[\frac{\sin \lambda}{2(K_2 + K_4 + K_3)} \cos 2(\bar{\nu}_0 + \bar{\omega}_{M0} + \bar{\Omega}_{M0}) + \right.$$

$$\left. + \frac{\cos \lambda \sin I}{K_3} \cos \bar{\Omega}_{M0} - \frac{\sin I}{2K_2 + 2K_4 + K_3} \cos(2\bar{\nu}_0 + 2\bar{\omega}_{M0} + \bar{\Omega}_{M0}) \right] -$$

$$- \tfrac{1}{2}\kappa a_0 \left[- \tfrac{1}{2}\omega_1 \sin 2\lambda \left(\frac{\cos(2\bar{\nu}_0 + 2\bar{\omega}_{M0} + 2\bar{\Omega}_{M0} + \Psi_0)}{2K_2 + 2K_4 + 2K_3 + \omega_1} - \right. \right.$$

$$\left. - \frac{\cos(2\bar{\nu}_0 + 2\bar{\omega}_{M0} + 2\bar{\Omega}_{M0} - \Psi_0)}{2K_2 + 2K_4 + 2K_3 - \omega_1} \right) -$$

$$- \omega_{11} \sin \lambda \left(\frac{\cos(2\bar{\nu}_0 + 2\bar{\omega}_{M0} + 2\bar{\Omega}_{M0} + \Psi_0)}{2K_2 + 2K_4 + 2K_3 + \omega_1} + \right.$$

$$\left. + \frac{\cos(2\bar{\nu}_0 + 2\bar{\omega}_{M0} + 2\bar{\Omega}_{M0} - \Psi_0)}{2K_2 + 2K_4 + 2K_3 - \omega_1} \right) -$$

$$- \omega_{11} \sin I \cos \lambda \left(\frac{\cos(\bar{\Omega}_{M0} + \Psi_0)}{K_3 + \omega_1} + \frac{\cos(\bar{\Omega}_{M0} - \Psi_0)}{K_3 - \omega_1} \right) -$$

$$- \omega_1 \sin I \cos 2\lambda \left(\frac{\cos(\bar{\Omega}_{M0} + \Psi_0)}{K_3 + \omega_1} - \frac{\cos(\bar{\Omega}_{M0} - \Psi_0)}{K_3 - \omega_1} \right) +$$

$$+ \frac{\sin I}{2K_2 + 2K_4 + K_3 + \omega_1} (\omega_{11} \cos(2\bar{\nu}_0 + 2\bar{\omega}_{M0} + \bar{\Omega}_{M0} + \Psi_0) +$$

$$+ \omega_1 \cos 2\lambda \sin(2\bar{\nu}_0 + 2\bar{\omega}_{M0} + \bar{\Omega}_{M0} + \Psi_0)) + \frac{\sin I}{2K_2 + 2K_4 + K_3 - \omega_1} \times$$

$$\times (\omega_{11} \cos(2\bar{\nu}_0 + 2\bar{\omega}_{M0} + \bar{\Omega}_{M0} - \Psi_0) -$$

$$\left. - \omega_1 \cos 2\lambda \sin(2\bar{\nu}_0 + 2\bar{\omega}_{M0} + \bar{\Omega}_{M0} - \Psi_0)) \right] -$$

$$- \frac{9}{4} \left(\frac{\bar{X}_{60}}{\bar{\bar{X}}_{60}} \right)^6 \left[\frac{1}{2} a_0 (1 + \kappa) \sin 2\lambda \right]^2 \cos 2\Psi_0 -$$

$$- Y_9 \left[\frac{\omega_{31} \cos \lambda - \omega_3}{K_1 + \omega_3} \cos(\bar{\alpha}_0 + \xi_0) + \frac{\omega_{31} \cos \lambda + \omega_3}{K_1 - \omega_3} \cos(\bar{\alpha}_0 - \xi_0) \right]$$

(For the case of no solar radiation pressure, the term containing χ becomes zero).

$$Y_1 = 3 \left(\frac{\bar{X}_{60}}{\bar{\bar{X}}_{60}} \right)^6 \left[\frac{1 - \omega_4}{1 - 2\omega_4} + \frac{3}{2} \left(\frac{1}{2} \omega_1^2 a_0^2 + \frac{1}{2} \omega_{11}^2 a_0^2 + D^2 \right) \right]$$

$$Y_2 = \frac{9}{4} \left(\frac{\bar{X}_{60}}{\bar{\bar{X}}_{60}} \right)^6 (\omega_{11}^2 - \omega_1^2) a_0^2$$

$$Y_3 = - 9 \left(\frac{\bar{X}_{60}}{\bar{\bar{X}}_{60}} \right)^6 \omega_{11} a_0 D$$

$$Y_4 = \frac{3}{2} \left[1 - \frac{3}{2} \left(\frac{\bar{X}_{60}}{\bar{\bar{X}}_{60}} \right)^6 (1 + \kappa) a_0 \varepsilon \sin 2\lambda \cos \Psi_0 \right]$$

$$Y_5 = \frac{1}{3} \frac{1}{1 - 2\omega_4} Y_4$$

$$Y_6 = \frac{9}{8} \left(\frac{\bar{X}_{60}}{\bar{\bar{X}}_{60}} \right)^6 (1 + \kappa) a_0 \varepsilon \sin 2\lambda$$

$$Y_7 = \frac{1}{3} \frac{1}{1 - 2\omega_4} Y_6$$

$$Y_8 = \frac{3}{2} a_0^2(\omega_1^2 + \omega_{11}^2) + 3D^2 + 1$$

$$Y_9 = \frac{-3}{8} r_c^2 \chi b_0$$

$$Y_{10} = \frac{-9}{16} \varepsilon \left(\frac{\bar{X}_{60}}{\bar{\bar{X}}_{60}}\right)^6 a_0^2(1 + \kappa)^2 \sin^2 2\lambda .$$

APPENDIX E

Satellite under the Influence of a Low Thrust Having a Fixed Angle with Its Orbital Radius Vector

For a satellite in an orbit of small eccentricity and subject to the influence of a low constant acceleration which is always having a constant angle with its orbital radius vector, the equations of motion are (see Figure E1.)

$$\ddot{r} - r\dot{\theta}^2 = \frac{-g_0 r_0^2}{r^2} + f \cos \Psi \tag{E1}$$

and

$$\frac{1}{r}\frac{d}{dt}(r^2\dot{\theta}) = f \sin \Psi , \tag{E2}$$

where r and θ are the polar coordinates, g_0 is the gravitational acceleration at $r = r_0$, f is the magnitude of a constant thrust acceleration, and Ψ is a known constant angle between vectors $\vec{f}$ and $\vec{r}$.

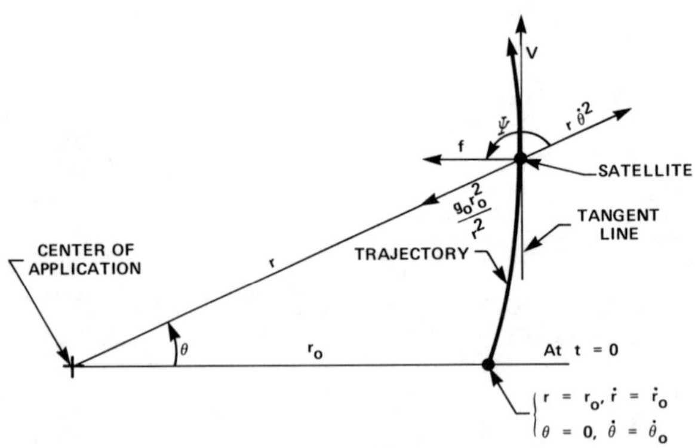

Fig. E1. Notations - low thrust power-flight trajectory.

If V is the velocity of the satellite, we have

$$V^2 = \dot{r}^2 + r^2\dot{\theta}^2 .$$

Letting

$$u = \frac{r_0}{r}, \quad h = \frac{r^2\dot{\theta}}{r_0^2\dot{\theta}_0} = \frac{\dot{\theta}}{u^2\dot{\theta}_0}, \quad C \equiv \frac{g_0}{r_0\dot{\theta}_0^2}, \quad \varepsilon = \frac{f}{g_0},$$

$$\sin \Psi \equiv M, \quad \cos \Psi \equiv N,$$

and

$$\dot{r} = -h\dot{\theta}_0 r_0 \frac{du}{d\theta},$$

$$\ddot{r} = -h\dot{\theta}_0 u^2 r_0 \left(h \frac{d^2u}{d\theta^2} + \frac{dh}{d\theta}\frac{du}{d\theta} \right),$$

$$V^2 = h^2 \dot{\theta}_0^2 r_0^2 \left[u^2 + \left(\frac{du}{d\theta}\right)^2 \right],$$

(E1) and (E2) become

$$h \frac{d^2u}{d\theta^2} + \frac{dh}{d\theta}\frac{du}{d\theta} + hu = \frac{C}{h} + \frac{\varepsilon}{u^3 h} CN \tag{E3}$$

and

$$\frac{dh}{d\theta} = \frac{\varepsilon}{u^2 h} M . \tag{E4}$$

Combining (E3) and (E4) results in

$$\frac{d^2u}{d\theta^2} + u = \frac{C}{h^2} + \frac{\varepsilon}{u^3h^2} CN + \frac{\varepsilon}{u^3h^2} CM \frac{du}{d\theta},$$

which can be replaced by a system of two first-order differential equations

$$\left.\begin{aligned} &\frac{du}{d\theta} = w \\ &\text{and} \\ &\frac{dw}{d\theta} = -u + \frac{C}{h^2} + \frac{\varepsilon}{u^3h^2} CN + \frac{\varepsilon}{u^3h^2} CM \frac{du}{d\theta} . \end{aligned}\right\} \tag{E5}$$

The first-order approximate solution to (E5) is obtained by assuming $\varepsilon = 0$ in (E4) and (E5), which results in

$$\left.\begin{aligned} &u = \frac{C}{h^2} + S\cos(\theta + \Phi) \\ &\text{and} \\ &w = -S\sin(\theta + \Phi) , \end{aligned}\right\} \tag{E6}$$

where h, S, and Φ are constants of integration.

Now if $\varepsilon \neq 0$, h, S, and Φ may be considered as functions of θ, the functional relationship between h and θ is given by (E4).

Differentiating (E6) with respect to θ and substituting the resulting equations into (E5) yield

$$\begin{aligned} \frac{du}{d\theta} &= -\frac{2C}{h^3}\frac{dh}{d\theta} + \cos(\theta + \Phi)\frac{dS}{d\theta} - S\left(1 + \frac{d\Phi}{d\theta}\right)\sin(\theta + \Phi) \\ &= w = -S\sin(\theta + \Phi) \end{aligned} \tag{E7}$$

and

$$\begin{aligned} \frac{dw}{d\theta} &= -\sin(\theta + \Phi)\frac{dS}{d\theta} - S\left(1 + \frac{d\Phi}{d\theta}\right)\cos(\theta + \Phi) \\ &= -u + \frac{C}{h^2} + \frac{\varepsilon}{u^3h^2} CN + \frac{\varepsilon}{u^3h^2} CM \frac{du}{d\theta} = -S\cos(\theta + \Phi) . \end{aligned} \tag{E8}$$

Solving (E7) and (E8) for $d\Phi/d\theta$ and $dS/d\theta$ gives

$$\begin{aligned} \frac{d\Phi}{d\theta} = &-\frac{2C}{h^3 S}\sin(\theta + \Phi)\frac{dh}{d\theta} - \frac{\varepsilon C\cos(\theta + \Phi)}{u^3h^2S} N - \\ &- \frac{\varepsilon C\cos(\theta + \Phi)}{u^3h^2S} M[-S\sin(\theta + \Phi)] \end{aligned} \tag{E9}$$

and

$$\frac{dS}{d\theta} = \frac{2C}{h^3}\cos(\theta + \Phi)\frac{dh}{d\theta} - \frac{\varepsilon C \sin(\theta + \Phi)}{u^3 h^2} N - $$
$$- \frac{\varepsilon C \sin(\theta + \Phi)}{u^3 h^2}[- S \sin(\theta + \Psi)] . \tag{E10}$$

Substituting (E5) and (E6) into (E4) yields

$$\frac{dh}{d\theta} = \frac{\varepsilon C h^7 M}{[C + Sh^2 \cos(\theta + \Phi)]^3} . \tag{E11}$$

Because of the low constant thrust acceleration, the application of the thrust will not change the eccentricity of the orbit appreciably; consequently, the orbit will remain nearly circular. If the first equation of (E6) is conceived as an equation of an ellipse, Sh^2/C is small for nearly circular orbits. Therefore, by expanding the right-hand side denominator of (E11) into a series and dropping out the terms containing $(Sh^2/C)^2$ and that of higher orders, (E11) gives

$$\frac{dh}{d\theta} = \varepsilon \frac{1}{C^2} h^7 M[1 - 3 \frac{Sh^2}{C} \cos(\theta + \Phi)] . \tag{E12}$$

By the same technique in obtaining (E12), Equations (E9) and (E10) become

$$\frac{d\Phi}{d\theta} = \frac{- 2\varepsilon M\bar{h}^4}{C\bar{S}} \sin(\theta + \bar{\Phi})[1 - 3 \frac{\bar{S}\bar{h}^2}{C} \cos(\theta + \bar{\Phi})] - \frac{\varepsilon N\bar{h}^4}{C^2\bar{S}} \cos(\theta + \bar{\Phi}) \times$$
$$\times [1 - 3 \frac{\bar{S}\bar{h}^2}{C} \cos(\theta + \bar{\Phi})] + \frac{\varepsilon M\bar{h}^4}{C^2} \sin(\theta + \bar{\Phi})\cos(\theta + \bar{\Phi}) \times$$
$$\times [1 - 3 \frac{\bar{S}\bar{h}^4}{C} \cos(\theta + \bar{\Phi})] \tag{E13}$$

and

$$\frac{dS}{d\theta} = \frac{2\varepsilon M\bar{h}^4}{C^2} \cos(\theta + \bar{\Phi})[1 - 3 \frac{\bar{S}\bar{h}^2}{C} \cos(\theta + \bar{\Phi})] - \frac{\varepsilon N\bar{h}^4}{C^2} \sin(\theta + \bar{\Phi}) \times$$
$$\times [1 - 3 \frac{\bar{S}\bar{h}^2}{C^2} \cos(\theta + \bar{\Phi})] + \frac{\varepsilon M\bar{S}\bar{h}^4}{C^2} \sin(\theta + \bar{\Phi}) \times$$
$$\times [1 - 3 \frac{\bar{S}\bar{h}^2}{C} \cos(\theta + \bar{\Phi})] , \tag{E14}$$

where (E12) has been employed.

Applying the asymptotic method (Bogoliubov and Mitropolsky, 1961) in nonlinear mechanics, (E12), (E13), and (E14) yield

$$\overline{\frac{dh}{d\theta}} = \varepsilon \frac{M\bar{h}^7}{C^2} ,$$

$$\overline{\frac{d\Phi}{d\theta}} = \frac{3}{2} \varepsilon N \frac{\bar{h}^6}{C^3} ,$$

and

$$\overline{\frac{dS}{d\theta}} = - 3\varepsilon M \frac{\bar{S}\bar{h}^6}{C^3} + \frac{1}{2} \varepsilon M \frac{\bar{S}\bar{h}^4}{C^2} .$$

Consequently,

$$\frac{1}{\bar{h}^6} = \frac{1}{\bar{h}_0^6} - \frac{6\varepsilon M}{C^2}(\theta - \theta_0) , \tag{E15}$$

$$\bar{\Phi} - \bar{\Phi}_0 = -\frac{N}{4CM}\ln\left[1 - \frac{6\varepsilon M\bar{h}_0^6}{C^2}(\theta - \theta_0)\right], \tag{E16}$$

and

$$\ln\left(\frac{\bar{S}}{\bar{S}_0}\right) = \frac{-1}{2C}\ln\left[1 - \frac{6\varepsilon M\bar{h}_0^6}{C^2}(\theta - \theta_0)\right] - \\ - \frac{1}{4\bar{h}_0^2}\left\{\left[1 - \frac{6\varepsilon M\bar{h}_0^6}{C^2}(\theta - \theta_0)\right]^{1/3} - 1\right\}, \tag{E17}$$

where θ_0 is the given initial condition, and $\bar{h}_0$, $\bar{\Phi}_0$, and $\bar{S}_0$ are the integration constants to be determined. Equations (E15)-(E17) are known as the first-order approximation to the problem. By applying the procedures stated in Section 2.5.4., the improved first-order approximate solution is obtained:

$$h = \bar{h} + \delta h \\ = \bar{h} - 3\varepsilon M\bar{h}^9 \frac{\bar{S}}{C^3}\sin(\theta + \bar{\Phi}), \tag{E18}$$

$$\Phi = \bar{\Phi} + \delta\Phi \\ = \bar{\Phi} + \frac{2\varepsilon M\bar{h}^4}{C\bar{S}}\cos(\theta + \bar{\Phi}) - \frac{\varepsilon N\bar{h}^4}{\bar{S}C^2}\sin(\theta + \bar{\Phi}) + \\ + \frac{3}{4}\frac{\varepsilon N\bar{S}\bar{h}^6}{C^3}\sin 2(\theta + \bar{\Phi}) - \frac{\varepsilon M\bar{h}^4}{2C^2}(3\bar{h}^2 + \tfrac{1}{2})\cos 2(\theta + \bar{\Phi}) + \\ + \frac{\varepsilon M\bar{S}\bar{h}^6}{C^3}\cos^3(\theta + \bar{\Phi}), \tag{E19}$$

and

$$S = \bar{S} + \delta S \\ = \bar{S} + \frac{2\varepsilon M\bar{h}^4}{C^2}\sin(\theta + \bar{\Phi}) + \frac{\varepsilon N\bar{h}^4}{C^2}\cos(\theta + \bar{\Phi}) - \frac{3}{4}\frac{\varepsilon N\bar{S}\bar{h}^4}{C^3}\cos 2(\theta + \bar{\Phi}) + \\ + \frac{\varepsilon M\bar{S}\bar{h}^4}{2C^2}\left[-3\frac{\bar{h}^2}{C} + \frac{1}{2}\right]\sin 2(\theta + \bar{\Phi}) - \frac{\varepsilon M\bar{S}^2\bar{h}^6}{C^3}\sin^3(\theta + \bar{\Phi}). \tag{E20}$$

Since $\dot{\theta} = hu^2\dot{\theta}_0$,

$$\frac{dt}{d\theta} = \frac{1}{hu^2\dot{\theta}_0} \\ = \frac{h^3}{C^2\dot{\theta}_0}\left[1 - 2\frac{Sh^2}{C}\cos(\theta + \psi) + 3\left(\frac{Sh^2}{C}\right)^2\cos^2(\theta + \psi) - \ldots\right],$$

and because of the smallness of Sh^2/C, we have

$$\overline{\frac{dt}{d\theta}} = \frac{\bar{h}^3}{C^2\dot{\theta}_0} = \frac{1}{C^2\dot{\theta}_0}\left[\frac{1}{\frac{1}{\bar{h}_0^6} + \frac{6\varepsilon M}{C^2}(\theta - \theta_0)}\right]^{1/2}.$$

An integration of the above equation yields

$$\bar{t} - \bar{t}_0 = \frac{1}{3\varepsilon M\dot{\theta}_0\bar{h}_0^3}\left\{-\left[1 - \frac{6\varepsilon M\bar{h}_0^6}{C^2}(\theta - \theta_0)\right]^{1/2} + 1\right\}, \tag{E21}$$

and the resulting improved first-order approximate solution to t is

$$t = \bar{t} + \delta t$$
$$= \bar{t} - \frac{2\bar{S}\bar{h}^5}{C^3\dot{\theta}_0} \sin(\theta + \bar{\Phi}), \tag{E22}$$

where $\bar{t}_0$ is an integration constant to be determined.

From the given initial conditions $r = r_0$, $\dot{r} = \dot{r}_0$, $\theta = \theta_0$, and $\dot{\theta} = \dot{\theta}_0$ at $t = 0$, the values of h_0, Φ_0, and S_0 can be obtained from the following relationships:

$$h_0 = 1 ,$$
$$\Phi_0 = \tan^{-1}\left[\frac{\dot{r}_0}{\dot{\theta}_0 r_0(1 - C)}\right] - \theta_0 ,$$

and

$$S_0 = \left[\left(\frac{\dot{r}_0}{\dot{\theta}_0 r_0}\right)^2 + (1 - C)^2\right] .$$

With these values and the relationships of (E18)-(E22), $\bar{h}_0$, $\bar{\Phi}_0$, $\bar{S}_0$, and $\bar{t}_0$ can be determined by iteration.

For the case of a pure, radial thrust acceleration, i.e., $M = 0$ and $N = \pm 1$, Equations (E9)-(E11) yield

$$\bar{h} = \bar{h}_0 ,$$
$$\bar{\Phi} - \bar{\Phi}_0 = \pm \frac{3}{2}\varepsilon\frac{\bar{h}_0^6}{C^3}(\theta - \theta_0) ,$$
$$\bar{S} = \bar{S}_0 ,$$

and

$$\bar{t} - \bar{t}_0 = \frac{\bar{h}_0^3}{C^2\dot{\theta}_0}\left[1 - \frac{3}{2}\left(\frac{\bar{S}_0\bar{h}_0^2}{C}\right)^2\right](\theta - \theta_0) ,$$

which have been obtained previously by Zee in a case including the effect of the Earth oblateness (Zee, 1969).

References

Bogoliubov, H. and Mitropolsky, J.; Asymptotic Method in the Theory of Nonlinear Oscillations, Hindustan Publishing Corporation, Delhi-6, India, 1961.

Zee, C.H.; 'Trajectories of Satellites under the Combined Influences of Low Radial Thrust and Earth Oblateness', Astronautica Acta **14**, 289-298 (1969).

APPENDIX F

Fourth-Order Earth Gravitational Potential Model Constants Used in Numerical Integration Computer Program

The fourth-order gravitational potential model constants used in the numerical integration digital computer program of the Communications Satellite Corporation are:

$$J_2 = 1082.7 \times 10^{-6}$$
$$J_{22} = -1.8191 \times 10^{-6} \qquad \lambda_{22} = -14.91737°$$
$$J_3 = -2.56 \times 10^{-6}$$
$$J_{31} = -2.13413 \times 10^{-6} \qquad \lambda_{31} = 8.62369°$$
$$J_{33} = -2.21820 \times 10^{-7} \qquad \lambda_{33} = 21.06797°$$
$$J_4 = -1.58 \times 10^{-6}$$
$$J_{42} = -1.65469 \times 10^{-7} \qquad \lambda_{42} = 31.71748°$$
$$J_{44} = -5.02195 \times 10^{-9} \qquad \lambda_{44} = 25.66314°$$

where the Earth Gravitational Potential Model is defined in Appendix A.

INDEX